GLACIER FLUCTUATIONS AND CLIMATIC CHANGE

GLACIOLOGY AND QUATERNARY GEOLOGY

Series Editor:

C. R. BENTLEY
University of Wisconsin-Madison,
Department of Geology and Geophysics,
Madison, Wisconsin, U.S.A.

The titles published in this series are listed at the end of this volume.

GLACIER FLUCTUATIONS AND CLIMATIC CHANGE

Proceedings of the Symposium on Glacier Fluctuations and Climatic Change, held in Amsterdam, 1-5 June 1987

Edited by

J. OERLEMANS

Institute of Meteorology and Oceanography, University of Utrecht, The Netherlands

KLUWER ACADEMIC PUBLISHERS

DORDRECHT / BOSTON / LONDON

Library of Congress Cataloging in Publication Data

Workshop on Glacier Fluctuations and Climatic Change (1987 :
Amsterdam, Netherlands)
Glacier fluctuations and climatic change : proceedings of the
Workshop on Glacier Fluctuations and Climatic Change, held in
Amsterdam, 1-5 June 1987 / edited by J. Oerlemans.
p. cm. -- (Glaciology and quaternary geology)
ISBN 0-7923-0110-2
1. Glaciers--Congresses. 2. Climatic changes--Congresses.
I. Oerlemans, J. (Johannes), 1950- . II. Title. III. Series.
QE576.W67 1987
551.3'12--dc19 88-34142

ISBN 0-7923-0110-2

Published by Kluwer Academic Publishers,
P.O. Box 17, 3300 AA Dordrecht, The Netherlands.

Kluwer Academic Publishers incorporates
the publishing programmes of
D. Reidel, Martinus Nijhoff, Dr W. Junk and MTP Press.

Sold and distributed in the U.S.A. and Canada
by Kluwer Academic Publishers,
101 Philip Drive, Norwell, MA 02061, U.S.A.

In all other countries, sold and distributed
by Kluwer Academic Publishers Group,
P.O. Box 322, 3300 AH Dordrecht, The Netherlands.

Printed in The Netherlands

CONTENTS

PREFACE

This book forms the proceedings of an international conference on Glacier Fluctuations and Climatic Change, held in Amsterdam, at the Royal Netherlands Academy of Sciences, 1 - 5 June 1987. The conference was organized by the Institute of Meteorology and Oceanography (University of Utrecht), and endorsed by the Netherlands section of the INQUA-commission.

The purpose of the meeting was to bring together scientists from various disciplines that, in one way or another, study glacier fluctuations. About 90 scientists participated in the meeting. Disciplines like geology, palynology, geomorphology, meteorology and glaciology were represented, and the 45 contributions gave a remarkably good overview of how many entries there are in studying glacier fluctuations. The differences in approach showed up clearly, leading to many lively discussions, in particular between modellers and people working in the field. I think we all learned a lot !

The production of this volume has taken more time than foreseen, due to the unexpectedly large number of papers that were submitted. The review procedure was time consuming, but has certainly contributed to the quality of the papers. I am indebted to all reviewers for their help in this matter.

I like to express my gratitude to the Ministry of Housing, Physical Planning and Environment for financial support in the organization of the conference. Further financial and administrative help was provided by the University of Utrecht. The pleasant and efficient organization was to a large extent due to the efforts of Marjolijn Verhoeven, Eveline Plesman and Pieter Thijssen.

My special thanks go to Marjolijn Verhoeven, who did most of the typing and lay-out of this book. I am also grateful to Eveline Plesman and Christina de Jong, who assisted in the last stage of preparation.

I hope that this volume will prove to be of value to anyone studying the fluctuations of glaciers.

Johannes Oerlemans

Utrecht, September 1988

EPISODIC PALAEOZOIC GLACIATION IN THE CAPE-KAROO BASIN, SOUTH AFRICA

J.N.J. Visser
Geology Department
P.O. Box 339, Bloemfontein 9300
South Africa

ABSTRACT

Late Ordovician (Caradoc to Ashgill) glacial sediments, having a maximum thickness of 150 m and being deposited over a period of about 12 Ma during a low-latitude temperate glaciation, are interbedded in quartz arenites of the Cape Supergroup. Marine sandstone and stratified diamictite deposited during an interglacial, and soft sediment pavements separate basal massive arenaceous diamictite from upper argillaceous diamictite. Permo-Carboniferous (Stephanian to Asselian) glacial deposits, having a maximum thickness of 800 m, at the base of the thick Karoo mudrock-sandstone sequence, unconformably overlies the Cape Supergroup and pre-Cape basement. Glacial sedimentation occurred over a period of about 19 Ma during a high-latitude temperate glaciation. About 90 per cent of the glacial sequence consists of massive argillaceous diamictite formed by lodgement, melt-out and debris rain processes on a marine shelf. Two units consisting of interglacial sediments are interbedded in the massive diamictites.

The two macro-scale glaciation periods, following no systematic pattern in southwestern Gondwana, occurred during a cool to cold climatic period lasting for about 270 Ma from the mid-Cambrian to the mid-Permian in the Cape-Karoo basin. The main reason for this period of below average global temperature, was the proximity of the basin during most of the time to the south pole. The high incidence of plate tectonics during this period affected the distribution of land masses and epicontinental seas, and mountain building on southwestern Gondwana as well as the orientation of the plate rotational axis. A combination of these factors probably triggered the two glaciations. Meso-scale warmer cycles with a periodicity of 8 to 9 Ma during the Permo-Carboniferous can be correlated with third-order Vail cycles and are attributed to plate tectonics.

J. Oerlemans (ed.), Glacier Fluctuations and Climatic Change, 1–12.

1. INTRODUCTION

Early Palaeozoic glacial deposits in the same stratigraphic context are found in southern and northern Africa and South America whereas late Palaeozoic glacial strata occur in southern Africa, South America, Antarctica, Madagascar, India, Australia and on the Falkland Island (Crowell, 1983). These regions formed part of Gondwana during the Palaeozoic and Mesozoic and the glaciations thus represent regional events.

This study which deals with time-stratigraphic glacier fluctuations in the Cape-Karoo basin, has the following objectives: (i) to describe episodic ice advance and retreat on different time scales during the period 260 to 530 Ma ago, and (ii) to briefly suggest possible causes for the cold periods.

It is absolutely necessary to have reliable ages for the glacial and interglacial deposits in order to reconstruct glacier fluctuations during the Palaeozoic. In the Cape-Karoo basin fossils are present in the mudrocks interbedded with or overlying the glacials, but unfortunately the Permo-Carboniferous diamictite has an unconformable base. Sedimentation rates were thus applied to calculate the duration of glacial deposition. Although this approach is debatable, using different rates, based on Quaternary glacial sedimentation, for subglacial till ($0.025 \text{ m}/10^3$ a), glaciomarine shelf diamicton and mud ($0.5 \text{ m}/10^3$ a) and fjord diamicton, mud and outwash sediment ($1 \text{ m}/10^3$ a) comparable ages to those based on palaeontological data, were obtained.

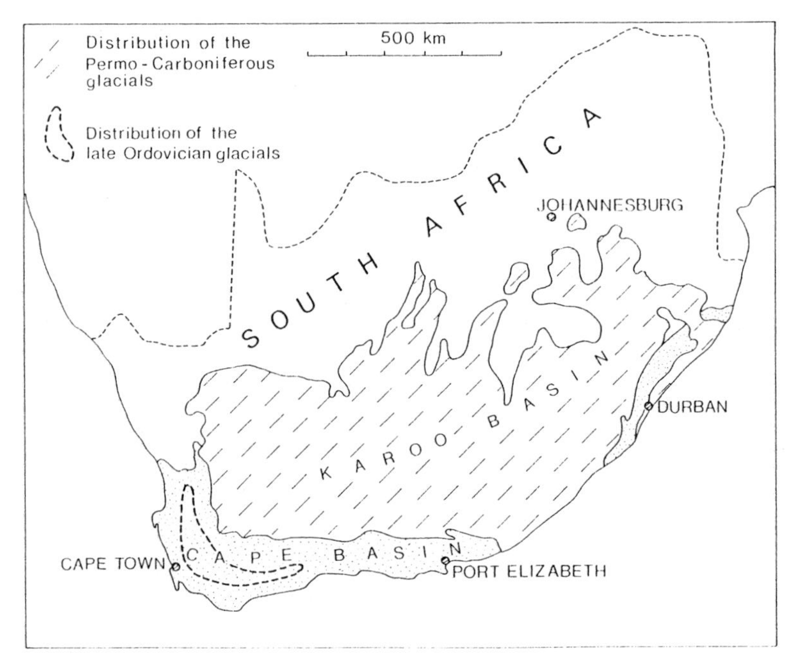

Figure 1. Distribution of the late Ordovician and Permo-Carboniferous diamictites in the Cape and Karoo basins respectively.

2. CAPE-KAROO BASIN

The Cape-Karoo basin covers an area of about 1,000,000 km^2 in South Africa (Fig. 1). This, however, represents only a fraction of the original basin which extended over thousands of kilometres in southwestern Gondwana. The sedimentary package which varies from late Cambrian to early Jurassic in age, has a maximum thickness of about 14 km and consists of a Cape Supergroup at the base and the unconformable Karoo Sequence at the top (Fig. 2). The Cape Supergroup contains shallow marine and deltaic sandstone, mudstone and shale of which the material was mainly derived from the north (Rust, 1973). The late Ordovician glacial beds (known as the Pakhuis Formation) occur together with fossiliferous mudrocks interbedded in a thick quartz arenite sequence (Fig. 2).

The Karoo Sequence consists of a glacigene basal diamictite formation (known as the Dwyka Formation), conformably overlain by a thick succession of shallow marine, lacustrine and fluvial mudrocks and sandstones. Sediments were derived from the north, east and south during the late Carboniferous to early Permian, but the late Permian to Mesozoic fluvial basin had a centripetal drainage. The glacigene formation unconformably overlies Cape Supergroup rocks and pre-Cape basement. At the upper contact diamictite grades into fossiliferous, carbonaceous mudstone and shale.

The Cape-Karoo sequence was deposited on a stable to unstable shelf bordering a continental interior in southwestern Gondwana. On the south and southwest the palaeo-Pacific ocean defined the shelf margin (Visser, 1987).

3. LATE ORDOVICIAN GLACIATION

The glacial deposits (Pakhuis Formation) which outcrop only in a small area in the southwestern part of the basin (Fig. 1), have been described in detail by Visser (1962), Rust (1973 and 1981) and Blignault (1981). Their glacial origin is substantiated by the presence of soft sediment pavements, striated stones in the diamictite and dropstone argillite. The sequence which has a cumulative thickness of 150 m, consists primarily of the basal Sneeukop arenaceous diamictite and the argillaceous Kobe diamictite at the top. The Sneeukop diamictite commonly lies in syndepositional folds with a maximum vertical dimension of 75 m, at the top of the Peninsula Formation (Fig. 3). The folding formed either by density loading of the unlithified sand by the ice (Rust, 1973) or by drag of the soft sediment by the overriding ice (Blignault, 1981). The thin Oshoek marine sandstone unconformably follows on the Sneeukop diamictite and is in turn conformably overlain by the Kobe diamictite in the north. Soft sediment pavements occur on top of the Oshoek sandstone and directional measurements of the grooves and flutes suggest ice flow from the north.

Similarity in composition between the Peninsula sandstone and the Sneeukop diamictite and the deformation of the subice sediments suggest deposition of a basal till. The Oshoek sandstone formed on a beach

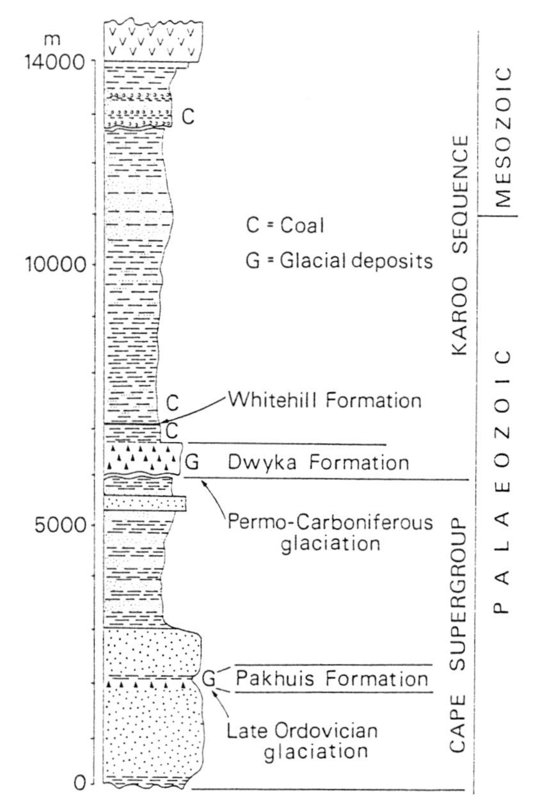

Figure 2. A generalized stratigraphy of the Cape Supergroup and Karoo Sequence in the Cape-Karoo basin.

during a blanket transgression (Blignault, 1981) indicating complete ice retreat (interglacial) from the basin. During the next depositional phase the ice front was grounded in a shallow sea so that the Kobe diamictite represents grounding line melt-out, suspension settling and debris rain deposits. The presence of predominantly extrabasinal material in the Kobe diamictite indicates glaciation over a much larger area and a more vigorous advance of the ice lobe into the basin. Rust (1981) estimated an areal extent of at least 200,000 km^2 for the ice lobe.

The trilobites, brachiopods and possible land plants in the overlying Cedarberg Formation either suggest a late Ashgill to early Llandovery (Gray et al., 1986) or an early Ashgill (Cocks and Fortey, 1986) age. The latter age is more acceptable in view of the trace fossils occurring below the glacials (Potgieter and Oelofsen, 1983, Cocks and Fortey, 1986), the age of the trilobite fauna in the Bokkeveld Group (Cooper, 1982), and the extremely low depositional rate of super-mature quartz arenites.

According to calculations deposition of the diamictite beds lasted approx. 12 Ma which suggests a late Caradoc to early Ashgill age for the glaciation. These deposits are thus of similar age (440 Ma) as the North African diamictites (Deynoux, 1985). A temperate glacial environment with the northern basin margin at palaeolatitudes 25° to 35°S (Smith et al., 1981) is suggested. Hambrey (1985) recognised two overlapping glacial epochs during the late Ordovician. The difference in character between the Sneeukop and Kobe diamictites may indicate deposition during Hambrey's Gander Bay and Tamadjert epochs respectively.

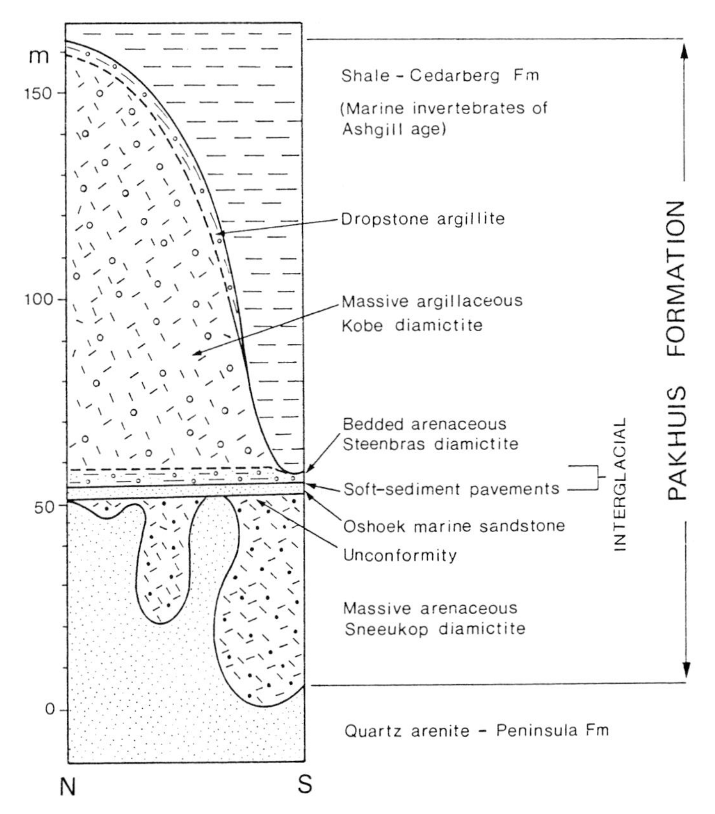

Figure 3. Stratigraphy of the glacigene late Ordovician Pakhuis Formation.

4. PERMO-CARBONIFEROUS GLACIATION

General descriptions of the glacigene Dwyka Formation which outcrops along the basin margin and underlies an area of about 600,000 km^2 (Fig. 1), are given by Crowell and Frakes (1972), Von Brunn and Stratten (1981) and Visser (1983). Visser (1983) recognized a valley facies in the continental interior and an 800 m thick platform facies which gives a more complete record of the glaciation. The glacial origin of the Dwyka Formation is substantiated by the presence of bedrock, boulder and soft sediment pavements, striated stones in the diamictite, dropstone argillite, glacial erosion surfaces, and distantly derived (> 500 km), large (> 1 m in diameter) clasts.

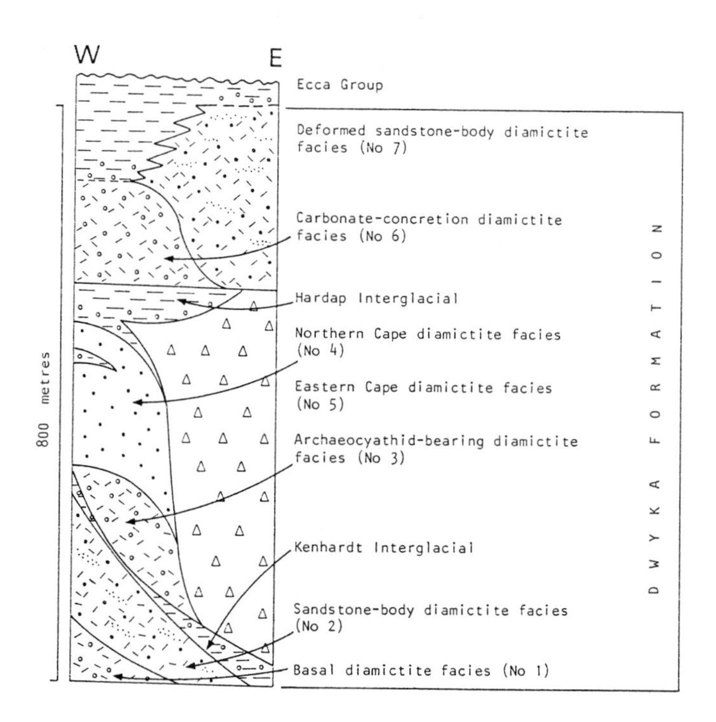

Figure 4. A schematic sequence of distinct diamictite facies groups in the Dwyka Formation of the southern Karoo. The interglacial deposits consist of shale, mudstone, sandstone, conglomerate, dropstone argillite and stratified diamictite. Units not exactly to scale.

The platform sequence consists of 7 conceptually distinct diamictite facies groups (Fig. 4). Near the unstable margin of the ice sheet in the west facies groups 1, 2, 3, 4 and 6, and in the east (closer to the pole) facies groups 5 and 7 are present. Facies groups 1 and 3 represent subglacial tills deposited during active glacial advance over land. Facies groups 4 and 5 formed by a combination of ice grounding zone melt-out and debris rain in a marine environment, whereas facies group 6 is entirely the product of undermelting of floating ice. Facies groups 2 and 7 represent a mixture of lodgement and melt-out tills, debris rain diamicton and glacial outwash deposited primarily during deglaciation.

Mudstone, black shale, sandstone with occasional trace fossils, conglomerate, stratified diamictite and dropstone argillite were deposited during the Kenhardt Interglacial. This sequence which has a maximum thickness of 30 m in the west and about 50 m in the east, can be traced for about 900 km across the basin (Fig. 4). Its absence between facies groups 2 and 3 in places is attributed to glacial erosion.

Mudstone, black shale, sandstone, dropstone argillite and stratified diamictite were deposited during the Hardap Interglacial. The sequence which has a thickness of 35 to 55 m in the west, can be traced for about 400 km eastwards before it interfingers with the massive diamictite of facies group 5 (Fig. 4). Invertebrate fossils are present in the sequence to the north of the Karoo Basin.

The invertebrate and microfossils in the Hardap Interglacial sequence and at the base of the overlying Ecca Group have an early Permian (Sakmarian?) age (McLachlan and Anderson, 1975; Martin, 1981). Palynological studies of mudrocks from the Kenhardt Interglacial sequence in the southern Karoo show the Permian-Carboniferous boundary to be more or less at that stratigraphic level (Anderson, 1977).

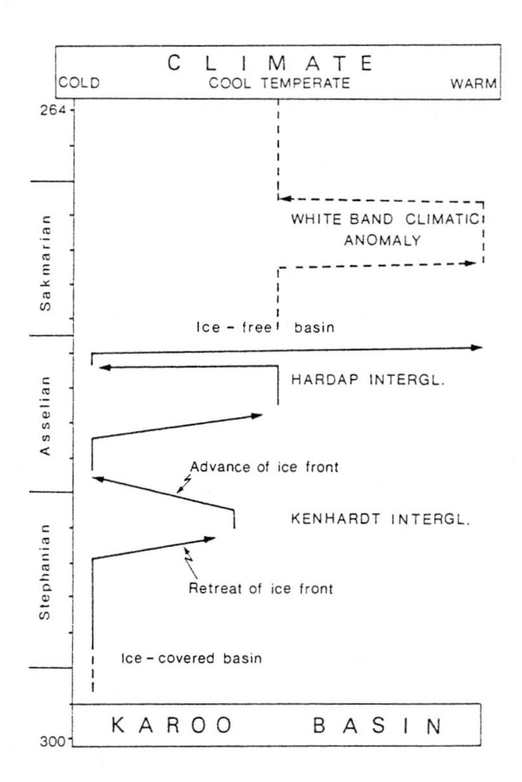

Figure 5. Meso-scale climatic cycles superimposed on the Permo-Carboniferous cool to cold climate. Intergl. = interglacial.

According to calculations deposition of the beds lasted approximately 19 Ma which suggests a Stephanian to Asselian age for the glaciation. A temperate glacial environment with the basin margins at palaeolatitudes 60° to 85°S (Smith et al., 1981) is suggested. The Kenhardt Interglacial occurred towards the end of the Stephanian, whereas the Hardap Interglacial was approximately at the middle of the Asselian (Fig. 5). After ice retreat the climate remained cool, except during deposition of the Whitehill (White Band) Formation (Fig. 2) when it suddenly ameliorated and water temperatures became favourable for aquatic reptiles (Mesosaurus) to invade the Karoo Basin. This climatic anomaly shows approximately the same periodicity as the two interglacials (Fig. 5).

5. MACRO- AND MESOSCALE CLIMATIC FLUCTUATIONS

The duration of the Palaeozoic glaciations is an unknown factor in establishing glacier fluctuations. Crowell (1983) estimated that the Ordovician-Silurian glaciation lasted about 30 Ma, but for the Cape-Karoo basin this figure apparently is too high and one of 15 to 20 Ma is perhaps more acceptable. During the Permo-Carboniferous glaciation the problem is aggravated as an erosional period of about 30 Ma preceded glacial deposition. Bell (1981) attributed the slumping in the upper sandstones of the Witteberg Group to glacial activity inferring a glaciation of between 60 and 70 Ma, whereas Crowell (1983) suggested a glacial period lasting about 110 Ma for Gondwana. Visser (1987) estimated that polar conditions in the basin and adjoining highlands became evident between 310 and 320 Ma giving rise to a glaciation lasting between 35 and 40 Ma.

During the Palaeozoic two glaciations which lasted 15 to 20 Ma and 35 to 40 Ma, respectively, and which were interrupted by 125 Ma of generally cool to cold conditions, occurred in the Cape-Karoo basin (Fig. 6). A cool to cold climate also probably preceded the late Ordovician glaciation. Since the middle Silurian to the middle Permian the Cape-Karoo basin was at high latitudes (> 60°S) which supports the postulated climatic conditions and it was only for a brief period during the middle Devonian when, presumably, warmer conditions prevailed (Fig. 6). This long (about 270 Ma in the Cape-Karoo basin), predominantly cool to cold period with occasional polar conditions, corresponds with the "Gondwanan glacial era" of Chumakov (1981). This was also the period when southwestern Gondwana drifted across the south magnetic pole and the possibility of small permanent ice caps over mountainous regions during this interval cannot be ignored. By a combination of factors these small ice caps rapidly expanded during the late Ordovician and Permo-Carboniferous causing the glaciations in the Cape-Karoo basin.

There is wide consensus that global climatic conditions during the Palaeozoic were not consistent (Frakes, 1979; John, 1979; Spjeldnaes, 1981, Copper, 1986). Below average global temperatures apparently prevailed during the Ordovician and early part of the Silurian (up to ± 430 Ma) and from the Devonian-Carboniferous boundary to the early Permian when there was a sudden increase in global temperature. During

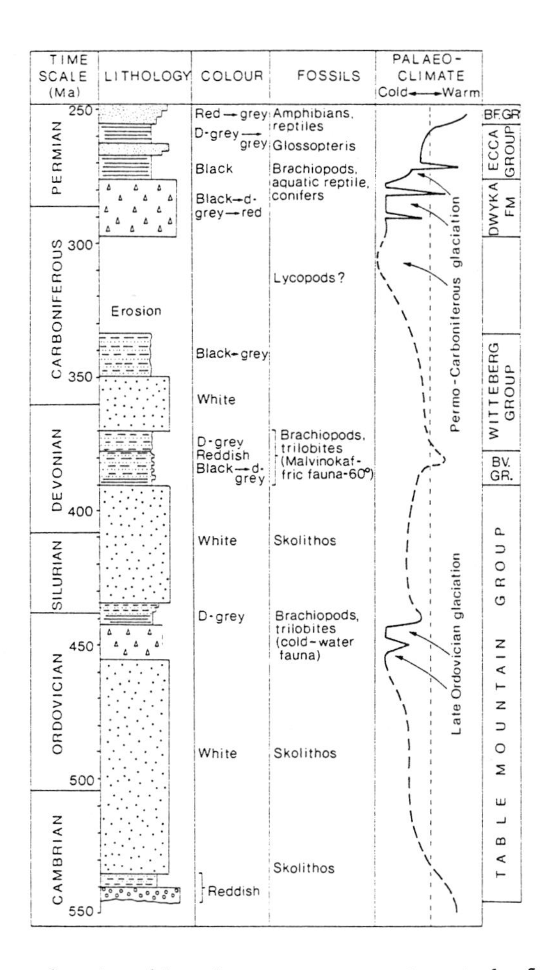

Figure 6. A climatic curve constructed for the Palaeozoic succession in the Cape-Karoo basin. BV = Bokkeveld, BF = Beaufort, GR = Group, FM = Formation. Time-scale after Harland et al. (1982).

such globally cool intervals climatic zones shifted equatorwards, latitudinal gradients were steep (Spjeldnaes, 1981), and conditions became favourable for ice sheet growth and its expansion into intermediate latitudes. However, when global temperatures were above average, like during the late Silurian and Devonian (up to the end of the Frasnian) climatic zones shifted polewards, but continents clustered at the pole, like southwestern Gondwana during this period, would have experienced cool (but ice-free) climatic conditions. The above line of argument corresponds with the "ice-age preparation" theory of Fairbridge (1973) which involved physiogeographic changes at the earth's surface.

The latter may be linked with periods of increasing or diminishing hotspot activity resulting in different rates of plate tectonics and heat flux to the earth's surface (Nicolysen, 1985).

The glacier fluctuations during the Gondwanan glacial era in the Cape-Karoo basin apparently did not follow a systematic pattern. The temperate, low-latitude, late Ordovician glaciation is attributed to a major equatorward shift of climatic zones and the positioning of the south pole over an extensive upland area in Africa. Although the centre of glaciation was located over the Sahara (Deynoux, 1985) distal ice lobes, probably influenced by a high topography and proximity of shallow seas, flowed into the Cape Basin. The growth of the extensive Permo-Carboniferous ice sheet can be attributed to a combination of physiogeographical factors. The pole was located over the highlands of Antarctica, subduction of the oceanic palaeo-Pacific plate beneath the Gondwana plate created an alpine mountain range, whereas the plate tectonic reorientation of the palaeo-Pacific - southwestern Gondwana margin to north-northeast - south-southwest allowed westerly winds to carry moisture deep into the interior. A major fall in sea-level during the early Carboniferous (Vail et al., 1977) also increased the albedo enhancing lower temperatures.

Meso-scale glacier fluctuations during the Permo-Carboniferous (Fig. 5) suggest a periodicity of 8 to 9 Ma. This corresponds with the third-order Vail cycle (Vail et al., 1977) which is attributed to geotectonic (changes in the rate of sea-floor spreading) causes.

In the Cape-Karoo basin glacier fluctuations occurred on different time-scales. On a macro-scale ice fluctuations apparently show correlation with globally cooler and warmer periods which can possible be attributed to the earth's internal mechanisms. Physiogeographical factors and pole positioning controlled ice sheet formation superimposed on the globally cool period. On a meso-scale an apparent periodicity in glacier fluctuation is noticeable and can be related to sea level changes and plate tectonics. Thus both macro- and meso-scale glacier fluctuations are caused by an integrated system of parameters controlled from within the earth itself.

REFERENCES

Anderson, J.M. 1977. The biostratigraphy of the Permian and Triassic. Pt 3, A review of Gondwana Permian palynology with particular reference to the northern Karoo Basin, South Africa. Mem. bot. Surv. South Africa 41, 188 pp.

Bell, C.M. 1981. Soft-sediment deformation of sandstone related to the Dwyka glaciation in South Africa. Sedimentology 28, 321-329.

Blignault, H.J. 1981. Ice sheet deformation in the Table Mountain Group, Western Cape. Ann. Univ. Stell., Ser. A1 3, 1-66.

Chumakov, N.M. 1981. Upper Proterozoic glaciogene rocks and their stratigraphic significance. Precambrian Res. 15, 373-395.

Cocks, L.R.M. and Fortey, R.A. 1986. New evidence on the South African lower Palaeozoic: age and fossils reviewed. Geol. Mag. 123, 437-444.

Cooper, M.R. 1982. A revision of the Devonian (Emsian-Eifelian) trilobita from the Bokkeveld Group of South Africa. Ann.S. Afr. Mus. 89, 1-174.

Copper, P. 1986. Frasnian/Famennian mass extinction and cold-water oceans. Geology 14, 835-839.

Crowell, J.C. 1983. Ice ages recorded on Gondwanan continents. Trans. Geol. Soc. S. Afr. 86, 237-262.

Crowell, J.C. and Frakes, L.A. 1972. Late Paleozoic glaciation: Part V, Karroo Basin, South Africa. Bull. Geol. Soc. Amer. 83, 2887-2912.

Deynoux, M. 1985. Les glaciations du Sahara. La Recherche 16 (169), 986-997.

Fairbridge, R.W. 1973. Glaciation and plate migration. In: D.H. Tarling and S.K. Runcorn (eds.), "Implications of continental drift to the Earth Sciences". Academic Press, London, 503-515.

Frakes, L.A. 1979. "Climates throughout geolocial time". Elsevier, Amsterdam, 310 pp.

Gray, J., Theron, J.N. and Boucout, A.J. 1986. Age of the Cedarberg Formation, South Africa and early land plant evolution. Geol. Mag. 123, 445-454.

Hambrey, M.J. 1985. The late Ordovician - early Silurian glacial period. Paleogeogr., Palaeoclimatol., Palaeoecol. 51, 273-289.

Harland, W.B., Cox, A.V., Llewellyn, P.G., Pickton, C.A.G., Smith, A.G. and Walters, R. 1982. "A Geological time scale". Cambridge University Press, Cambridge, 131 pp.

John, B.S. 1979. "The winters of the world". David and Charles, London, 256 pp.

Martin, H. 1981. The late Palaeozoic Dwyka Group of the south Kalahari Basin in Namibia and Botswana and the subglacial valleys of the Kaokoveld in Namibia. In: M.J. Hambrey and W.B. Harland (eds.), "Earth's pre-Pleistocene glacial record". Cambridge University Press, Cambridge, 61-66.

McLachlan, I.R. and Anderson, A.M. 1975. The age and stratigraphic relationship of the glacial sediments in southern Africa. In: K.S.W. Campbell (ed.), "Gondwana Geology". Australia National University Press, Canberra, 415-422.

Nicolaysen, L.O. 1985. On the physical basis for the extended Wilson cycle, in which most continents coalesce and then disperse again. Trans. Geol. Soc. S. Afr. 88, 562-580.

Potgieter, C.D. and Oelofsen, B.W. 1983. Cruziana acacensis - the first Silurian index-trace fossil from southern Africa. Trans. Geol. Soc. S. Afr. 86, 51-54.

Rust, I.C. 1973. The evolution of the Paleozoic Cape Basin, southern margin of Africa. In: A.E.M. Nairn and F.G. Stehli (eds.), "The ocean basins and margins, vol I." Plenum Publishing Corp., New York, 247-276.

Rust, I.C. 1981. Early Paleozoic Pakhuis tillite, South Afria. In: M.J. Hambrey and W.B. Harland (eds.), "Earth's pre-Pleistocene glacial record". Cambridge University Press, Cambridge, 113-116.

Smith, A.G., Hurley, A.M. and Briden J.C. 1981. "Phanerozoic paleocontinental world maps". Cambridge University Press, Cambridge, 102 pp.

Spjeldnaes, N. 1981. Lower Palaeozoic palaeoclimatology. In: C.H. Holland (ed.), "Lower Palaeozoic of the Middle East, Eastern and Southern Africa, and Antarctica". John Wiley and Sons, New York, 199-256.

Vail, P.R., Mitchum, R.M. and Thompson, S. 1977. Seismic stratigraphy and global changes of sea level, part 4: Global cycles of relative changes of sea level. In: C.E. Payton (ed.), "Seismic stratigraphy - Application to hydrocarbon exploration". Mem. Amer. Assoc. Petroleum Geologists 26, 83-97.

Visser, J.N.J. 1962. Die voorkoms en oorsprong van die tillietband in die Serie Tafelberg. Unpubl. M.Sc. thesis, Univ. O.F.S., Bloemfontein, 58 pp.

Visser, J.N.J. 1983. Glacial-marine sedimentation in the late Paleozoic Karoo Basin, Southern Africa. In: B.F. Molnia (ed.), "Glacial-marine sedimentation". Plenum Publishing Corp., New York, 667-701.

Visser, J.N.J. 1987. The Palaeogeography of part of southwestern Gondwana during the Permo-Carboniferous glaciation. Palaeography, Palaeoclimatolgy, Palaeoecology 61, 205-219.

Von Brunn, V. and Stratten, T. 1981. Late Palaeozoic tillites of the Karoo Basin of South Africa. In: M.J. Hambrey and W.B. Harland (eds.), "Earth's pre-Pleistocene glacial record". Cambridge University Press, Cambridge, 71-79.

QUATERNARY GLACIATIONS AND PALAEOCLIMATE OF MOUNT KENYA, EAST AFRICA

W.C. Mahaney, R.W. Barendregt[1] and W. Vortisch[2]
Dept. of Geography, Atkinson College
York University, 4700 Keele Street
North York Ontario, Canada M3J 1P3

ABSTRACT

Quaternary glacial and periglacial deposits, in several drainages on Mount Kenya, have been dated using relative (RD) and absolute dating (AD) methods to assist in refining the glacial chronology and in reconstructing palaeoclimate. Topographic position of deposits, weathering features, degree of soil development, and radiocarbon assist in assigning ages to Quaternary deposits in the Afroalpine and bamboo forest-Hagenia woodland zones. AD and RD methods provide reasonably precise ages for late Pleistocene and Holocene deposits; however, early and middle Pleistocene deposits proved difficult to date with precision. Palaeosols provide information on palaeoclimate and allow palaeoenvironmental reconstruction to and beyond the last interglacial > 100,000 yrs BP.

In the higher valleys sequences of younger glacial deposits mapped between 4500 and 4800 m a.s.l., belong to the late Holocene (<1000 yrs BP). Moraines, formerly mapped by other workers as mid-Holocene glacial deposits, are considered to date from recession of valley ice during the late glacial (~12,000 yrs BP). Ice advance during the last glacial (Liki Glaciation) presumably began < 100,000 yrs BP, ending ~15,000 yrs BP, and reaching to 3200 m a.s.l. Only minimum ages of ~15,000 yrs BP are available from bog-bottom sediments recovered from cores taken in the lower valleys around the margin of glaciation. In Teleki Valley, a sequence of pre-Liki tills, loesses, and tephras provides information on the nature and character of environmental fluctuations over the period > 100,000 yrs BP to > 730,000 yrs BP. These glaciations produced a succession of four tills older than the last glaciation, from youngest to oldest: Teleki, Naro Moru, Lake Ellis, and Gorges drifts. The Teleki glaciation occurred prior to deposition of post-Teleki loess; Teleki

[1] Dept. of Geography, University of Lethbridge, Lethbridge, Alberta, Canada T1K 3M4
[2] University of Marburg, Inst. of Geology and Paleontology, D-3550 Marburg/ Lahn, W. Germany

J. Oerlemans (ed.), Glacier Fluctuations and Climatic Change, 13–35.

till, high in ferromagnetic minerals, and almost free of quartz, contains more highly weathered feldspars and altered glass than is commonly found in younger Liki palaeosols. The compound palaeosol formed in these sediments is the product of weathering during and after the Teleki/Liki interglaciation.

The older Naro Moru, Lake Ellis and Gorges tills have variable ages that can be established by topographic position, K/Ar ages on outwash, and palaeomagnetism. The oldest tills, dating from the Gorges and Lake Ellis glaciations, have strong reversed remanent magnetism placing them in the Matuyama Chron. Younger drift of Naro Moru age has stong normal remanent magnetism placing it in the lower to middle Brunhes Chron.

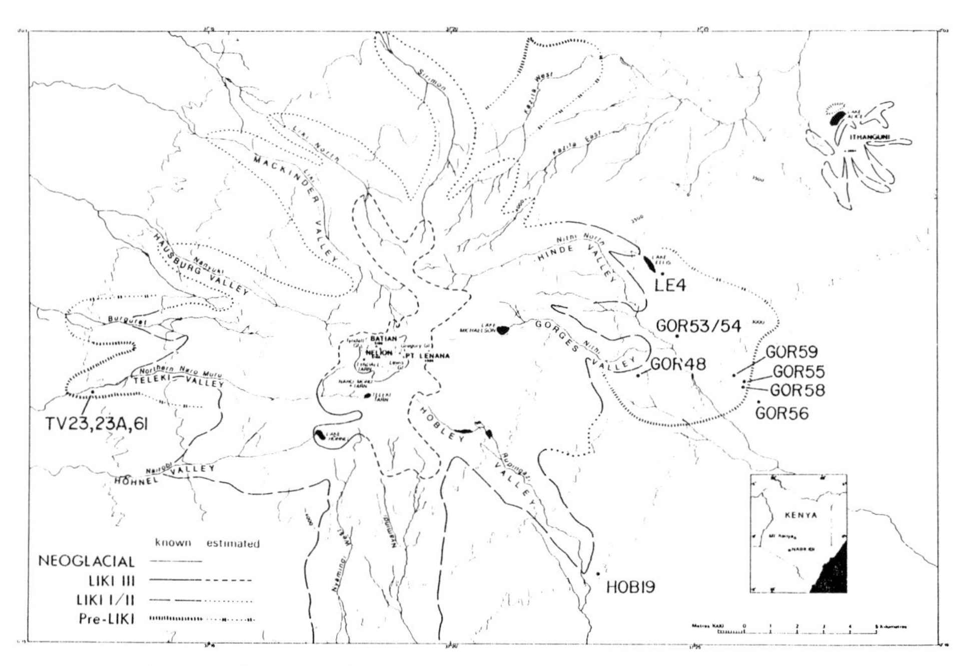

Figure 1. Map showing the spatial extent of Liki and pre-Liki glaciations on Mount Kenya.

1. INTRODUCTION

Five Pleistocene glaciations and two Neoglacial advances are known on Mount Kenya in East Africa (Figs. 1 and 2). The two oldest glaciations occurred prior to the Matuyama/Brunhes boundary (730,000 yrs) extending to ~2850 m on the eastern side of the mountain. The oldest of these deposits, emplaced during the Gorges glaciation, are covered with thin loess sheets which have strong reversed remanent magnetism adjacent to till, becoming normal upwards. Till emplaced during the Lake Ellis glaciation has stong reversed remanent magnetism; however, overlying loess is sometimes reversed and other times normal. Tills documenting the Naro Moru glaciation are found on the western, eastern and southern

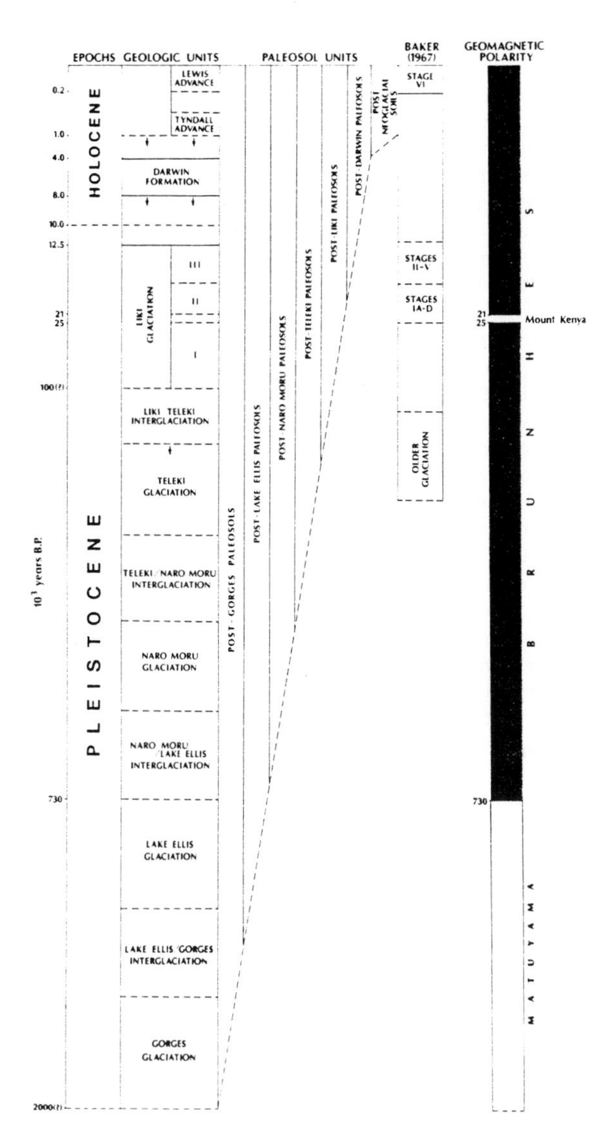

Figure 2. Stratigraphy on Mount Kenya for pre-Teleki, Teleki, Liki and Neoglacial deposits. Unit boundaries are based on maximum and minimum dates from nonglacial sediments. Dashes denote boundaries for which there is no radiometric control. Arrows indicate directions in which boundaries are likely to shift when new dates become available. Geomagnetic polarity events are shown where they provide dating control.

flanks of the mountain between ~2950 and 3150 m, however, they tend to occur at lower elevations on the western flank. The strong normal remanent magnetism in these drifts suggest they date from the lower Brunhes Chron. Clast composition in this older group of deposits assists in unravelling the erosional history of the mountain from the removal, first, of the roof rocks of basaltic composition (Gorges glaciation), and later of nepheline syenite and porphyritic phonolite, from the volcanic plug. Clasts of the plug rock are encountered in drifts emplaced during the Lake Ellis glaciation. Higher percentages of intrusive rock in Naro Moru drifts indicate that erosion of the plug rock was well underway by the early Brunhes Chron. The palaeosols in these older glacial deposits occur both as relict (surface) and buried entities. Collectively they provide important evidence of the totality of weathering (and hence palaeoclimate) over varying time periods during the early and middle Pleistocene.

Following the Naro Moru glaciation, loess with normal remanent magnetism was emplaced and weathered prior to the emplacement of Teleki till (considered equivalent with ocean core stage 6). This till is named from its type locality in Teleki Valley in the wetter Hagenia woodland; on the drier eastern side of the mountain, Teleki till is less weathered and contains thinner surface palaeosols. However, a detailed study of the relative weathering in the sand (250-63μm) fractions of a section (TV23) comprising post-Teleki loess, Teleki till, post-Naro Moru loess plus tephra, and Naro Moru till in Teleki Valley showed that the post-Naro Moru palaeosol was more highly weathered than the overlying post-Teleki unit (Vortisch, Mahaney and Fecher, 1987); the main differences in weathering were attributed to changes in palaeoclimate.

While relatively little is known of the beginning of the Liki I (Würm, Wisconsinan) glaciation, considerably more information is available to document the beginning and the end of the Liki II and III substages. Liki II is known to have started just prior to 25,000 yrs BP when slope instability led to burial of palaeosols in lower Teleki Valley (Mahaney, 1988a). In overlying slope wash deposits a magnetic excursion ^{14}C dated at between ~21,000 and 24,000 yrs BP, is possibly correlative with the Mono Lake excursion of North America (Barendregt and Mahaney, 1988). Moraines deposited during the Liki glaciation extend to ~3200 m, but are best displayed on the western and eastern flanks. Because they lack datable organic materials it is impossible to determine the maximum ages of Liki I or II moraines. However, ^{14}C dates on bog bottom sediments in lower Teleki Valley place the end of the Liki II substage at ~15,000 yrs BP (Mahaney, 1988a). In all valleys ice receded relatively rapidly from 3200 m to circa 4000 m, between circa 15,000 yrs BP and circa 12,500 yrs BP (Mahaney, 1985, 1988a); thereafter during the late glacial several recessional moraines were constructed during stillstand events (Mahaney, 1987a,b). These glaciers either disappeared completely or retreated to their present positions in cirques by the end of the Pleistocene (Fig. 3).

The Holocene on Mount Kenya was comparatively warmer and wetter than today. Contrary to the opinions of some workers (Karlén, 1985; Johansson and Holmgren, 1986; Perrott, 1982), there is no evidence for ice advances in the early and middle Holocene (Mahaney, 1985, 1987a,b). Towards the end of the Holocene, during the Neoglacial, ice advances documented by the Tyndall advance (~1000 yrs BP), and Lewis advance (~100 yrs BP) (Mahaney, 1985) show that ice reached circa 4500 m and remained in local cirque basins.

In this paper we intend to discuss the glacial stratigraphy, palaeoclimatic reconstructions that are possible from terminus elevations, and the newly reconstructed palaeomagnetic chronology.

Figure 3. Central Peaks area from the south showing the ~4100 m moraines in Teleki Valley, Neoglacial moraines below Lewis and Tyndall glaciers, and Batian and Nelian.

2. METHODS

Palaeosol descriptions follow the Soil Survey Staff (1951, 1975) and Birkeland (1984). Samples for radiometric and thermoluminescence age determiniations were handled with great care (TL samples were collected in the dark, following procedures established by Lamothe et al., in Mahaney, 1984). Those samples for which amino acid dates are available were collected according to procedures outlined in Mahaney et al. (1986). Palaeomagnetic samples were analyzed following the standard procedures outlined by Barendregt, (in Mahaney, 1984), and Barendregt and Mahaney (1988), and Fisher (1953).

3. GLACIAL SEQUENCE

3.1 Gorges glaciation

Moraines of Gorges age on Mount Kenya represent the basal till and occur on the eastern flank as low as circa 2850 m. Because they are composed of basaltic clasts without any nepheline syenite and/or porphyritic phonolite they are considered to represent the earliest stage of glacial deposition following the removal of surface lavas on the Mount Kenya volcano. The plug rocks of the volcano have a K/Ar age of 2.64 m y (Baker, 1967) which indicates the lavas eroded by the earliest glaciation probably postdate the Pleistocene/Pliocene boundary. Indeed beyond the limit of Gorges drift, bedrock composed principally of tuff displays well weathered palaeosols 2.5 m thick with rather extensive moorland-like surface A horizons. The parent material (and lower subsoil C horizons) in this palaeosol yield a strong normal remanent magnetism with a reversed overprint indicating the tuff may have been emplaced some time in the upper Gauss Chron (> 2.4 m y) or in the Olduvai subchron.

The magnitude of the Gorges glaciation was considerable given that the mountain was probably circa 1000 m higher than today - similar to the present-day Mount Kilimanjaro - when the Gorges Glaciers filled major valleys on the eastern flank with drift. On the western and southern flanks of the mountain it appears that Gorges drift was removed by younger glacial episodes that emplaced Naro Moru and Teleki tills. Deeply weathered palaeosols, formed in Gorges drifts, attest to the amount of weathering since deposition in the early Quaternary. As with other pre-Liki tills, Gorges drifts have high percentages of ferromagnetic minerals, especially magnetite (Vortisch, Mahaney and Fecher, 1987), which make them ideally suited for palaeomagnetic measurements. All the sites sampled thus far have yielded strong reversed remanent magnetism, often with normal overprints, suggesting to us that these tills were emplaced prior to the Matuyama/Brunhes boundary (730,000 yrs).

Loess caps on these palaeosols also have strong reversed remanent magnetism suggesting emplacement prior to 730,000 yrs ago. Because we avoided the upper loess (root zone affected by bioturbation) it is impossible to assess the degree to which younger airfall sediments might have been added to these palaeosols during the Brunhes Chron. While it is not the intent here to discuss the palaeosols at length, it is important to note that these older palaeosols often have thick (30-50 cm) A horizons, with strong (10YR 2/1 and 1/1) colors, and high (20-30%) organic matter content. Because it is possible to demonstrate some movement of clay, organic matter, and organically complexed Fe and Al in these palaeosols it is possible to assign them to the Alfisol order (Soil Survey Staff, 1975). The thick 'moorland-like' Ah horizons in those palaeosols probably reflect depression of vegetation belts during the last glaciation as well as low microbial numbers in the ericaceous and Hagenia belt palaeosols (Mahaney and Boyer, 1986). These thick "moorland-like" A horizons may be relict features that remain due to the inability of microorganisms to rapidly decompose organic matter (Mahaney and Boyer, 1986).

3.2 Lake Ellis glaciation

The second glaciation on Mount Kenya is known mainly from the Lake Ellis and Hobley Valley areas (Barendregt and Mahaney, 1988). This glaciation emplaced drift composed of clasts of basalt, kenyte and phonolite giving rise to surface and buried palaeosols belonging to the Alfisol order, and displaying relatively thick profiles consisting largely of hematite, gibbsite and halloysite in about the same quantity as in the older post-Gorges palaeosols.

Palaeomagnetic analyses of the Lake Ellis and Hobley Valley tills yielded strong reversed remanent magnetism indicating an age older than 730,000 yrs (Barendregt and Mahaney, 1988). The relatively thin till bodies (~5 m), and closeness to the steeper slopes at ~3100 m on the eastern flank, suggest that while a significant volume of rock was removed during the Lake Ellis Glaciation, relatively little remained behind to form moraines. The presence of clasts of different composition indicate that valley glaciers were excavating lava cover as well as rock from the volcanic plug in the Central Peaks area.

Loess overlying Lake Ellis till has about the same thickness as that overlying Gorges drift (circa 0.5 m) and gives a stongly reversed remanent magnetism. Because these sites are generally higher than those on Gorges drift they are beyond the range of timberline fluctuations and occur on flat-lying plateau areas that are imperfectly drained. As a result they have cemented subsoils (C horizons) where clasts and matrices alike are coated with thick skins of MnO_2 (Derbyshire and Mahaney, 1988). In some cases the upper parts of loess caps in these palaeosols give a strong normal remanent magnetism indicating that some airfall deposition of fine-grained material occurred during the Brunhes Chron.

3.3 Naro Moru glaciation

In lower Teleki Valley (western flank) and in Gorges Valley (eastern flank; Mahaney, 1988a), tills of Naro Moru age occur in juxtaposition to older and younger glacial deposits. In some cases they lie buried beneath extensive slope wash deposits and overlie well-weathered soils in residual regolith. As with Lake Ellis drifts they contain clasts of nepheline syenite and porphyritic phonolite indicating that this was the second glaciation to reach the volcanic plug (presently at an elevation of circa 5100 m). Because these moraines are found as low as 2950 m in Teleki Valley and somewhat higher (3100 m) on the eastern flank we may postulate a certain asymmetry in the form of the ice cap covering the Central Peaks during Naro Moru time. Moreover, this is the first glaciation for which there is any evidence on the northern slopes (in the Kazita West Drainage, Fig. 1)

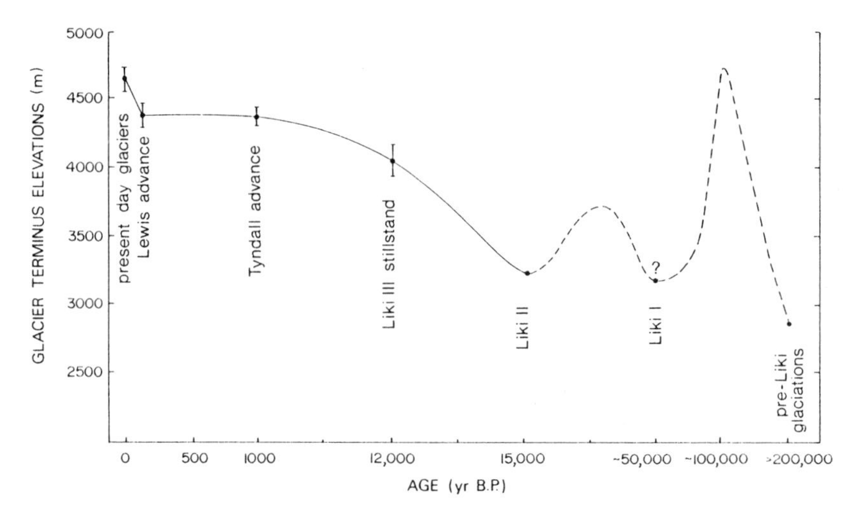

Figure 4. Glacier terminii altitude fluctuations over the last ~200,000 yrs using the mean altitudes of lateral moraines - Curve is based on ^{14}C and TL (solid line) and relative age indicators (dashed line).

Palaeosols in Naro Moru drifts have variable degrees of development depending on whether they occur on the wetter western or drier eastern flanks. On the wetter western flank they are Tropudalfs, relatively thick and well-developed, with a high degree of feldspar dissolution (Vortisch, Mahaney and Fecher, 1987). On the eastern flank they are Fragiboralf's (Soil Survey Staff, 1975) or soils with cemented pans in the subsurface. Loess caps in these palaeosols vary in thickness from 0.3 to 0.6 m, and in some cases, discrete layers of tephra alternate with loess (Mahaney, 1988a; Vortisch, Mahaney and Fecher, 1987).

Palaeomagnetic measurements in till and loess bodies show always normal remanent magnetism indicating that both the till and loess were emplaced during the Brunhes Chron (< 730,000 yrs). Since the tephras associated with the Naro Moru drifts are similar in age with tuff described by Charsley (1987) in the Nanyuki Formation it is likely that the Naro Moru till was emplaced prior to 320,000 ± 20,000 yrs (Charsley, 1988).

3.4 Teleki glaciation

The youngest of the pre-Liki tills is known from sites in Teleki, Gorges and Ruguti drainages on Mount Kenya, and on Ithanguni (Derbyshire and Mahaney, 1988). These tills usually contain a mix of clasts consisting of phonolite, porphyritic phonolite and nepheline syenite. In Teleki Valley at the type section (TV23; Vortisch, Mahaney and Fecher, 1987) the post-Teleki palaeosol is a well weathered Alfisol that bears the full extent of weathering from the last interglacial period on Mount Kenya, as well as the deposition of loess followed by weathering during the Liki glaciation and postglacial period. All palaeosols of post-Teleki age and older contain some halloysite, kaolinite, and gibbsite as the main weathering product (Vortisch, Mahaney and Fecher, 1987); in general the amounts of these clay minerals increase in older palaeosols provided they are exposed at the surface to continual subaerial weathering. For the formation of gibbsite it is important that the palaeosols are relatively permeable and capable of aggressive leaching (Vortisch, Mahaney and Fecher, 1987). Buried palaeosols record, for the most part, weathering that occurred while they were exposed to subaerial processes.

Because the Teleki glaciation was slightly stronger than the full ice age (Liki) advance, tills of Teleki age are prevalent only in a few drainages. Increases in percentages of clasts from the central volcanic plug and decreases in basaltic clasts, indicate that erosion of the main volcanic roof rocks was nearly complete by Teleki time (ocean core stage 6).

3.5 Liki Glaciation

The onset of the Liki Glaciation is probably equivalent with ocean core stage 4 beginning circa 100,000 yrs BP. Most major valleys on Mount Kenya have end and lateral moraine complexes formed during Liki I and II time at circa 3200 m and higher. Moraines are divided into I and II on the basis of topographic position, weathering parameters, and palaeosol cover (Mahaney, 1979, 1988a; Vortisch, Mahaney and Fecher, 1987). Liki II is equivalent to ocean core stage 2, and in some valleys (e.g. Teleki, Hausberg) it is the only vestige of the Liki glaciation.

Attempts at dating bog-bottom sediments from Liki II moraine surfaces gave minimum ages of ~15,000 yrs BP (Mahaney, 1988a). Attempts at dating these moraines directly have been fruitless owing to the scarcity of organic materials in them. However, the onset of the Liki II glaciation occurred circa 25,000 yrs BP, as documented from buried palaeosols in lower Teleki Valley. At site TV61 (Barendregt and Mahaney, 1988) a series of buried palaeosols (Fig. 5) provide evidence for burial of a thick "moorland-like" A horizon sequence circa 0.5 m thick at circa 25,000 yrs BP. The pollen from this buried palaeosol indicates a cooler climate than today but warmer than the palaeoclimate that existed during most of the Liki II substage.

Following the end of the Liki II substage ice receded upvalley at a fast rate so that by 12,500 yrs BP most valley glaciers had their terminii at circa 4000 m (Fig. 1; Mahaney 1987a, 1988a,b). The available pollen sequence for the period 15,000-12,000 yrs BP (Coetzee, 1967; Mahaney, 1985) shows a progression from colder to warmer to cold temperatures during the late glacial (Liki III) substage. After the end of the Liki III recessional stillstand, ice in most valley either disappeared completely, or receded into major cirques giving a distribution similar to the present (Fig. 3). There is no evidence for early to middle Holocene ice advances despite the interpretations of 4 cores from Naro Moru Tarn (Johansson and Holmgren, 1985), Thompson Tarn (Perrott, 1982) and Hausberg and Oblong tarns (Karlén, 1985). Data gathered from 33 cores recovered from major bogs and lakes in many valleys shows conclusively that the ~4000 m end moraines have later Pleistocene ages (Mahaney, 1987a,b).

Figure 5. Palaeomagnetic sampling stations in a slope wash deposit at site TV61, Teleki Valley, Mount Kenya (Fig. 1 for location).

3.6 Neoglaciation

Following the climatic optimum on Mount Kenya - circa 2000 yrs BP - pollen evidence indicates colder and drier climate (pollen zone Z). During the last thousand years the record shows two ice advances (Mahaney, 1985): first, the Tyndall advance at approx. 1000 yrs BP, and second, the Lewis advance which is undated but considered equivalent to the classical Little Ice Age.

The Tyndall advance is known from moraines, outwash, and debris flows in front of the Lewis and Tyndall glaciers (Mahaney, 1987, 1988a,b) which radiocarbon date at circa 900-950 yrs BP. Because these deposits are not found in other cirques on Mount Kenya it is likely that ice resurgence was confined to the southwestern flank of the Central peaks area. Soils on the Tyndall deposits are thin Entisols with A/Cox profiles making them significantly different from the palaeosols on nearby Liki III moraines, which are Inceptisols with cambic (color) B horizons and greater depth. Lichens also provide a means of separating Liki III deposits from Tyndall deposits, as lichen cover in general, and maximum diameters of *Rhizocarpon geographicum* are radically different. Lichen cover (including the genera *Usnea*, *Umbilicaria*, *Buellia*, and *Rhizocarpon*) ranges from 50 to ~80 per cent on the older Liki III deposits, while on Tyndall-age surfaces it is usually 30-50 per cent.

The Lewis advance, which was still evident when J.W. Gregory visited the mountain in 1893 (Gregory, 1900) emplaced about 90 per cent of Neoglacial surficial materials that are presently exposed in the Central Peak area (Fig. 3). Moraines are bouldery, often with an open network of stones lacking matrix materials, and practically devoid of soil cover. Where present, soils form only 10% of the surface cover, and to a depth of circa 10-15 cm - C/Cu horizon sequences (Mahaney, 1979, 1988a).

4. PALEOCLIMATIC RECONSTRUCTION

Palaeoclimate can be recontructed from the terminus elevations of moraines of various ages. For present-day glaciers ice fronts are at 4650 m ±100 m (Fig. 4), which is about 250 m higher than the front positions during the two-stage Neoglacial. From the landform evidence (Mahaney, 1985), the earliest Neoglacial moraines, emplaced during the Tyndall advance lie farther out on the cirque floors, and approximately at the same altitude as moraines emplaced during the Lewis advance (classical Little Ice Age). Because the older Neoglacial advance occurred only on the southwestern flank of the mountain, we consider that microclimatic changes, induced primarily by changes in localized cloud cover (perhaps produced by local anabatic winds), might explain the local advances. The palaeoclimatic change that produced the Lewis advance was widespread, affecting every cirque on the mountain, including three cirques on the eastern flank.

During the late glacial the terminus elevations were at ~4100 ±150 m (Fig. 4) as determined from recessional moraine positions. Even though some of these moraines are considered to be of Holocene age, there is no

Figure 6. Palaeomagnetic sampling stations in outwash at site GOR54, Gorges Valley, Mount Kenya (Fig. 1 for location).

evidence to support moraine emplacement over preweathered surfaces, nor is there any evidence from the moraines in support of a middle Holocene age (Mahaney, 1987a). The spatial extent of late glacial (Liki III) ice as shown in Fig. 1 is considered to result from a halt in the recession and decay of the main Liki valley glaciers. The time of this stillstand is fixed at ~12,500 yrs BP (Mahaney, 1988a,b) plus or minus approximately 1000 yrs. Palaeoclimate as reconstructed from pollen evidence (Coetzee, 1967; Mahaney, 1985) confirms a colder and drier climate during the late glacial, although not as cold as during the Liki II glaciation.

Liki II equilibrium line altitudes are difficult to work out for all valleys owing to the lack of well defined lateral moraines. Taking the terminii of several major glaciers the equilibrium line altitudes were undoubtedly higher than 3200 m (mean snout positions shown in Fig. 4), and probably at ~3700 m. Liki I terminus elevations, while slightly lower than Liki II end moraines, are considered to have formed from glaciers which had essentially the same equilibrium line altitudes. The available pollen evidence from Sacred Lake, on the northeastern flank of Mount Kenya (Coetzee, 1967), shows colder and drier climate during the Liki II period and warmer climate during the deposition of older deposits considered equivalent in age with the Kalambo Interstadial (>25,000 yrs BP; Coetzee, 1967).

Pre-Liki glaciations produced terminus elevations as low as 2850 m (somewhat lower on the eastern flanks of the mountain) that reflect the major source of moisture derived from the Indian Ocean monsoon. Just as the northern and northeastern sides of the mountain are dry today, during the pre-Liki glaciations very little drift was emplaced on the northern flank. Even during the Liki glaciation ice terminii in the Sirimon Drainage reached only to ~3400 m and ice in the Kazita Valley was no lower than ~3500 m.

There is considerable evidence for an ice cap on Mount Kenya during the Liki glaciation including polished bedrock pavement on Point Lenana, extensive lateral moraines to ~4100 m on the eastern flank, as well as striae on bedrock interfluves at ~3600-3800 m between the Gorges and Hobley valleys. Because the Two Tarn area in the upper Burguret Drainage was ice free during the Liki glaciation, the Teleki and Burguret glaciers appear to have been separated so that they operated more or less as individual valley glaciers. This gives rise to the hypothesis that the ice cap flowed mainly to the north, east and south with the strongest flow to the east and southeast - rather thicker on the east and southeastern sides and somewhat thinner on part of the western and northern sides.

5. PALEOMAGNETIC CHRONOLOGY

5.1 General remarks

Palaeomagnetism has proven to be a valuable tool for the differentiation of sediments which were deposited during the early Pleistocene, and those which were deposited during the middle and late Pleistocene. The boundary between these two groups of sediments falls between 735 and 790 ka (Johnson, 1982) and separates the normally magnetized Brunhes Chron from the reversely magnetized Matuyama Chron. Although polarity subchrons lasting from 20 to 200 ka occur within the Brunhes and Matuyama, most sequences of glacial deposits reveal stratigraphic evidence of multiple glacial and interglacial events spanning long periods of time. The glacial deposits sampled on Mount Kenya appear to fall into two major groups, each representing a substantial part of a polarity chron. The periods of time represented by these deposits far exceeds the estimated temporal extent of known excursions (subchrons).

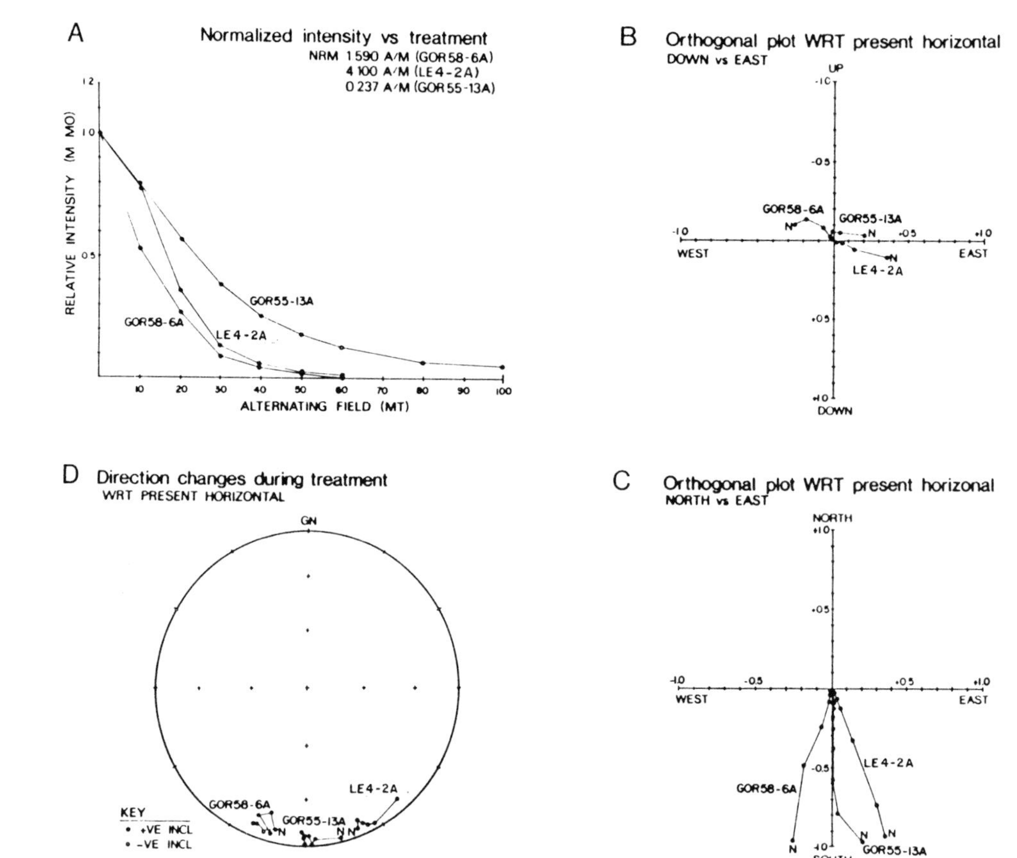

Figure 7. Typical demagnetization curves for samples GOR58-6A, LE4-2A and GOR 55-13A. [A] Ratio of normalized intensity (M/MO) of natural remanent magnetization (N.R.M.) to residual moment after progressive alternating field demagnetization (a.f.d.) at levels of 10, 20, 30, 40, 50, 60, 80 and 100 millitesla (mT). Magnetization values have been normalized to the highest intensity and are thus unitless. Zijderveld plots showing orthogonal projections of the changing resultant magnetization vector with respect to present horizontal: Down vs East [B], North vs East [C], and a stereoplot showing directional changes during treatment [D]. The samples are reversely magnetized.

The deposits found at lower elevations (~2850 m) were laid down by stronger glaciations (the Gorges and Lake Ellis glaciations) and reveal a reversed magnetic polarity while deposits found at the higher elevations (~3000–3200 m) represent weaker glaciations and reveal normal polarities.

These palaeomagnetic results provide the first indication of the age of the Gorges and Lake Ellis glaciations. Being reversely magnetized, they must be older than 735–790 ka, which is the age of the Brunhes–Matuyama boundary.

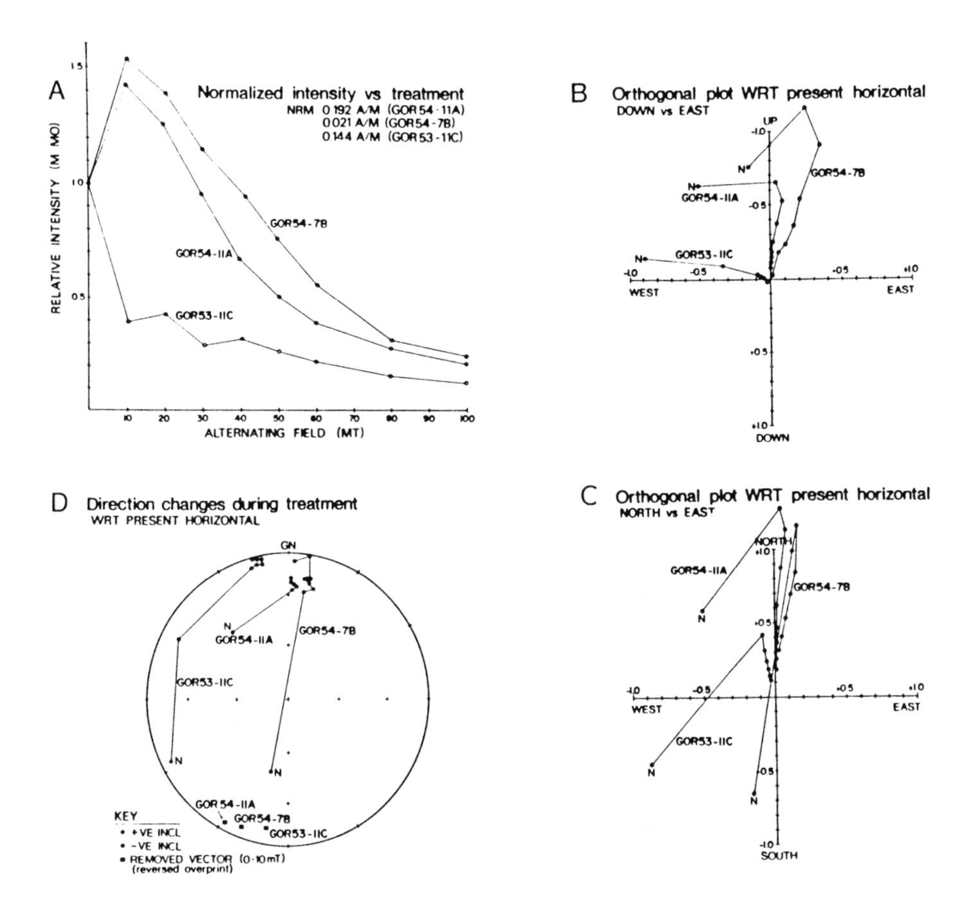

Figure 8. Typical demagnetization curves for samples GOR54-7B, and GOR53-11C. Description of [A] to [D] as in Fig. 3. The samples are normally magnetized and show a reversed overprint that is completely removed with 10mT demagnetization.

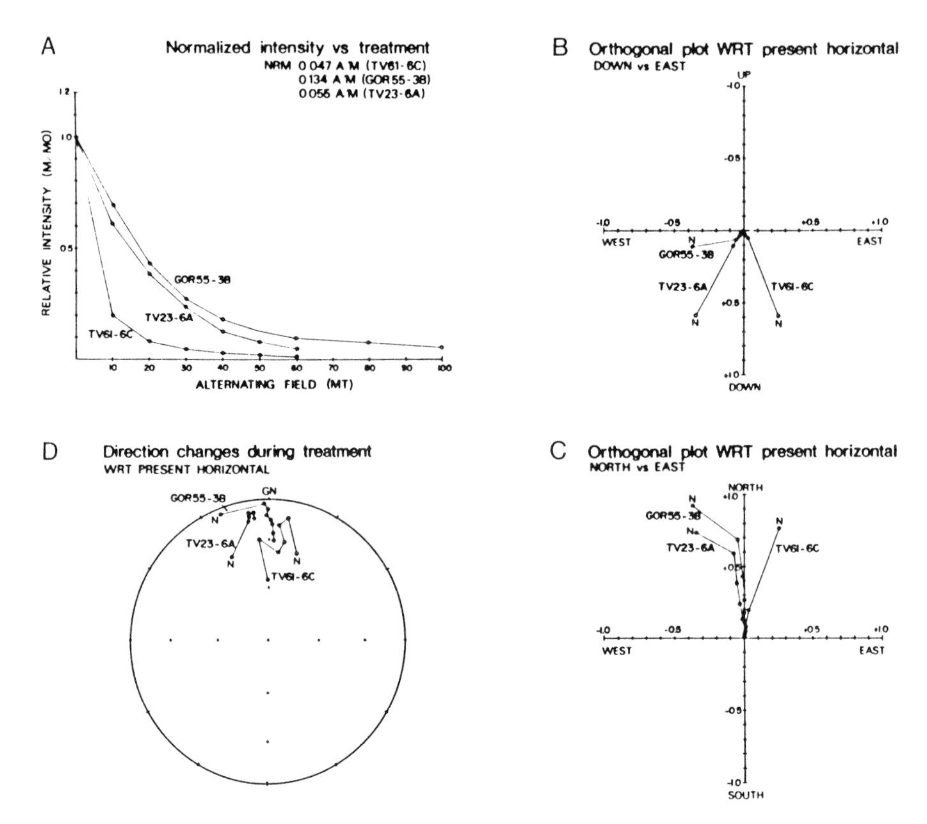

Figure 9. Typical demagnetization curves for samples TV61-6C, GOR55-3B and TV23-6A. Description of [A] to [D] as in Fig. 3. The samples are normally magnetized.

5.2 Sampling sites

During the summers of 1984 and 1986, twelve sites (Figs. 5 and 6, for example), which contained suitable materials for palaeomagnetic sampling, were located and sampled. A total of 432 samples were collected and analyzed. As the method of sampling was restricted to the use of separately oriented plastic cubes, only soft sediments were used. Sediments substantially influenced by root systems (or with krotovinas) or containing abundant coarse clastic material were avoided. The sampling sites were located in three of the main valleys: Gorges Valley (7 sites), Teleki Valley (3 sites), and Hobley Valley (1 site).

As these samples are composed of soft unconsolidated sediments with some moisture, it is impossible to thermally demagnetize them. Pilot specimens were selected from each of the sites and subjected to stepwise alternating field (a.f.) demagnetization at levels of 10, 20, 30, 40,

Table I

Table 1 Stratigraphy and Remanent Magnetization Directions[a]

Site[e]	Thickness (meters)	Lithology[c]	Age[d]	Polarity	N	M	D	I	K	α95	"A" Pole[b]	
											Lat.	Long.
TV23	4.4	sequence of tills and loesses	Late Pleistocene(?)	normal (Brunhes)	39	0.04	353	+08	71	03	+82	-021
TV23A	3.2	pebbly till overlain with loess	Late Pleistocene(?)	normal (Brunhes)	16	0.01	006	+00	11	12	+84	+124
TV61	0.2	loess	Late Pleistocene(?)	excursion(?) (Brunhes)	08	0.02	160	+40	02	63	-60	+076
	6.0	sequence of weathered bedrock, till, loess, colluvium, and slopewash	Late Pleistocene(?)	normal (Brunhes)	56	0.01	004	+16	150	03	+81	+064
LE4	1.3	stony till, weathered; covered with outwash and loess	Early Pleistocene(?)	reversed (Matuyama)	18	2.83	162	+10	27	07	-72	+112
GOR48	1.4	pebbly till, weathered, partly cemented	Early Pleistocene(?)	normal (Olduvai?)	33	4.25	002	-01	16	07	+88	+131
GOR55	0.9	loess	Late Pleistocene(?)	normal (Brunhes)	13	0.13	[illegible]57	+02	186	03	+82	-031
	1.0	pebbly till, weathered	Early Pleistocene(?)	reversed (Matuyama)	07	0.16	179	-05	112	06	-87	-168
GOR58	1.7	pebbly till; overlain with loess	Early Pleistocene(?)	reversed (Matuyama)	09	3.5	173	+08	07	20	-82	+100
GOR59	1.8	pebbly till; weathered, partially cemented	Early Pleistocene(?)	normal (Olduvai?)	21	0.03	352	-03	12	10	+82	-060
HOB19	0.5	loess	Late Pleistocene(?)	normal (Brunhes)	08	5.12	358	-23	132	5	+78	-135
	0.8	pebbly till, weathered, partially cemented	Early Pleistocene(?)	reversed (Matuyama)	04	0.67	188	-07	221	6	-81	-076
GOR53	1.8	fine-grained colluvium and slopewash	Late Pliocene/ Early Pleistocene(?)	normal (Gauss)	64	0.02	360	+04	78	02	+88	+034
				with reversed overprint OVERPRINT (Matuyama)	26	1.00	226	-01	62	9	-44	-054
GOR54	1.6	sandy outwash	Late Pliocene/ Early Pleistocene(?)	normal (Gauss)	32	0.03	001	-09	09	09	+86	-154
				with reversed overprint OVERPRINT (Matuyama)	23	0.24	193	+01	17	10	-77	-052
GOR56	1.8	coarse-grained tuff	Late Pleistocene(?)	normal (Gauss)	17	0.10	352	-06	25	07	+82	-074

Footnotes to Table I:

(a) Mean location of Mt. Kenya sampling sites: -0.3° Latitude; 37.4° Longitude.
(b) "A" Pole as defined by Irving, 1964.
(c) Since the authors were primarily interested in the age of glaciations, most of the samples wer collected from till facies.
(d) Late Pleistocene ages are based on polarity data, ^{14}C, thermoluminescence, amino acid racemization ratios, degree of weathering and soil formation, and topographic position.
Early Pleistocene ages are based on polarity data, K/Ar, degree of weathering and soil formation, and topographical position.
(e) Sites exhibiting both normal and reversed polarities are listed as such (TV61, GOR55, and HOB19).

Legend to Table I:

N = number of samples per site
M = average intensity of magnetization (Am^{-1})
D = average declination (degrees)
I = average inclination (degrees)
K = precision parameter (Fisher, 1953)
$\alpha 95$ = radius of the 95% confidence circle about the mean direction (degrees)

50, 60, 80 and 100 mT. Qualitative assessment of changes in magnetic vector directions and magnetization values was used in the selection of appropriate cleaning fields. Typical demagnetization plots are shown in Figs. 7-9. On the basis of the low median destructive field of these samples, it would appear that detrital magnetite is the predominant remanence carrier [magnetite is one of the principal components of the tills on Mount Kenya (Vortisch, Mahaney and Fecher, 1987)].

5.3 Correlation of the stratigraphy and palaeomagnetic data.

A generalized cross-section showing the major geologic units (Fig. 2) depicts the proposed stratigraphic correlations of the sampling sites. Table I summarizes the lithologies of the sites sampled and provides the remanent magnetization directions of these sites. Table II and Fig. 10 provide a summary of mean pole positions for all sites.

Of the twelve sites sampled, five were normally magnetized two were reversely magnetized, two showed reversely magnetized sediments overlain by normally magnetized sediments and two showed normal magnetization with reversed overprinting. One site revealed a normal sequence containing a possible excursion (Table I). For the most part, the tills found at lower elevations record a reversed polarity of the earth's field. Exceptions are sites GOR 48 and GOR 59. It is thought that these tills may have been deposited during one of the Matuyama subchrons, when the field was normal. Which one is difficult to say, but since the Olduvai lasted some 200 ka the probability of locating till deposited during this subchron is far greater than the other likely candidates (Jaramillo or Réunion) which lasted only 20 ka or so.

Table II. Mean Poles (Mount Kenya)

	N	M	D	I	K	α95	Pole Lat.	Pole Long.
[c]Model A								
Mean N [all]	10	1.00	359	-1	49	7	88	297
[a]Mean R [all]	7	1.20	183	7	9	21	-85	349
Mean N & R	17	1.07	360	-2	18	9	89	189
[b]Mean R/	3	1.20	176	-1	25	25	-86	138
Mean N & R/	13	1.03	358	-1	46	6	88	305
[c]Model B								
Mean N [all]	10	--	--	--	117	5	88	300
[a]Mean R [all]	7	--	--	--	12	19	-85	359
Mean N & R	17	--	--	--	26	7	88	216
[b]Mean R/	3	--	--	--	35	21	-81	136
Mean N & R/	13	--	--	--	89	4	88	306

Where

N = number of sites
M = average intensity of magnetization (AM^{-1})
D = average declination (degrees)
I = average inclination (degrees)
K = precision parameter
α95 = radius of the 95% confidence circle about the mean direction (in degrees)

[a]Mean R includes excursion and α95 > 15 degrees.
[b]Mean R/ excludes excursion and α95 > 15 degrees.
[c]Model A Pole and Model B Pole as defined by Irving (1964).

Two sites (GOR 53 and GOR 54) provide evidence for the earliest glaciation to affect Mount Kenya. These deposits record a normal polarity with a reversed overprint, suggesting deposition during the late Gauss Chron with an early Matuyama overprinting. Alternatively, they may have been deposited during the Olduvai subchron. The nature of the deposits, namely fine-grained colluvium and slope wash, and sandy outwash, suggest that the ice was nearby, but was not directly depositing materials at these sites. Based on pedologic and topographic criteria, these sites occur at or just beyond the limits of glaciation and may well record the first glaciation on Mount Kenya. Previous workers have suggested that climatic deterioration, leading to glaciation would have been felt first and foremost in high alpine areas. If this assumption is correct, an early glaciation on Mount Kenya occurring at the time of the Gauss/Matuyama boundary is conceivable.

Further work will be carried out to attempt a refinement of the palaeomagnetic data collected thus far. Additional sampling near site TV61 and elsewhere will hopefully provide information leading to verification or rejection of a possible excursion found in sediments dated around 24,000 yrs BP.

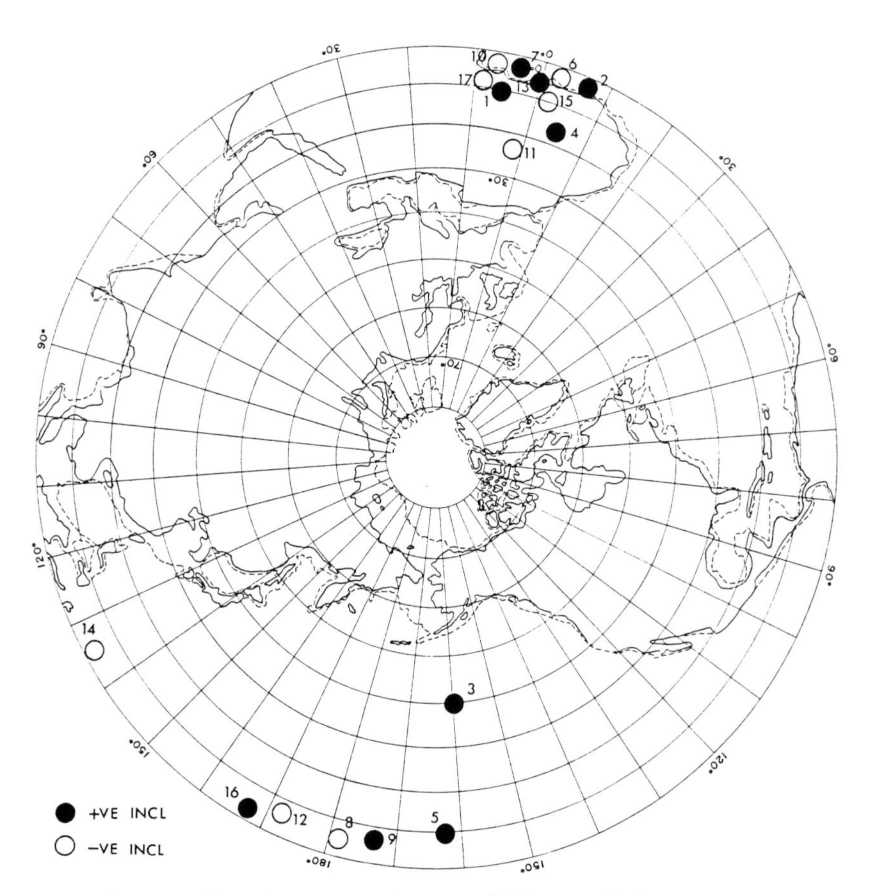

Figure 10. Summary of mean "A" and "B" poles for Mount Kenya sediments.

Legend -	1 - TV23	7 - GOR58
	2 - TV23A	8 - GOR59
	3 - TV61	9 - HOB19
	4 - LE4	10 - GOR53
	5 - GOR48	11 - GOR54
	6 - GOR55	12 - GOR56

6. CONCLUSIONS

Five Pleistocene and two late Holocene glacial depositional units are known on Mount Kenya. These stratigraphic units have been dated using a variety of relative and absolute methods which provide reasonably precise ages. In the high valleys, above 4400 m, glacial deposits of Neoglacial age belong to the late Holocene; no middle Holocene tills are known on Mount Kenya. Ice advances during the Liki glaciation (= Wisconsinan) presumably began < 100,000 yrs BP, ending ~15,000 yrs BP and reached 3200 m a.s.l. Tills older than ocean core stage 5 include Teleki and Naro Moru drifts with strong normal remanent magnetism, and Lake Ellis and Gorges deposits with strong reversed remanent magnetism. Tills have high amounts of ferromagnetic minerals making them excellent carriers of remanent magnetism. Palaeosols formed in these older glacial deposits occur as both relict (surface) and buried entities which provide important data on weathering and palaeoclimate over varying time intervals from the early to late Pleistocene.

ACKNOWLEDGEMENTS

This research was funded by grants from the National Geographic Society, Natural Sciences and Engineering Research Council of Canada, German Science Foundation and York University. This research was carried out with the permission of the office of the President, Republic of Kenya and with the cooperation of Mountain National Parks. We gratefully acknowledge the assistance of Tim Charsley (U.K. Geological Survey) especially with regard to discussions on the extent of Gorges drifts on Mount Kenya. An anonymous reviewer provided several helpful and constructive criticisms of the original manuscript.

REFERENCES

Baker, B.H. 1967. Geology of the Mount Kenya Area. Geological report no. 79, 78 pp.

Barendregt, R.W. 1984. Using paleomagnetic remanence and magnetic susceptibility data for the differentiation, relative correlation and absolute dating of Quaternary sediments. In: W.C. Mahaney (ed.) "Quaternary dating methods". Amsterdam, Elsevier, 101-122.

Barendregt, R.W. and Mahaney, W.C. 1988. Reconnaissance paleomagnetic determinations in Liki and pre-Liki sediments on Mount Kenya, East Africa. In: W.C. Mahaney (ed.), "Quaternary and environmental research on East African mountains". Rotterdam, Balkema, 355-366.

Birkeland, P.W. 1984. Soils and Geomorphology. N.Y., Oxford Univ. Press, 372 pp.

Charsley, T.J. 1988. Composition and age of older outwash deposits along the northwestern flank of Mount Kenya. In: W.C. Mahaney (ed.), "Quaternary and environment research on East African mountains". Rotterdam, Balkema, 165-174.

Coetzee, J.A. 1967. Pollen analytical sties in East and soutern Africa. Palaeoecology of Africa 3, 146 pp.

Derbyshire, E. and Mahaney, W.C. 1988. A preliminary scanning electron microscope study of five diamictons from Mount Kenya. In: W.C. Mahaney (ed.), "Quaternary and environmental research on East African mountains". Rotterdam, Balkema, 155-164.

Fisher, R.A. 1953. Dispersion on sphere. Proc. of the Royal Society, London A217, 295-305.

Gregory, J.W. 1900. The geology of Mount Kenya. Quart. Jour. Geol. Soc. 56, 205-222.

Irving, E. 1964. Paleomagnetism and its applications to geological and geophysical problems. N.Y., Wiley, 399 pp.

Johansson, L. and Holmgren, K. 1985. Dating of a moraine on Mount Kenya. Geografiska Annaler 67A, 123-128.

Johnson, R.G. 1982. Brunhes-Matuyama magnetic reversal dated at 730,000 yrs BP by marine-astronomical correlations. Quaternary Research 17, 135-147.

Karlén, W. 1985. Glacier and climate fluctuations on Mount Kenya, East Africa. Zeitschrift für Gletscherkunde und Glazialgeologie 21, 195-201.

Lamoth, M., Dreimanis, A.D., Morency, M. and Raukas, A. 1984. In: W.C. Mahaney (ed.), "Quaternary dating methods". Amsterdam, Elsevier, 153-170.

Mahaney, W.C. 1979. Reconnaissance quaternary stratigraphy of Mount Kenya, East Africa. In: E.M. van Zinderen-Bakker and J.A. Coetzee (eds.) "Palaeoecology of Africa" 10, 163-170.

Mahaney, W.C. 1985. Late glacial and Holocene paleoclimate of Mount Kenya, East Africa. Zeitschrif für Gletscherkunde und Glazialgeologie 21, 230-211.

Mahaney, W.C. 1987a. Reinterpretation of dated moraines at 4000 m in the Mount Kenya Afroalpine area. Palaeogeography, Palaeoclimatology, Palaeoecology 60, 47-57.

Mahaney, W.C. 1987b. Dating of a moraine on Mount Kenya: Discussion. Geografiska Annaler 69A (2), 359-363.

Mahaney, W.C. 1988a. Quaternary glacial geology of Mount Kenya. In: W.C. Mahaney (ed.) "Quaternary and environment research on East African Mountains". Rotterdam, Balkema, 121-140.

Mahaney, W.C. 1988b. Glacial advances in the middle Holocene on Mount Kenya: fact or fiction. In: W.C. Mahaney (ed.), "Quaternary and environment research on East African Mountains". Rotterdam, Balkema, 141-154.

Mahaney, W.C. and Boyer, M.G. 1986. Microflora distributions in paleosols: a method for calculating the validity of radiocarbon-dated surfaces. Soil science 142 (2), 100-107.

Mahaney, W.C., Boyer, M.G. and Rutter, N.W., 1986. Evaluation of amino acid composition as a geochronometer in buried soils on Mount Kenya, East Africa. Géographie Physique et Quaternaire 42 (2), 171-183.

Perrott, R.A. 1982. A high altitude pollen diagram from Mount Kenya: its implications in the history of glaciation. In: E.M. van Zinderen-Bakker and J.A. Coetzee (eds.), "Palaeoecology of Africa" 14, 77-85.

Soil Survey Staff 1951. Soil Survey Manual. Washington U.S. Govt. Printing Office, 503 pp.

Soil Survey Staff 1975. Soil Taxonomy, Agriculture Handbook 436. Washington U.S.D.A., 754 pp.

Vortisch, W.B., Mahaney, W.C. and Fecher, K. 1987. Lithology and weathering in a paleosol sequence on Mount Kenya, East Africa. Geologica et Paleontologica 21, 245-255.

TEPHROCHRONOLOGICAL STUDIES OF HOLOCENE GLACIER FLUCTUATIONS IN SOUTH ICELAND

A.J. Dugmore[1]
Dept. of Geography, University of Aberdeen
Aberdeen AB9 2UF, Scotland

ABSTRACT

Stratigraphic studies of tephra layers interbedded with soils have been used to date accurately Holocene glacier fluctuations in southern Iceland. 132 sections up to 11 m deep, and containing up to 78 tephra layers, were logged to a resolution of 0.25 cm. The chronological framework was completed with 15 radiocarbon dates, and by examining the association of the tephra stratigraphy with moraines representing former ice margins, a chronology of Holocene glacier fluctuations was constructed. The forelands of five non-surging glaciers were studied: Seljavallajökull, Gigjökull and Steinholtsjökull (outlets of Eyjafjallajökkull) and Sólheimajökull and Klifurárjökull (outlets of Mýrdalsjökull).

This study has shown for the first time that a large ice mass existed in mid-Holocene Iceland, because after 7000 BP and before 4500 BP Sólheimajökull extended up to 5 km beyond its present limits. Major advances also culminated before 3100 BP, and between 1400-1200 BP. In the tenth century AD the glacier was also longer than during the Little Ice Age (1600-1900 AD). In contrast, Klifurárjökull and all the outlets of Eyjafjallajökull reached a maximum Holocene extent during the Little Ice Age. The anomalous behaviour of Sólheimajökull is probably a result of catchment changes caused by the development of the Mýrdalsjökull ice cap over the last 5000 years. This study highlights the great value of apparently anomalous glacier behaviour to studies of climatic change, but it draws attention to the caution that should be used when interpreting the climatic significance of patterns of glacier fluctuations deduced from glacial deposits.

1. INTRODUCTION

Knowledge of Holocene glacier fluctuations in Iceland is important because the island is strategically located in a region where climate is strongly influenced by movements of the oceanic Polar Front.

[1] Now at the Dept. of Geography, University of Edinburgh, Drummond Street, Edinburgh EH8 9XP, U.K.

J. Oerlemans (ed.), Glacier Fluctuations and Climatic Change, 37–55.

Consequently, even comparatively minor variations in the position of this important feature of the global climatic system are likely to be reflected in significant fluctuations of Icelandic glaciers. At present, however, very little is known about the pattern of Holocene glacier fluctuations in Iceland before the Little Ice Age (1600-1900 AD). It is commonly supposed that the major icecaps disappeared c. 8000 BP not to reform until c. 2500 BP, at which time it has been suggested that responsive mountain glaciers were able to react rapidly to deteriorating climate and reach a Holocene maximum extent (Thórarinsson, 1956, 1964; Björnsson, 1979). Pre-Little Ice Age moraines exist around Öraefajökull (Thórarinsson, 1956) and Snaefell (Thórarinsson, 1964), but at present the formation times of these moraines have not been bracketed by close limiting dates. The lowlying outlet glaciers of the Icelandic ice caps are assumed to have reached a Holocene maximum extent in recent centuries; an idea which is supported by firm stratigraphic evidence from Hagafellsjökull Eystri (Thórarinsson, 1966), Skalafellsjökull and Eybakkajökull (Sharp, 1982; Sharp and Dugmore, 1985). In order to test these ideas, and consequently improve knowledge of pre-Little Ice Age glacier history, geomorphological and stratigraphic investigations were undertaken in the forelands of five non-surging glaciers in south Iceland.

2. STUDY AREA

A geographically diverse group of five glaciers were selected for study around the Eyjafjallajökull and Mýrdalsjökull ice caps (Fig. 1). The glacial equilibrium line altitude (ELA) on Eyjafjallajökull and neighbouring parts of Mýrdalsjökull lies at c. 1100 m.

Gigjökull and Steinholtsjökull both originate at about 1600 m close to the summit of Eyjafjallajökull, and both descend about 1300 m to reach the northern foot of the mountain. On Gigjökull the area around the ELA is narrow and confined by rockwalls; in contrast on Steinholtsjökull the ELA crosses a wide part of the glacier. The average gradients of both Gigjökull and Steinholtsjökull are about 1:4.5, though Gigjökull has a rather uneven long-profile. Where its upper reaches flow across the summit of Eyjafjallajökull the gradient is less than 1:25, but central parts slope at over 1:2.5. Gigjökull terminates at the edge of a sandur where a large terminal moraine up to 50-70 m high impounds a small pro-glacial lake. In contrast Steinholtsjökull terminates in a narrow valley, and several terminal moraines, just a few metres high mark recent positions of the ice margin. Landforms on the valley floor in front of Steinholtsjökull have been extensively modified as a result of a rockfall that swept down the glacier in 1967, but this event did not affect the ridge where this study was undertaken (Kjartansson, 1967).

The lobate glacier Seljavallajökull also originates close to the summit of Eyjafjallajökull, but this glacier terminates midway down the southern flanks of the mountain at an altitude of 700-800 m (Fig. 1). The surface gradient of Seljavallajökull is steeper than the two northern glaciers and averages 1:3.75. Below the present ice margin, series of subdued terminal moraines lie within a clearly defined limit that marks the recent maximum southern extent of the ice cap.

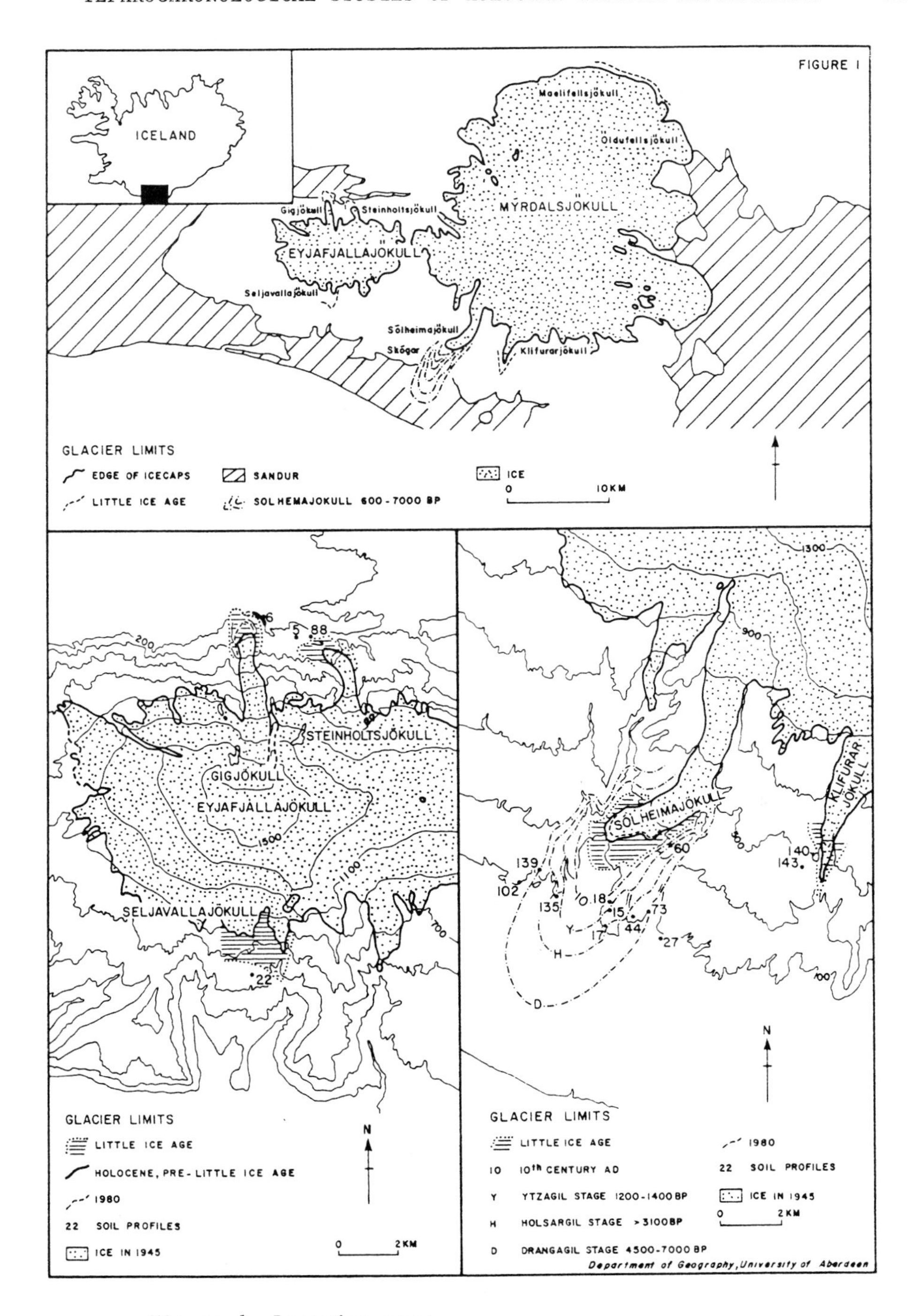

Figure 1. Location maps

Sólheimajökull and Klifurárjökull are both southern outlets of Mýrdalsjökull (Fig. 1). This ice cap covers the comparatively low lying massif formed by the volcano Katla; the highest parts of the bedrock rise to over 1300 m, but most lies below 1000 m (Björnsson, 1978). Sólheimajökull is over 15 km long and has a longitudinal profile characterized by long shallow gradients, typically 1:25. Moraines occur on the broad valley floor in front of Sólheimajökull where they form arcuate ridges rarely more than a few metres high. Discontinuous glacial deposits cover extensive areas on the uplands in front of Sólheimajökull, and discrete, long, low morainic ridges occur to the south and east of the glacier. Klifurárjökull is about 9 km long and has an average slope of c. 1:16. It terminates in a gorge, and in contrast to Sólheimajökull it is surrounded by only a restricted area of recent moraines.

3. METHODS

Geomorphological studies were to establish the location of former Holocene ice marginal positions and produce a spatial framework for the stratigraphical studies. The identification and accurate mapping of lateral and terminal moraines, and other ice marginal landforms was therefore of great importance. Special attention was also paid to the morphostratigraphy created by jökulhlaups, because these catastrophic flood events have the power to modify, or completely destroy pre-existing landforms and sedimentary sequences (Jónsson, 1982). As a result of its relation to particular ice margins, jökulhlaup morphostratigraphy may also supplement the incomplete data contained in the morainic record, and its very existence may indicate that glaciers formerly existed even when little direct evidence remains. Detailed geomorphological maps of the five pro-glacial areas considered in this paper are presented in Dugmore (1987).

The ages of glacial landforms and jökulhlaup deposits were determined through the application of tephrochronology, a dating technique based on the identification, correlation and dating of tephra layers (Thórarinsson, 1981). Tephra is a Greek work that, following the proposal by Thórarinsson in 1944, has been generally adopted as a collective term for all pyroclasts from volcanic eruptions, including both airfall and pyroclastic flow material. This usage complements rather than replaces terms such as ignimbrite, welded tuff, pumice, ash, lapilli or volcanic bombs that are used to designate specific types of tephra produced by distinctive types of eruptions (Thórarinsson, 1974). The tephra stratigraphy in complete Holocene soil profiles at sites outside the area affected by any conceivable recent glacial activity was studied in detail (Dugmore, 1987). Tephra layers of historical age (post c. 900 AD) were identified by reference to the works of Thórarinsson (1967, 1975), Larsen (1979, 1981, 1984), and Einarsson et al. (1980). Key pre-historic tephra layers were dated through the application of radiocarbon analyses (Table I). The complete Holocene tephra stratigraphy was traced between closely spaced profiles into the areas affected by recent glacial activity, and, by examining the association of the tephra stratigraphy with moraines and tills, a chronology of glacier fluctuations was constructed.

Table I. Radiocarbon dates on pre-historic tephra layers, and profiles 22 and 44.

Sample no.	Date (^{14}C yr BP)	$\delta^{13}C$ Value ‰	Material	Overlying tephra layer
SSR-2802 (a)	7480 ± 60	-27.0	fine peat (< 1 mm)	layer S
SSR-2802 (b)	7530 ± 60	-26.6	coarse peat (> 1 mm)	layer S
SSR-2802 (c)	7030 ± 60	-28.5	cellulose (twigs)	layer S
SSR-2803 (a)	1740 ± 50	-28.1	fine peat (< 1 mm)	layer La
SSR-2803 (b)	1500 ± 70	-28.7	coarse peat (> 1 mm)	layer La
SSR-2803 (c)	1880 ± 60	-27.1	cellulose	layer La
SSR-2804	2260 ± 60	-29.5	twigs/bark	layer L
SSR-2805	2660 ± 60	-29.6	fine peat (< 1 mm)	layer Un
SSR-2809	3480 ± 60	-28.1	twigs/bark	layer K
SSR-2801 (a)	7380 ± 60	-27.2	fine peat (< 1 mm)	(profile 44)
SSR-2801 (b)	7450 ± 60	-26.3	coarse peat (> 1 mm)	(profile 44)
SSR-2801 (c)	7210 ± 70	-27.1	cellulose (twigs)	(profile 44)
SSR-2806	2820 ± 50	-28.9	peat	(profile 22)
SSR-2807	4620 ± 50	-28.0	peat	(profile 22)
SSR-2808	5930 ± 50	-28.6	peat	(profile 22)

4 RESULTS

4.1 Seljavallajökull

South of Eyjafjallajökull the stratigraphy is relatively simple. Deep soils over 6000 years old extend from sea level to over 600 m without the incorporation of any non aeolian deposits (profile 22, Fig. 2). At about 600 m the stratigraphy of the superficial deposits changes abruptly. Above the well-developed moraine at this height the hillsides are mantled in a bouldery till extending about 1 km from the present glacier, and only a few tens of metres from the position of the ice margin in 1930 (Eythórsson, 1931). No loessial soils occur above this till. The historical and geomorpholocial evidence indicates that this till has been exposed very recently, and profile 22 (Fig. 2) indicates that Seljavallajökull has not extended beyond this till deposit at any time in the last 6000 years.

4.2 Steinholtsjökull and Gigjökull

The geomorphology of the pro-glacial areas to the north side of Eyjafjallajökull is more complicated than that of the southern side, but the stratigraphic record of Holocene glacier fluctuations is effectively the same. At Steinholtsjökull terminal moraines lie only a few dozen metres beyond the limits of the glacier recorded by the 1932 Geodetic survey (Norand, 1944). Immediately outside the limits of the recent till shallow loessial soils have developed within the channels that extend downslope from the outermost moraines. A typical stratigraphical section in shown in profile 88 (Fig. 2). These soils contain a maximum of just four tephra layers. The uppermost was found to be the coarse grey-brown pumice produced by the 1947 AD eruption of Hekla. Below that is the fine black ash produced by Katla in 1918 AD; and below that lies the fine white tephra that originated between 1821-1823 AD during the only historical eruption of Eyjafjallajökull. Toward the base of the deepest loessial deposits within these channels another layer of fine black tephra is present. This ash was almost certainly produced by an eighteenth century eruption of Katla, probably that of 1755 AD. These shallow profiles indicate that immediately outwith the area of bouldery till there is a zone where patches of loessial soil have been accumulating since the early eighteenth century.

Chronic contemporary erosion has removed most of the soils outside the recent moraine limits, but some still survive, and a typical section is presented in profile 5 (Fig. 2). Over 30 tephra layers are present in this early Holocene soil, the basal beds of which are probably more than 7000 years old. A careful examination of the hundreds of square metres of eroding silt surfaces around profile 5 (Fig. 2) failed to find any till stones or other large clasts. As a result it seems that despite the loss of much of the Holocene stratigraphic record from this area it is most unlikely that these silts have ever been overridden by Steinholtsjökull.

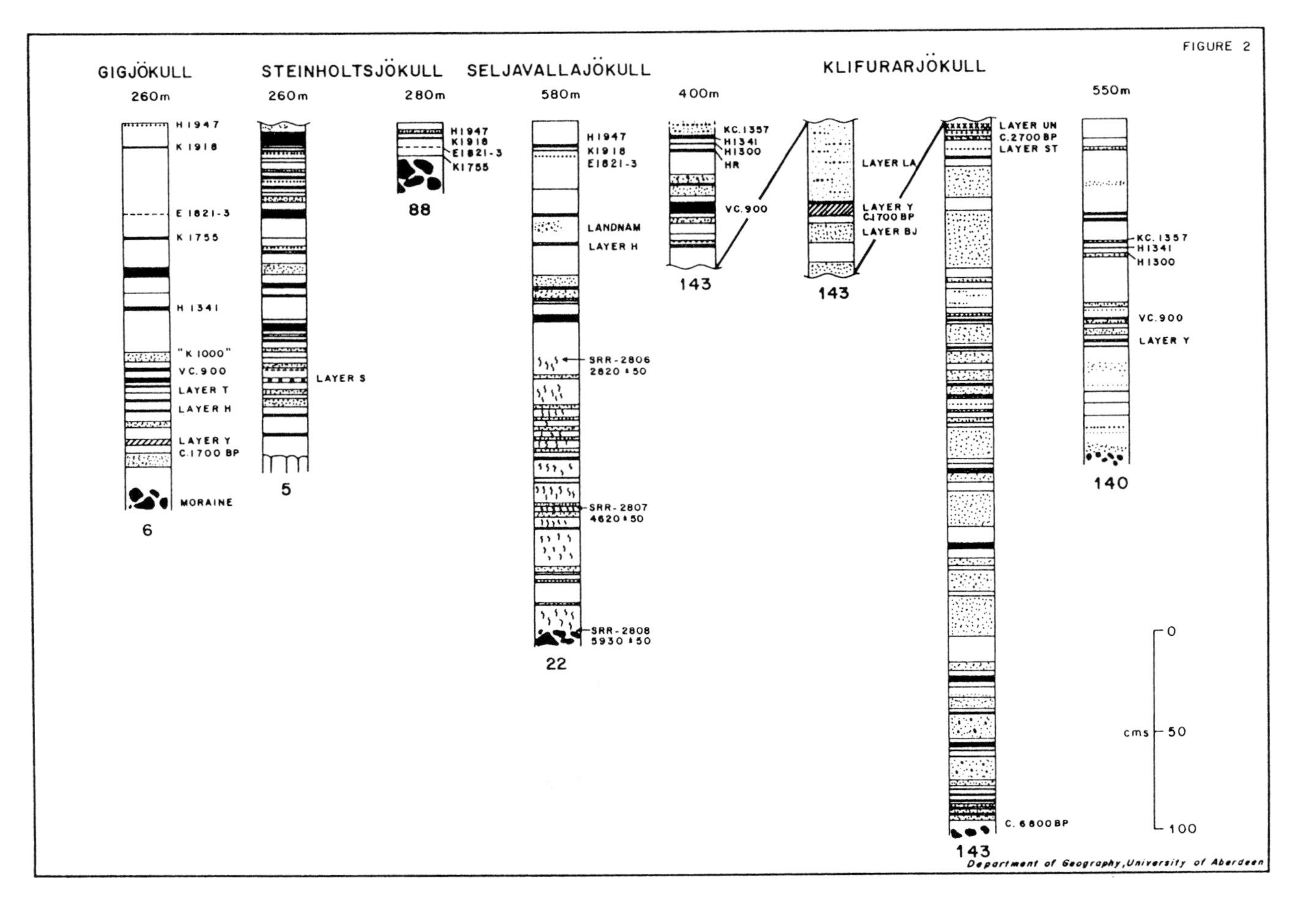

Figure 2. Soil profiles in the study area. Locations in Fig. 1, legend in Fig. 5.

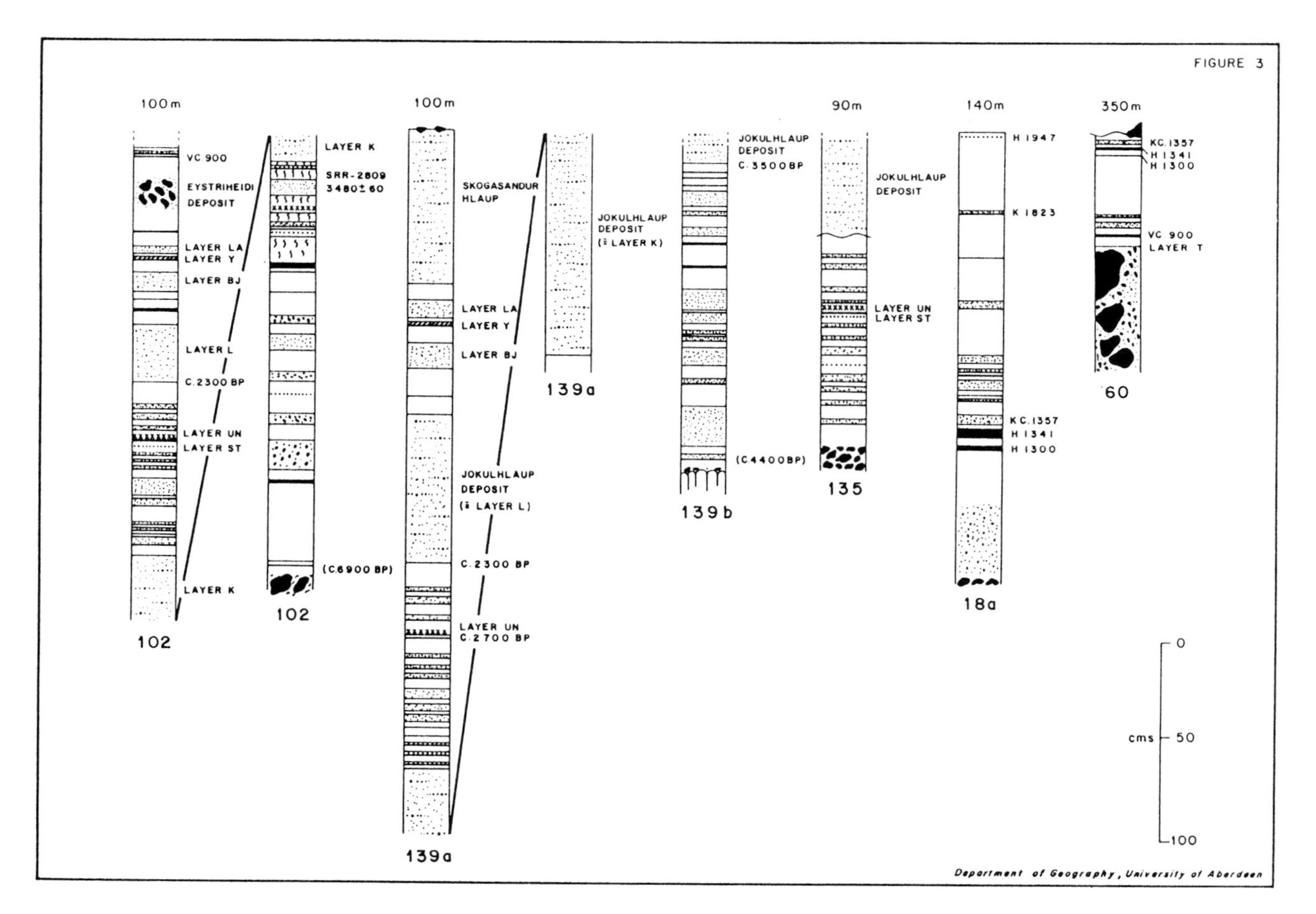

Figure 3. Soil profiles around Solheimajokull. Locations in Fig. 1, legend in Fig. 5

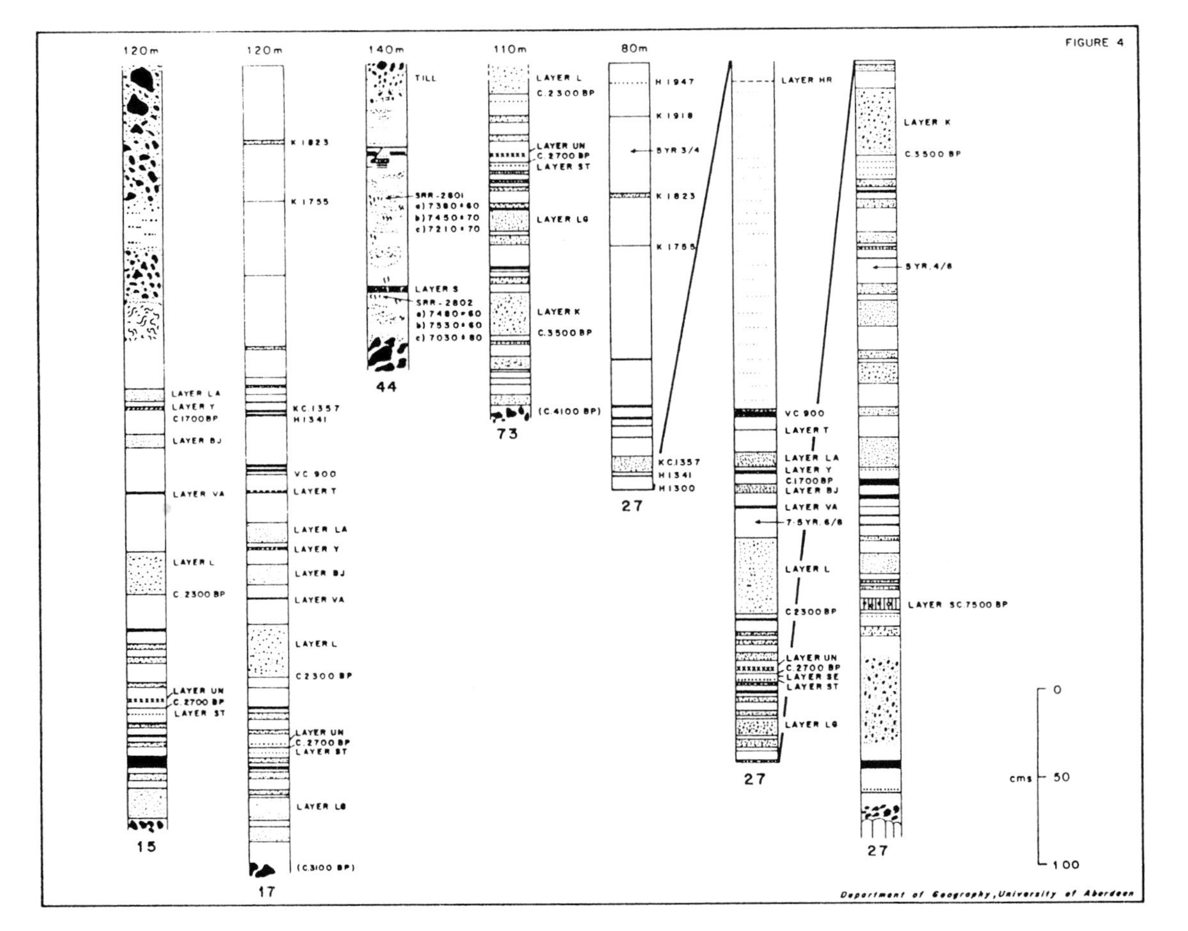

Figure 4. Soil profiles around Solheimajokull. Locations in Fig. 1, legend in Fig. 5

KEY TO STRATIGRAPHIC SYMBOLS

TEPHRA LAYERS, AIRFALL DEPOSITS

	Particle Size	Thickness of Layer
WHITE		
	<0.2mm	0.25cm-0.5cm
	<0.2mm	>0.5cm, Thickness Proportional to Scale
	>0.2mm <2mm	0.25cm-0.5cm
	>0.2mm <2mm	>0.5cm, Thickness Proportional to Scale
YELLOW		
	<0.2mm	>0.5cm, Thickness Proportional to Scale
DARK (Olive-Brown, Grey-Brown, Dark Grey-Brown, Blue-Grey, Black)		
	<0.2mm	0.25cm-0.5cm
	<0.2mm	>0.5cm, Thickness Proportional to Scale
	>0.2mm <2mm	0.25cm-0.5cm
	>0.2mm <2mm	>0.5cm, Thickness Proportional to Scale
	>2mm<20mm	0.25cm-0.5cm
	>2mm<20mm	>0.5cm, Thickness Proportional to Scale

Water-sorted Tephra: Thickness Proportional to Scale

Needle-grained Tephra Layer (Pelees Hair), Layer Thickness <0.5cm

SRR-2809
3480±60 : Stratigraphic Location of Radiocarbon Date (14C years BP)

H1341 : Origin and Age of Tephra

: Tephra Produced by Hekla Eruption in 1341AD

H : Hekla
K : Katla
E : Eyjafjallajokull
V : Veidivotn

7.5YR6/8 ➤ 'Munsell' soil colour

Loessial Soil — Generally Dark Reddish Brown (5YR3/4) to Yellowish Red (5YR4/6)

Peat/Organic Silt

GRAVEL 0.2mm 20mm

GRAVEL 20mm

Bedded Gravels

TILL

JOKULHLAUP DEPOSIT (Tephra-Rich)

BEDROCK

TOP OF SECTION AT GROUND LEVEL

Ground Surface Formed by Erosional Unconformity

Incomplete Diagram

Erosional Unconformity

Iron Pan

Shear Plane

Heterogenous Mix Of Loess And Tephra (Particle Size 0.2mm)

Altitude of Section(±10m) 550m

500m

SPLIT SECTION

ISOCHRONOUS HORIZON

Similar Facies Not Necessarily Contemporaneous

Section Number 95 78

Figure 5. Key to stratigraphic symbols.

The stratigraphy, therefore, indicates that the pattern of Holocene glacier fluctuations around Steinholtsjökull is similar to that of the south of Eyjafjallajökull; in both regions the immediate proglacial areas have been affected by ice very recently; but beyond these limits there is no evidence of any previous Holocene glacier activity.

The stratigraphic evidence around Gigjökull is somewhat less decisive than that around Steinholtsjökull. Historical records and photographic evidence show that Gigjökull extended to the proximal slopes of the great terminal moraine in the early twentieth century (Eythórsson, 1931). A soil section on the distal slopes of this moraine proves, however, that a part of this landform pre-dates c. 1700 BP. In profile 6 (Fig. 2) it can be seen that an uninterrupted sequence of 33 loess and tephra beds has accumulated over the north-eastern part of the moraine during the last 1700 years. The underlying till must, therefore, have been deposited at least 1700 years ago, but it is not yet possible to determine the maximum age of this till, other than to say that it does not pre-date late-glacial times.

4.3 Sólheimajökull

In contrast to the glaciers around Eyjafjallajökull, Sólheimajökull was found to have created a number of widely spaced pre-Little Ice Age moraines. The outermost limits to these fluctuations are indicated by the profiles 102 (Fig. 3), and 27 (Fig. 4), because the stratigraphy shows that aeolian material has accumulated at these sites without interruption for 7-8000 years. Just within these limits, however, a till was discovered which is more than 4500 years old (profiles 73 (Fig. 4) and 139 (Fig. 3)), and which overlies 7000 year old peat, soil and tephra deposits (profile 44, Fig. 4). These traces of a previously unrecorded mid-Holocene glacier advance occur up to 2 km beyond the most extensive limits of recent centuries, but their lateral location implies that the snout advanced even further to the south (see Fig. 1). This advance is here called the Drangagil after a local stream. Within the Drangagil limits a second limit may be defined by moraines that mark the glacier-ward extent of a 3500 years old tephra. Within this limit, here called the Hólsárgil, all the soils are less than c. 3100 years old (profile 135 (Fig. 2) and 17 (Fig. 4)). A third limit is defined by lateral moraines over 3 km long that overlie soils formed c. 3100-1400 years ago (profile 15, Fig. 4), yet are themselves overlain by soils up to 1200 years old (profile 60, Fig. 3). Jökulhlaups in the early historic period indicated the probable position of the ice margin in the tenth century AD (Maizels and Dugmore, 1985; Dugmore, 1987). This position also lies outside the limits of the Sólheimajökull in the last 300 years (Thórarinsson, 1943).

4.4 Klifurárjökull

The moraines around the snout seem to have been formed within the last few centuries. Vegetation cover is sparse, there are no overlying soils and despite the very active peri-glacial processes affecting this high exposed area, geliflution has yet to degrade the till surfaces. Outside the moraine limit there is an abrupt change in the nature of the landscape. The hillslopes are mantled with smooth lag gravel deposits, and in sheltered locations these gravels are covered with deep aeolian deposits. Within 100 m of the moraine limit, loess and tephra have been accumulating without interruption for over 2000 years (profile 140, Fig. 2), and, less than 500 m from the moraines, in the closest well-sheltered location to the western side of the glacier, aeolian sediments have been accumulating for over 6800 years (profile 143, Fig. 2). This pattern is similar to that found around Eyjafjallajökull, but radically different to the long record of glacier fluctuations found around Sólheimajökull.

4.5 Summary of results

In the study area at least two tills were deposited after 7000 BP, but before 3100 BP; the older of these was probably deposited between 4500 BP and 7000 BP, and is here called the Drangagil. The younger till formed before 3100 BP, and is here called Hólsárgil. A third pre-historic till was deposited between 1200-1400 BP and is here called the Ytzagil. These three pre-historic tills were identified only around Sólheimajökull. No conclusive stratigraphical evidence was found to support the idea that an extensive glacier expansion occurred soon after 2500 BP (cf. Thórarinsson, 1964). The glacier fluctuations in the study area are summarized in Fig. 6.

5. DISCUSSION

The pattern of glacier variations discovered in the study area was quite different from that which was expected. From previous knowledge three probable patterns of variation could have been predicted. It was assumed that Mýrdalsjökull melted away during the early Holocene and reformed during a brief period of severe climatic deterioration when the regional ELA was lower than that of today. During this initial phase of neoglaciation climatically sensitive mountain glaciers were assumed to have reached a Holocene maximum, whereas ice caps were not because of their longer response times (Thórarinsson, 1956, 1964); consequently early Neoglacial moraines were expected around the mountain glaciers of Eyjafjallajökull but not around the low-lying ice cap outlets of Mýrdalsjökull. There were, however, two alternative patterns of glacier variation that might also have been predicted from previous knowledge. Mýrdalsjökull could have responded to a lowering of ELA with sufficient speed to match the fluctuations of Eyjafjallajökull; alternatively, both ice caps could have responded slowly. Consequently early Neoglacial moraines could also have been predicted to lie outwith the Little Ice Age limits of both ice caps, or neither.

The discovery in this work that extensive pre-Little Ice Age moraines occurred around a Mýrdalsjökull outlet, but not around Eyjafjallajökull, was the one pattern that, according to predictions from previous knowledge, was most improbable. Any lowering of the snow line that enabled the low-lying Sólheimajökull to reach a Holocene maximum should also have enabled the neighbouring mountain glaciers of Eyjafjallajökull to reach a maximum. It is possible that part of the Holocene sequence may be missing from around Mýrdalsjökull because of delayed response due to the time necessary for a low-lying massif to accumulate a substantial ice mass. On the other hand, none of the Holocene sequence around the low-lying ice cap should be missing from the alpine glaciers of Eyjafjöll since the glaciers would have an immediate response to the lowering snowline altitude. Following these arguments, the presence of extensive pre-Little Ice Age moraines around Sólheimajökull and not around Eyjafjallajökull would seem to be very unlikely, but that is exactly what was found (Fig. 1). Moreover, the earliest of these glacier advances, which brought a major outlet glacier virtually to sea level, occurred at a time when the island is supposed to have been almost completely deglaciated (Einarsson, 1978). These unexpected findings can be explained by a model of ice divide migration (Dugmore, 1987). Bedrock divides lie up to 7 km to the north of the ice divides that define the present catchment of Sólheimajökull. The model predicts that when Mýrdalsjökull is thin and unable to form a true ice cap, and the direction of ice flow is determined by the underlying topography, Sólheimajökull will have a very large catchment and extent. When Mýrdalsjökull grows larger, ice divides move south away from the underlying bedrock divides, Sólheimajökull's catchment is reduced, and the extent of the glacier is restricted (Dugmore, 1987) (Fig. 6). Alternative possible explanations for the unusual chronology of glacier fluctuations, such as gross topographic change and variations in vulcanic heat flow, have been considered elsewhere and dismissed as insufficient (Dugmore, 1987). As a result the considerable extension of Sólheimajökull in mid-Holocene, late pre-historic and Mediaeval times must be seen as a local anomaly related more to conditions peculiar to its catchment rather than regional climatic change.

The dating of the Drangagil till (4500-7000 BP) indicates that Neoglaciation in Iceland had begun by the mid-Holocene. The assumption that no major, low-lying ice masses could have existed at that time has now been shown to be incorrect. Major ice masses did exist because Sólheimajökull extended virtually to sea level. The ELA of this period were, however, no lower than those of the Little Ice Age, and so the overall area of glacier ice in the island was probably less than the maximum of recent centuries. There is no evidence in the study area that glaciers were more extensive during the third millenium BP than during the later part of the Little Ice Age. This lack appears to be puzzling in view of the botanical evidence of climatic deterioration at this time (Einarsson, 1961, 1963), and especially since the botanical evidence of climatic deterioration in southern Iceland c. 1300 BP (Haraldsson, 1981) does coincide with the Ytzagil advances of Sólheimajökull. Mýrdalsjökull certainly did exist during the early "Sub-Atlantic" period because c. 2300 BP the Layer "L" eruption resulted in a jökulhlaup. The model of ice divide migration can explain this apparent anomaly.

Let us assume that the vegetation changes of the early "Sub-Atlantic" period were produced by a climatic deterioration that approached the severity of the Little Ice Age; given these conditions

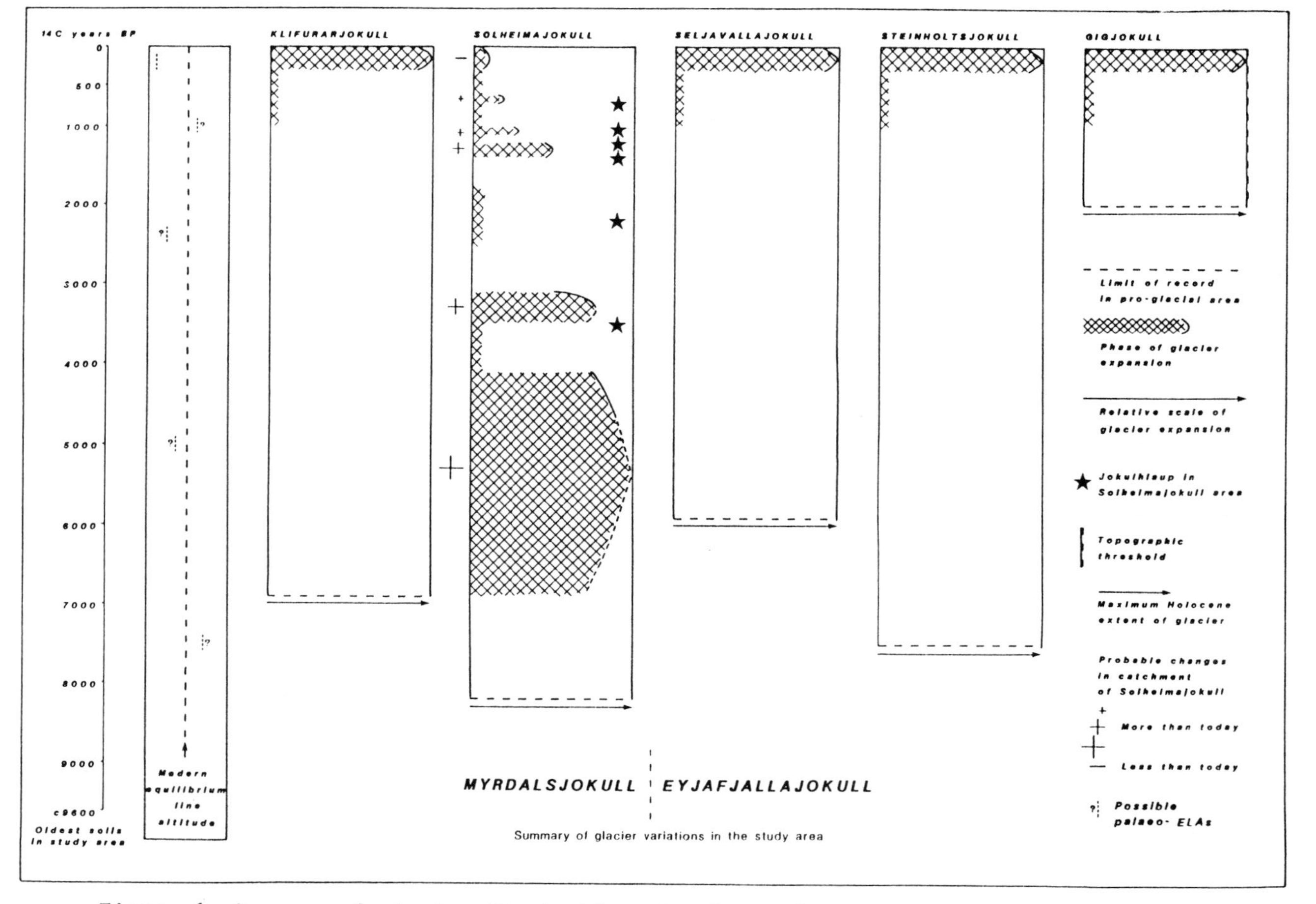

Figure 6. Summary of glacier fluctuations in the study area.

the model would predict that the development of Mýrdalsjökull would also approach that of the Little Ice Age. The extensive catchments of mid-Holocene Sólheimajökull (Drangagil and Hólsárgil periods) would be reduced. Consequently, despite a lower ELA, Sólheimajökull would be less extensive as the ablation zone shrank to balance the reduced input from the smaller accumulation zone. During the early part of the "Sub-Atlantic" the lower part of Sólheimajökull could have been up to c. 7 km^2 larger than during the Little Ice Age; it could not have exceeded this area, because Sólheimajökull did not exceed the Ytzagil limits at any time between 1700 and 3100 BP (profile 15, Fig. 4). Given a major climatic deterioration c. 2500 BP, other glaciers in the study area would also have expanded, but a "Sub-Atlantic" deterioration is unlikely to have exceeded that of the late Little Ice Age as no morainic record of this event can be found in the pro-glacial areas of Klifurárjökull, Steinholtsjökull, or Seljavallajökull. At Gigjökull, however, local anomalies have probably led to the preservation of pre-Little Ice Age moraines.

It has been suggested that pre-historic Holocene moraines exist in front of climatically sensitive mountain glaciers that are able to respond rapidly to climatic deterioration (Thórarinsson, 1956, 1964). The absence of these moraines in front of other Icelandic glaciers is explained as being a result of longer response times that prevent these glaciers from rapidly reaching climatic equilibria. Assuming that a major climatic deterioration did occur c. 2500 BP, the author would agree with Thórarinsson and suggest that part of the Gigjökull moraine could have been formed at that time. In contrast, however, a different explanation is proposed for its survival. The pre-historic moraine may have survived not because it is the result of the most profound climatic deterioration, but because of another local topographic anomaly.

At Gigjökull the soil profile 6 (Fig. 2) shows that the crest of the moraine has not been overrun by the glacier during historical time, yet data from Eythórsson (1931) shows that the glacier was close to the crest in 1930. Consequently, the moraine has clearly formed at a position repeatedly occupied by the glacier margin. This position is probably particularly stable as a result of local topographic controls (cf. Mercer, 1961). Gigjökull has a high-level, low-gradient accumulation area, the central sections are narrow and particularly steep, and it terminates at the edge of a sandur. During a climatic oscillation, as soon as the ELA falls below the summit plateau a glacier will form and probably reach the edge of the sandur. As conditions deteriorate the ELA will descend particularly steep, narrow sections of the glacier. Consequently, the falling ELA will produce only comparatively small increases in the accumulation area. At the same time, the glacier snout will have reached the sandur, where increases in lateral extent produce a dramatic increase in the ablation area. As a result the sandur edge probably represents a particularly stable topographic location, which is reached rapidly during the initial stages of a climatic deterioration, but beyond which the glaciers may only advance during full glacial conditions. During the Holocene the critical climatic threshold conditions that enable Gigjökull to reach the sandur have probably crossed on a number of occasions. The threshold which would enable the glacier to advance across the sandur has not, and as a result a large composite moraine has been formed. It is therefore proposed that the pre-historic terminal moraine of Gigjökull exists because local topographic conditions inhibited the late Little Ice Age advances of the glaciers, and so prevented the moraines formed earlier in the Holocene from being completely destroyed.

The pre-historic part of the Gigjökull moraine is more than 2000 years old, and it probably post-dates the Budi stage. If it dates from the Neoglacial period as opposed to the Late-glacial then it might have been formed between 2000-7000 BP. The absence of the c. 2700 year old tephra Layer "Un" from the soils on the moraine suggests that it might have been formed about c. 2500 BP; the moraine could, however, pre-date the "Sub-Atlantic" and the observed stratigraphy could be the result of extensive soil erosion.

A discussion of the status of the early "Sub-Atlantic" period must also consider other moraines in Iceland attributed to this period. These fall into two groups; those around Öraefajökull (Thórarinsson, 1956) and the one on Snaefell (Thórarinsson, 1964). Locationally and morphologically the Öraefi moraines are similar to the terminal moraine at Gigjökull. They all occur within 2.5 km of the sandur edge, they are all massive features 35-113 m high, and they all represent ice limits that are coincident with or very close to glacier margins of the last century (Thórarinsson, 1956; Eythórsson, 1931). The glaciers are morphologically similar as they have extensive high altitude accumulation zones, steep and narrow central reaches that contain the present ELA, and they all terminate at the edge of the sandur. The Öraefi glaciers lie in the same climatic region as Gigjökull (Einarsson, 1976), and Öraefajökull and Eyjafjallajökull are similar types of volcano (Williams et al., 1983). Consequently, the explanation for the origins of the pre-historic moraines of Öraefi is likely to be the same as that for the Gigjökull moraine. In this paper the preferred explanation is that all these composite moraines represent topographically controlled threshold locations.

The only other pre-historic Holocene moraine known to exist in Iceland lies on Snaefell in east Iceland (Thórarinsson, 1964). It is here proposed that the occurrence of this moraine outside the Little Ice Age glacier limits of the Snaefell glacier is a result of the development of Vatnajökull. It has been argued that Vatnajökull did not develop to its present form until late mediaeval times (Tómasson and Vilmundardóttir, 1967). The modern Vatnajökull forms a great ice wall that has profound effects on the local climate, particularly to the north where a precipitation shadow affects a large area of the central highlands (Einarsson, 1976). Since this precipitation shadow effect produces a dramatic elevation of the snowline (Björnsson, 1979), the development of Vatnajökull has probably exerted a primary control on glacier fluctuations within the affected area and Snaefell is one of the mountains in this zone. At times when the Vatnajökull ice wall is reduced, the barrier to the northern penetration of precipitation bearing winds falls, so greater precipitation in the eastern central highlands lowers the ELA, and results in glacier expansion. As the climate deteriorates, Vatnajökull grows, and the mountain glaciers to the north become starved of moisture and fail to exceed their earlier limits. Consequently, the greater extent of pre-Little Ice Age glaciers on Snaefell seems to be an inevitable consequence of the mountain's location, rather than an indication of severe climatic deterioration in the mid-Holocene. If a maximum Holocene age could be attached to the formation of the pre-historic moraines on Snaefell it could give an indication of the onset of Neoglaciation in Iceland. It could show when conditions first began to deteriorate before the reformation of Vatnajökull and the development of the present precipitation shadow.

6. CONCLUSIONS

The data presented in this paper go some way towards reducing the apparent anomalies between the general pattern of Holocene glacier fluctuations in Iceland and those of other parts of the world (Grove, 1979), as the Icelandic record is now known to be longer and more complex than previously thought and glaciers probably existed throughout the later half of the Holocene. This extended history has only become apparent from a study of an unusual glacier where the catchment shrinks as the icecap from which it flows grows, and where as a result glacier extent is inhibited as climate deteriorates, and exaggerated as climate ameliorates.

Although in Iceland the behaviour of Sólheimajökull appears to be rare, if not unique, it is probable that many similar glaciers exist elsewhere in the world. An interesting example of ice divide migration determined from glacier surface studies (as opposed to pro-glacial studies employed in this paper) is provided by Waddington and Marriott (1986), who have indentified ice divide migration on Blue Glacier, Washington State, U.S.A. The identification and study of these oddities deserve special attention for two reasons; firstly to avoid unnecessary confusion if these records are compared with those of fixed catchment glaciers, and secondly because the forelands of these glaciers are likely to provide some of the longest and most complex records of glacier fluctuation, the interpretation of which, while not easy, may be extremely rewarding. This study also provides an additional warning against the over-reliance on glacier chronologies based on single glaciers, and reinforces the arguments in favour of regional studies, or at least several geographically diverse glaciers.

ACKNOWLEDGEMENTS

I would like to gratefully acknowledge the receipt of a U.K. N.E.R.C. Research Studentship at the Department of Geography, University of Aberdeen, supervised by Prof. D.E. Sugden and Dr. C.M. Clapperton. Dr. D.D. Harkness undertook the 14C analyses. The Icelandic National Research Council gave permission for the work to be carried out, and Dr. P.C. Buckland and A. Kirkbride gave valuable assistance in the field. Additional funding was provided by the Quaternary Research Association, and the Royal Society of London.

REFERENCES

Björnsson, H. 1978. Könnun á jöklum med rafsegulbylgjum (Radio-echo sounding of temperate glaciers). Náttúrufraedingurinn 47 (3-4), 184-194.

Björnsson, H. 1979. Glaciers in Iceland. Jökull 29, 74-80.

Dugmore, A.J. 1987. Holocene glaciers fluctuations around Eyjafjallajökull, south Iceland: A tephrochronological study. Unpubl. PhD. thesis, University of Aberdeen.

Einarsson, M.A. 1976. Vedurfar á Íslandi. Indunn: Reykjavik.

Einarsson, E.H., Larsen, G. and Thórarinsson, S. 1980. The Sólheimar tephra layer and the Katla eruption of ~ 1357 AD. Acta Naturalia Islandica 28, 1-24.

Einarsson, Th. 1961. Pollenanalytische Untersuchungen zur spät und postglazialen klimageschichte Islands. Sonderveroff. d. Geol. Inst., Univ. Köln 6, 52 pp.

Einarsson, Th. 1963. Pollen-analytical studies on the vegetation and climate history of Iceland in late and post-glacial times. In: A. Löve and D. Löve (eds.) "North Atlantic biota and their history". Pergamon, Oxford, 355-365.

Einarsson, Th. 1978. Jardfraedi. Mál og Menning: Reykjavik, 240 pp.

Eythórsson, J. 1931. On the present position of the glaciers in Iceland. Vísindafélag Íslendinga (Societas Scientiarum Islandica) 10, 1-35.

Grove, J.M. 1979. The glacier history of the Holocene. Progress in Physical Geography 3, 1-54.

Haraldsson, H. 1981. The Markarfljót sandur area, southern Iceland: Sedimentological, petrological and stratigraphic studes. Striae 15, 1-58.

Jónsson, J. 1982. Notes on the Katla volcano glacial debris flows. Jökull 32, 61-68.

Kjartansson, G. 1967. The Steinholtshlaup, central south Iceland on January 15th 1967. Jökull 17, 249-262.

Larsen, G. 1979. Um aldur Eldjárhrauna (Tephrochronological dating of the Elgjá lavas in southern Iceland). Náttúrfraedingurinn 49 (1), 1-26.

Larsen, G. 1981. Tephrochronology by microprobe glass analysis. In: S. Self and R.S.J. Sparks (eds.) "Tephra studies". D. Reidel, Dordrecht, 95-102.

Larsen, G. 1984. Recent volcanic history of the Veidivötn fissure swarm, southern Iceland;- an approach to volcanic risk assessment. In: N. Óskarsson and G. Larsen (eds.) "Volcano monitoring". Journal of Volcanology and Geothermal Research 22, 33-58.

Maizels, J.K. and Dugmore, A.J. 1985. Lichenometric dating and tephrochronology of sandur deposits, Sólheimajökull area, southern Iceland. Jökull 35, 69-78.

Mercer, J.H. 1961. The response of fjord glaciers to changes in the firn limit. Journal of Glaciology 3 (29), 850-858.

Norand, N.E. 1944. Islands Kortlaegning en Historisk Fremstilling. Ejnar Munksgaard: København.

Sharp, M.J. 1982. A comparison of the landforms and sedimentary sequences produced by surging and non-surging glaciers in Iceland. Unpubl. PhD. thesis, University of Aberdeen.

Sharp, M.J. and Dugmore, A.J. 1985. Holocene glacier fluctuations in East Iceland. Zeitschrift für Gletscherkunde un Glazialgeologie 21, 341-349.

Thórarinsson, S. 1943. Oscillations of the Iceland glaciers in the last 250 years. Geogr. Ann. 25, 1-54.

Thórarinsson, S. 1944. Tefrokronologiska studier på Island. Geogr. Ann. 26, 1-217.

Thórarinsson, S. 1956. On the variations of Svínafellsjökull, Skaftafellsjökull and Kviarjökull in Öraefi. Jökull 14, 67-75.

Thórarinsson, S. 1964. On the age of the terminal moraines of Brúarjökull and Hálsajökull. A tephrochronological study. Jökull 14, 67-75.

Thórarinsson, S. 1966. The age of the maximum post-glacial advance of Hagafellsjökull Eystri. Jökull 16, 207-210.

Thórarinsson, S. 1967. The eruptions of Hekla in historical times. A tephrochronological study. The Eruption of Hekla 1947-48, 1, 183 pp.

Thórarinsson, S. 1974. The terms Tephra and Tephrochronology. In: J.A. Westgate and C.M. Gold (eds.) "The World Bibliography and Index of Quaternary Tephrochronology". University of Alberta, 27-28.

Thórarinsson, S. 1975. Katla og annáll Kötlugosa. Árbók Fedafél. Íslands. 1975, 129-149.

Thórarinsson, S. 1981. Tephra studies and Tephrochronology. A historical review with special references to Iceland. In: S. Self and R.S.J. Sparks (eds.) " Tephra studies". Reidel, Dordrecht, 1-12.

Tómasson, H. and Vilmundardóttir, E. 1967. The lakes Storisjor and Langisjor. Jökull 17, 280-295.

Waddington, E.D. and Marriott, R.T. 1986. Ice divide migration at Blue Glacier, USA. Annals of Glaciology 8, 175-176.

Williams, R.S., Thórarinsson, S. and Morris, E.C. 1983. Geomorphic classification of Icelandic volcanoes. Jökull 33, 19-24.

RELATIONSHIP OF LAND TERMINATING AND FJORD GLACIERS TO HOLOCENE CLIMATIC CHANGE, SOUTH GEORGIA, ANTARCTICA

C.M. Clapperton, D.E. Sugden & M. Pelto[1]
Department of Geography
University of Aberdeen
Scotland

ABSTRACT

This paper uses ELA and mass balance gradient reconstructions to understand the climatic significance of Holocene glacier fluctuations in South Georgia. The position of the island between mid-latitude southern South America and coastal Antarctica makes it an important link in the chain of Holocene glacier histories in the Southern Hemisphere. A Neoglacial cooling of 1°C and a Little Ice Age cooling of 0.5°C reflect the increased frequency of southern air masses from the Weddell Sea which lowered temperatures but modified precipitation only slightly. Radiocarbon dating the most extensive Neoglacial advance is problematic because most data have been obtained only for the moraine of a calving fjord glacier. This, the Moraine Fjord glacier, is believed to have advanced 450-650 years later than land-terminating glaciers, since its progress down-fjord depended on the construction of a pinning moraine bank. The scale of late Holocene-Little Ice Age glacier advances in South Georgia was small by mid-latitude standards, possibly because of the relatively arid environment in the lee side of the island.

1. INTRODUCTION

This paper examines the climatic significance of Holocene glacier fluctuations in South Georgia, a sub-Antarctic island that provides a terrestrial stepping stone in the southern Atlantic Ocean between South America and Antarctica (Fig. 1). The island's location gives a unique opportunity to attempt a link between the glacial histories of mid-latitude ice fields in Patagonia and high-latitude glaciers in Antarctica. Such information may permit an important constraint on models of Southern Hemisphere climatic change. This study draws attention in particular to problems of interpreting climatic fluctuations from the moraines of former fjord glaciers.

[1] Dept. of Geology, University of Maine, U.S.A.

J. Oerlemans (ed.), Glacier Fluctuations and Climatic Change, 57–75.

2. CLIMATIC SETTING

The position of South Georgia, firmly embedded in the circum-Antarctic westerlies, implies that any global circulation changes may affect the island's climate in the same way as they would affect Patagonia and the Antarctic Peninsula. This is because climatic change in a Polar Front region reflects changes in mid- and high-latitudes synchronously (Weller, 1980). In particular, temperature fluctuations on decadal and centennial scales might be expected to follow a similar pattern in South Georgia as in adjacent land to the north and south. Analysis of weather charts showing monthly means of geostrophic winds and pressure patterns (Jenne et al., 1974) shows that the climate of South Georgia is dominated by the passage of cyclonic depressions which originate mainly on the downward limb of a fairly stable wave in the Antarctic Polar Front (Fig. 1). Because the longitudinal position of the limb is

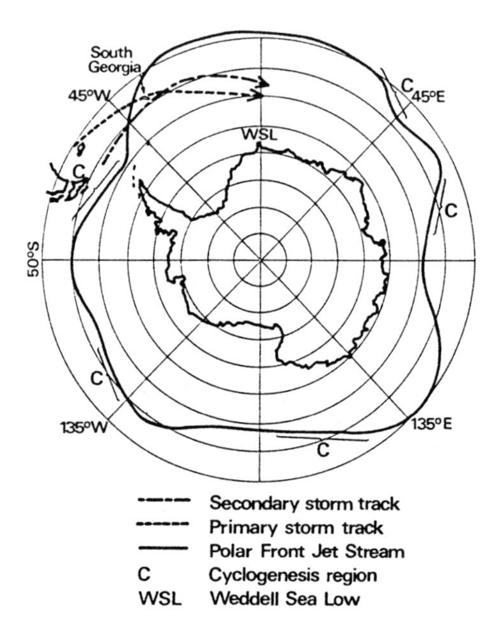

Figure 1. The location of South Georgia in relation to South America, the Antarctic continent, the Polar Front jetstream and storm tracks in the western South Atlantic.

controlled by large scale topography and ocean circulation in the region of the Bellingshausen, Scotia and Weddell Seas, it is a quasi-stable feature and has probably not changed significantly during the Quaternary (Manabe and Hahn, 1977; Schwerdtferger, 1970). The latitudinal position of the limb appears to vary slightly, however, with changes in the strength of the Weddell Sea Low pressure cell (Fig. 1). Fluctuations in the position and intensity of the Weddell Sea Low (W.S.L.) may reflect changes in the global radiation budget on the decadal scale and, on a longer time scale, are influenced by the combined extent of the Antarctic ice sheet, fringing ice shelves and the ice pack. For example, a reduction in ice extent reduces the intensity of the Low while an expansion pushes the Low northwards without weakening it (Schwerdtferger 1975).

The effect of variations in the W.S.L. on the climate of South Georgia may be predicted, thereby forming a basis for understanding glacier fluctuations on the island. A weakened W.S.L. will reduce the frequency with which cold and dry southerly air masses affect the island during summer months (December-January), and this permits a greater penetration of milder maritime air from more temperate latitudes in the north. The effect is to slightly increase annual precipitation and to significantly increase mean summer temperature, thereby causing glacier recession. An extension northwards of the W.S.L. permits a greater influence of Antarctic air masses on the island's climate reducing both precipitation and mean annual temperature; since ablation is the more dynamic term for glaciers in maritime climates, reduced temperatures cause glacier advance.

3. GEOMORPHOLOGICAL SETTING

South Georgia is a relatively small island c. 170 km long and c. 2-30 km wide, but has a mountainous backbone with several peaks rising more than 1800 m above sea level. Climatic statistics obtained since 1914 at Grytviken on the north-east coast indicate a mean annual temperature of 2°C, varying between -1.2°C in winter (June-August) and 4.5°C in summer (December-February); precipitation, averaging 1601 mm, is distributed regularly throughout the year, while mean wind speeds of 4.4 m s^{-1} are fairly consistent for much of the year. Some 58% of the island is covered in glaciers and many terminate at sea level, particularly on the 'windward' south-western side of the island. The north-eastern side is indented by fjords and intervening peninsulas which are generally lower than 650 m; these are either ice free or contain only small cirque glaciers and ice fields (Fig. 2). It is in such ice-free areas that there is good geomorphological evidence of late Quarternary glacier history (Clapperton, 1971; Sugden and Clapperton, 1977; Clapperton et al., 1978). Following the retreat of an ice cap which extended offshore during the last glacial maximum, glaciers withdrew close to their present limits before 9700 radiocarbon years ago. This conclusion is based on the presence of undisturbed peat of this age lying within glacial troughs several hundred metres from existing glacier snouts. Subsequently, there have been at least two main Holocene advances referred to here as the outer (Neoglacial) and inner (Little Ice Age) moraine series (Fig. 3). The outer series is distinguished from glacial deposits outside its limit by freshness of form, continuity, association with a relative sea level close to the present, relative lack of solifluction and lack of iron-staining on surface clasts (Clapperton and

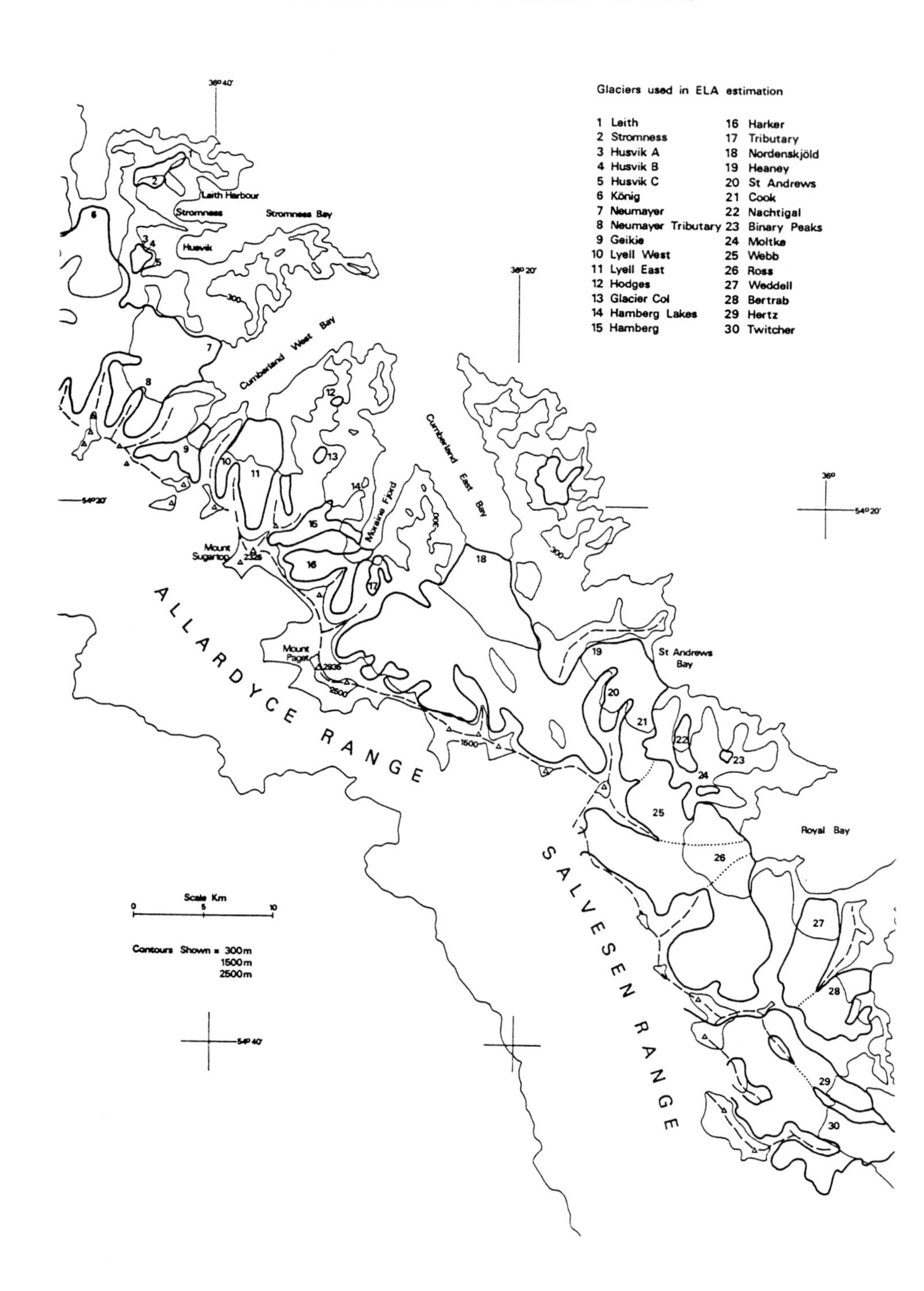
Glaciers used in ELA estimation
1 Leith
2 Stromness
3 Husvik A
4 Husvik B
5 Husvik C
6 König
7 Neumayer
8 Neumayer Tributary
9 Geikie
10 Lyell West
11 Lyell East
12 Hodges
13 Glacier Col
14 Hamberg Lakes
15 Hamberg
16 Harker
17 Tributary
18 Nordenskjöld
19 Heaney
20 St Andrews
21 Cook
22 Nachtigal
23 Binary Peaks
24 Moltke
25 Webb
26 Ross
27 Weddell
28 Bertrab
29 Hertz
30 Twitcher
36°40'
36°20'
36°
54°20'
54°40'
Leith Harbour
Stromness
Stromness Bay
Husvik
Cumberland West Bay
Cumberland East Bay
Moraine Fjord
Mount Sugartop
2325
Mount Paget
2935
2500
1500
300
St Andrews Bay
Royal Bay
ALLARDYCE RANGE
SALVESEN RANGE
Scale Km
0
5
10
Contours Shown = 300m
1500m
2500m

Figure 2. (previous page) Glaciers on the northeastern side of the mountain backbone of South Georgia used in the calculation of present day and past Equilibrium Line Altitudes (ELAs).

Sugden, INQUA Abstracts, 1987). Most fjord and land-terminating glaciers advanced only several hundred metres and produced moraine ridges less than c. 5 m high. Certain fjord glaciers, such as those entering Possession Bay, Antarctic Bay (both on the north coast) and Moraine Fjord, advanced up to 6.5 km, however, and deposited large multiple moraine complexes (Fig. 3). Radiocarbon dates associated with this stage imply that glaciers may have been less extensive than now at c. 2,230-3,330 (Gordon, 1987), but completed a readvance between c. 1,700 BP and 1,460 BP (Clapperton and Sugden, 1987). The inner series usually occurs within a few hundred metres of the present glacier margins. In contrast to the earlier Neoglacial moraines, the younger ones are poorly vegetated, blocky and, in the case of those closest to the glacier margins, ice-cored and unstable. Historical records and radiocarbon dating allow these moraines to be attributed to glacier advances of the Little Ice Age (Hayward, 1983; Timmis, 1986; Clapperton and Sugden, 1987). At face value, implications of the geomorphological evidence are that some glaciers in South Georgia advanced close to their maximum Holocene extent as recently as c. 1700 radiocarbon years BP and that glacier response varied considerably from place to place. Moreover, the Little Ice Age advance of most glaciers was only slightly smaller than the maximum Neoglacial advance.

4. PALAEOCLIMATIC QUESTIONS

Although the field evidence is far from complete in terms of detail and dating, it is sufficient to pose three important questions of climatic significance. First, what scale of climatic deterioration was necessary to bring the glaciers forward to their maximum Neoglacial extents? Second, why is there such a dramatic difference between the majority of glaciers which advanced only a few hundred metres and others such as that in Moraine Fjord which advanced 6.5 km? Third, why are the apparent dates of the Neoglacial maximum (1700-1460 BP) different to those established in Patagonia? In the latter area three Neoglacial advances have been recognized at 4600-4200 BP, 2700-2000 BP and AD 1600-1850 (Mercer, 1976, 1983; Clapperton, 1983). Not only does South Georgia bear no morainic evidence of the earlier advances, but there appears to be no Patagonian equivalent of the advance dated to 1700-1460 BP on South Georgia. This is particularly puzzling since Little Ice Age advances appear to have occurred synchronously in both regions during the last two hundred years.

This paper approaches these questions in the first instance by estimating present and past Equilibrium Line Altitudes (ELAs), which are then linked to predicted mass balance gradients; this helps to establish glacier discharges and to explore the role of calving in affecting glacier snout responses to climatic change.

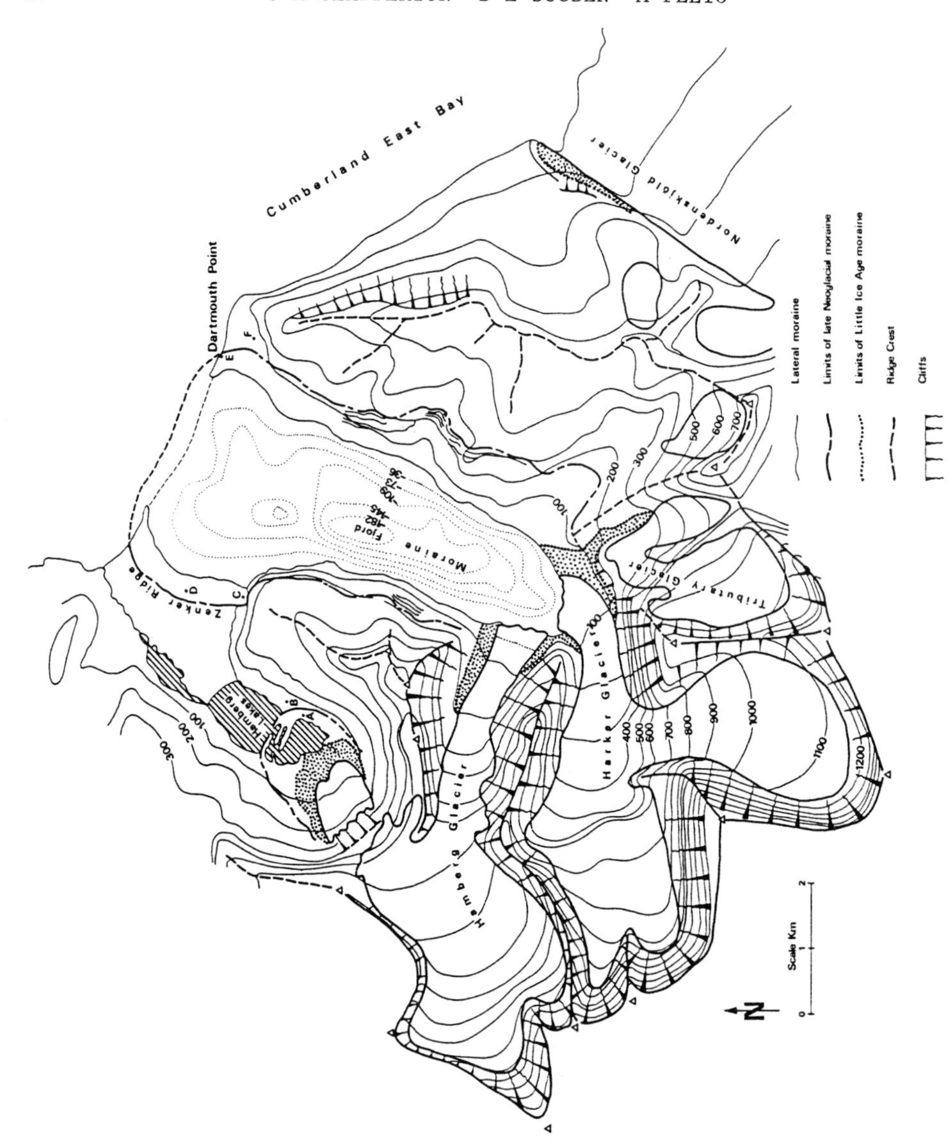

Figure 3. Holocene moraines in the Cumberland East Bay area. The Neoglacial moraine represents a much bigger advance of Moraine Fjord glacier than of Nordenskjöld glacier. The latter typical of the response of most South Georgia glaciers. Relevant radiocarbon dates are: A-950±50 (SRR-3002), B-2210±50 (SRR-3001), C-910±50 (SRR-2997), D-1460±60 (SRR-2998), E-1700±50 (SRR-2760a) and 2000±50 (SRR-2760b), F-1000±50 (SRR-2996).

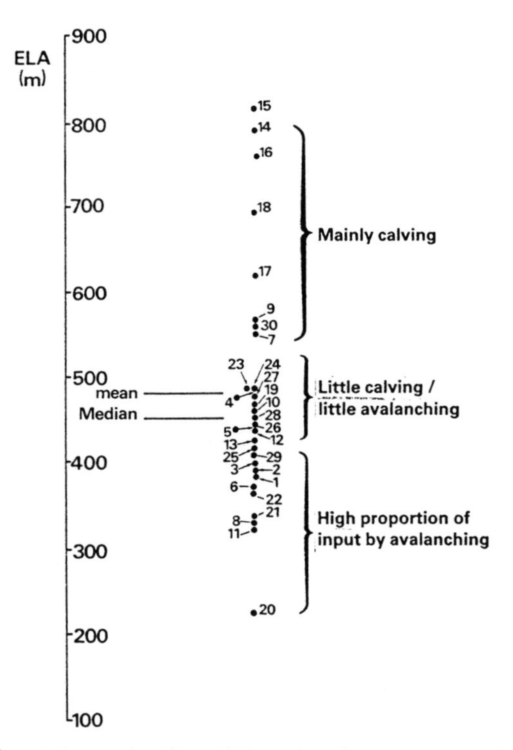

Figure 4. ELAs calculated by the linear mass balance gradient method showing the main groupings of glaciers as well as the overall mean and the median ELAs. Numbers refer to Fig. 2.

5. PRESENT DAY EQUILIBRIUM LINE ALTITUDES

There are no comprehensive estimates of ELAs in South Georgia and thus a simple method was used to calculated the ELAs on 30 glaciers in the Stromness-Iris Bay area on the northeastern side of the island (Fig. 2). The sample includes a wide range of glacier types and sizes, including compact cirque glaciers and valley glaciers nourished in high mountain couloirs. The method makes the assumption that the mass balance of the glacier is zero, i.e. that it is in equilibrium. It further assumes that the mass balance gradient is linearly related to altitude and thus the ELA, the elevation where total accumulation above balances total ablation below, can be calculated from knowledge of the altitudinal distribution of the glacier (L.D. Williams, pers. comm.). Braithwaite (1984) noted that this relationship, first pointed out by Kurowski in 1891, was rediscovered by Liestol (1967); it was successfully used by Sissons and Sutherland (1976). In this study the method assumes no ablation by calving. The basic data are derived from measurements on topographic maps; for 25 glaciers the 1:50,000 geomorphological maps of

Stromness Bay to Royal Bay were used (Clapperton, 1971; Clapperton and Sugden, 1980) and for 5 glaciers the 1:200,000 topographic map of South Georgia (D.O.S., 1958) provided adequate data. The altitude of the ELA (z_e) is found from the equation

$$z_e = \frac{\Sigma_{i=1}^{n} Z_i W_i}{\Sigma_{i=1}^{n} W_i} \quad (1)$$

where Z_i = the mean elevation of an area within a specified altitudinal range, W_i = the mean width of the glacier at the same altitude.

The results of the calculations are shown in Table I and Figure 4. The ranking in terms of altitude indentifies three groups within the population of 30 glaciers. A group of eight glaciers have high ELAs well above the mean and median values of 480 m and 450 m respectively. These glaciers all terminate in the sea and are subject to calving, a factor which is not incorporated in the model and would have the effect of producing higher than appropriate ELAs. A lower group of 11 glaciers with ELAs well below the median consists of glaciers which terminate aground or close to sea level but which receive considerable accumulation through avalanching; most are backed by high mountains rising steeply above the accumulation basins. As would be expected, the linear model employed would underestimate the ELA in such cases. The remaining group of 11 glaciers which clusters round the median value of 450 m is composed of glaciers where the mass balance is least affected by avalanching and calving. This is the group which would be expected to approximate most closely to a regional ELA. The median value for the regional ELA of 450 m is probably a better approximation of reality than the mean of 480 m, since the latter figure is biassed upwards by the inclusion of glaciers affected by large-scale calving.

There is good agreement between an estimated mean ELA of c. 450 m and field measurements. Accurate mass balance data are restricted to Hodges Glacier where measurements during 1972-1977 gave a mean value for the ELA of 448 m (Timmis, 1986). In a wider survey Smith (1960) estimated the 'firn limit' along the northeastern coast to be 460 m. There is also good agreement with estimates of the ELA through the construction of a theoretical mass balance gradient curve based on known budget gradients from other glaciers (Braithwaite, 1984; Pelto, 1987) and the associated prevailing climatic conditions. Figure 5 shows a reconstruction of the most probable mass balance curve and its likely limits. The reconstruction is based on the notion that a specific budget gradient is associated with each particular climatic regime. Here, it employs relationships between present day climatic statistics for Grytviken (1950-1983), radiosonde records (1950-1983) and satellite weather maps (1975-1983). The curve suggests a regional ELA of the order of 475 m. Since the reconstruction ignores the effect of avalanching, the estimate is likely to be slightly higher than exists in reality. Overall we conclude that the calculated present day ELA of around 450 m agrees well with field measurements and climatic predictions.

Table I. Present day ELAs for 30 glaciers on South Georgia, estimated from the linear mass balance equation and measurements from topographic maps. Glaciers are arranged from NW-SE.

Glacier	Altitude of Source (m)	Altitude of Terminus (m) (C=Calving)	Altitudinal Range (m)	ELA (m)
König	460	C	460	369
Leith	613	30	583	386
Stromness	610	122	488	394
Husvik A	488	300	188	402
Husvik B	518	300	218	483
Husvik C	500	300	200	442
Neumayer	1070	C	1070	543
Neumayer Tributary	915	C	915	335
Geikie	1160	C	1160	568
Lyell West	1040	C	1040	459
Lyell East	810	C	810	322
Hodges	600	212	388	440
Glacier Col	460	335	125	429
Hamberg Lakes	1390	C	1390	798
Hamberg	1390	C	1390	863
Harker	1375	C	1375	758
Tributary	890	220	670	623
Nordenskjöld	1525	C	1525	693
Heaney	900	0	900	465
St Andrews	430	0	430	223
Cook	750	0	750	337
Nachtigal	760	185	575	364
Moltke	540	410	130	485
Binary Peaks	610	360	250	485
Webb	590	C	590	416
Ross	900	C	900	449
Weddell	1070	0	1070	471
Bertrab	915	0	915	451
Hertz	1370	0	1370	413
Twitcher	1525	C	1525	560
Median ELA:				450
Mean ELA:				480

6. FORMER EQUILIBRIUM LINE ALTITUDES

In the case of those glaciers where clear lateral moraines mark the Neoglacial limits, it was possible to reconstruct the glacier morphology in detail and calculate the former ELA by means of the linear mass balance equation (1). The method of reconstruction is illustrated in

Fig. 6 for the case of Moraine Fjord glacier, formed by confluence of the main Harker and Hamberg glaciers and the smaller Tributary Glacier. Because of the relatively small scale of the topographical maps (1:50,000), estimations of former ELAs could be made only for glaciers that had extended more than c. 250 m beyond their present snouts. Consequently, data were obtained for 9 valley glaciers and 9 cirque glaciers (Table II). The two populations of glaciers were initially kept separate because of the wide range of basin morphologies. Mean and median values of ELA lowering for cirque glaciers are 50 m for the Little Ice Age and 81 m and 88 m for the Neoglacial; corresponding mean and median values for the valley glaciers are 43 m and 46 m and 98 m and

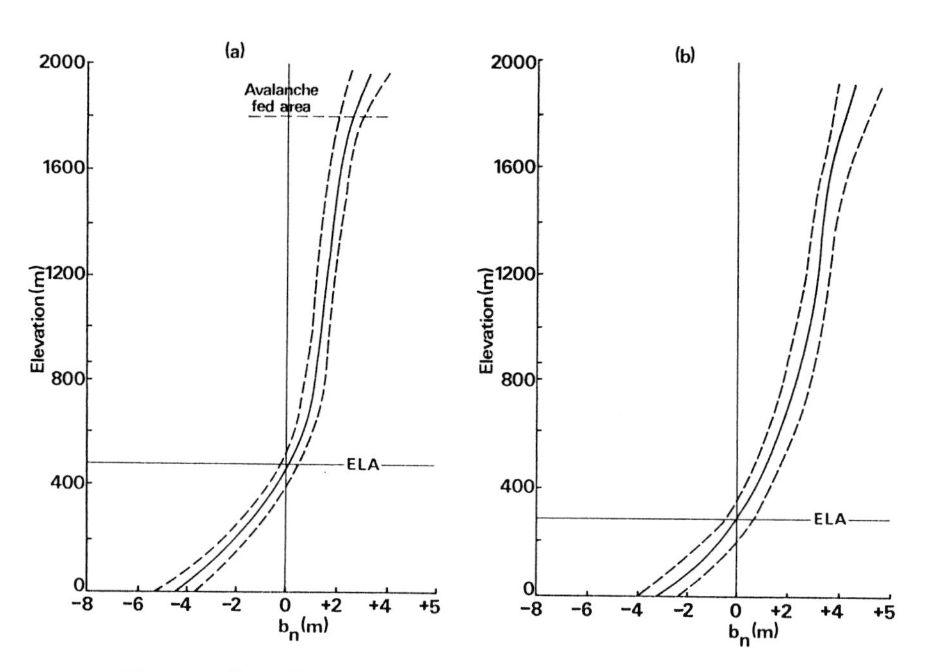

Figure 5. Theoretical predictions of the mass balance budget gradient (a) for present-day conditions and (b) for the Little Ice Age or similar Neoglacial cooling in the Cumberland Bay area. The best estimate is indicated by a solid line and limits by the pecked line. See text for derivation.

122 m respectively. These data suggest that a reasonable generalization for ELA depression in South Georgia during the Little Ice Age and Neoglacial periods was of the order of 50 m and 100 m respectively. If the apparent uniformity of the ELA lowering is real, then it is reasonable to infer that there has been little change in storm track directions and that annual precipitation has not changed significantly. In such circumstances the climatic change would be one of cooler

temperatures. Using radiosonde records from Grytviken and lapse rates

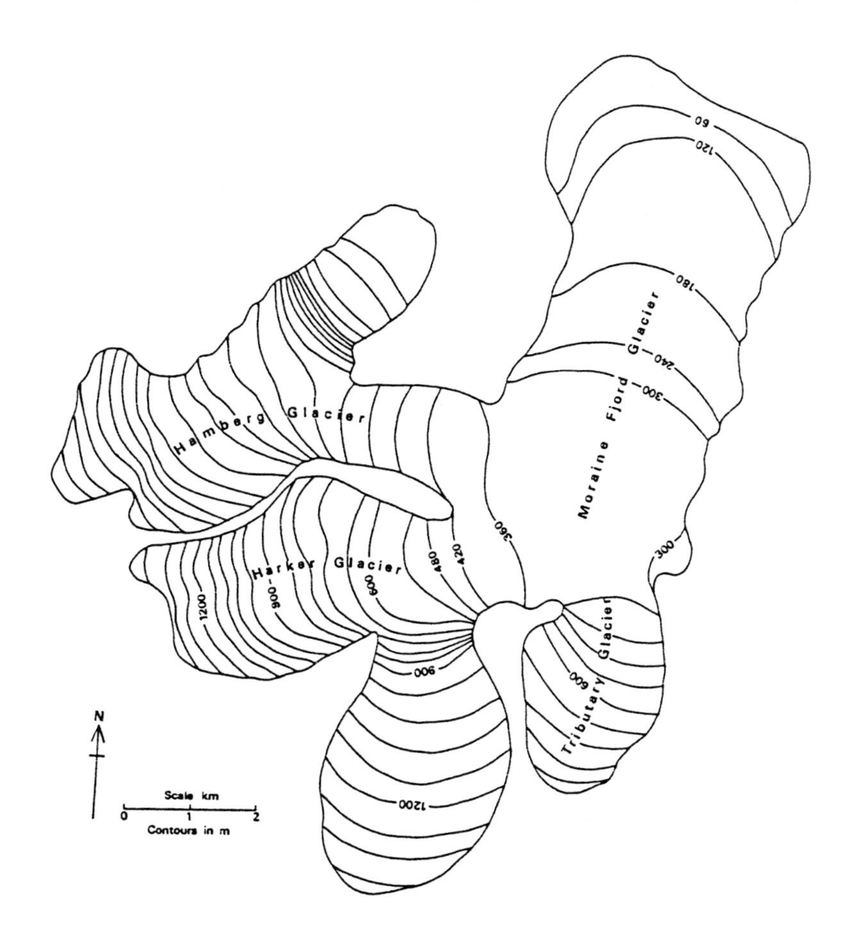

Figure 6. A reconstruction of the Neoglacial Moraine Fjord glacier based mainly on lateral moraines in the ablation zone.

from similar climates a lowering of 50 m and 100 m in the ELA would represent cooling of 0.5°C and 1.0°C for the Little Ice Age and Neoglacial periods respectively. Annual precipitation could have increased by c. 15 % but no more, otherwise the glacier response would not have been uniform.

The pattern of a 1°C Neoglacial cooling and a 0.5°C cooling in the Little Ice Age is well within average global estimates as, for example, outlined by Porter (1981, 1986). In the case of South Georgia the climatic conditions most likely to have caused both the Neoglacial and Little Ice Age cooling is an increase in air flow from the south associated with a northward movement of the Weddell Sea Low and a

Table II. Little Ice Age and Neoglacial ELAs for 9 valley glaciers (A) and 9 cirque glaciers (B) in South Georgia; estimated from the linear mass balance equation.

A: Valley glaciers

Glacier	Present ELA (m)	Little Ice Age ELA	Lowering	Neoglacial ELA	Lowering
Geikie	568	526	42	451	117
Lyell West	459	418	41	356	103
Lyell East	322	280	42	261	61
Hamberg Lakes	798	733	65	657	141
Hamberg	863	815	48	560	303
Harker	758	715	43	560	198
Tributary	623	573	50	560	63
Heaney	465	439	26	424	41
Nachtigal	364	306	58	286	78
			Mean = 46		Mean = 122
			Median = 43		Median = 98

B: Cirque glaciers

Glacier	Present ELA (m)	Little Ice Age ELA	Lowering	Neoglacial ELA	Lowering
Leith	386	346	40	295	91
Stromness	394	335	59	282	112
Husvik A	402	331	71	313	89
Husvik B	483	432	51	402	81
Husvik C	442	370	72	283	159
Hodges	440	390	50	374	66
Glacier Col	429	385	44	358	71
Moltke	485	456	29	426	59
Binary Peaks	485	449	36	422	63
			Mean = 50		Mean = 88
			Median = 50		Median = 81

corresponding decrease in warmer air masses from the west and north. This conclusion is supported by correlation of the mass balance changes of Hodges Glacier and the twentieth century climatic record at Grytviken. Timmis (1986) has shown that temperature was the main controlling factor and that periods of positive mass balance in the years 1921/22 to 1950/51 correlated with periods of more southerly air flow while negative mass balances during 1951/52 to 1980/81 correlated with more westerly and northerly air flow.

7. EXAGGERATED RESPONSE AND POSSIBLE ANOMOLOUS AGE OF THE MORAINE FJORD NEOGLACIAL GLACIER

The remaining question is, why did a Neoglacial climatic deterioration cause an apparently anomalous response from Moraine Fjord glacier? This kind of problem was first addressed by Mercer (1961) and recently discussed by Mann (1986) who drew attention to Alaskan fjord glaciers that may have fluctuated asynchronously with one another and with climate. Mann suggested that possible causes for anomalous fjord glacier responses to climatic change involve the instabilities inherent in a calving terminus, the interaction of fjord geometry with the extent of the calving terminus, and the intersection of a low-lying snowline by a thickening glacier. A mean Neoglacial ELA lowering of c. 100 m would have evoked a response from all glaciers in South Georgia, but the advance of that entering Moraine Fjord was much more expansive than most other glaciers on the island. The ability of a fjord glacier to advance, however, depends on the formation of a moraine bank that protects it from large calving loss (Brown et al., 1982). Without a bank, tidewater glaciers respond to ice accumulation by advancing only a short distance into deeper water where loss through calving increases. The relationship between water depth and calving flux is so sensitive that in practice it is uncommon for a glacier to advance into water which is more that 150 m deep.

The bathymetry of Moraine Fjord is shown in Fig. 3 and the crucial feature is a basin with a depth in excess of 180 m. It is possible to calculate whether the amount of ice being discharged during the Neoglacial advance was sufficient to keep up with predicted calving rates and, if not, what the maximum depth of the moraine bank must have been to allow an advance. The method used to calculate discharge was to superimpose a mass balance budget gradient curve onto the area/altitude distribution of the reconstructed glacier. The mass balance gradient curve is shown in Fig. 5b and is derived from general climatic principles as in the case of the present day curve, but with an assumed cooling of c. 1°C. Calculations reveal that the volume of ice at the snout available for calving was 33×10^6 m^3. The volume lost by calving must therefore have been less than this to permit the glacier to advance.

The calving flux was estimated by reference to an empirical relationship showing the mean velocity of calving ($\overline{V}_c$) to mean water depth ($\overline{h}_w$). Figure 7 shows the relationship found both by Brown et al. (1982) and by Pelto from studies of Alaskan glaciers. The Brown et al. (1982) relationship may over-estimate the calving velocity, however, because it is based mainly on retreating glaciers and glacier velocity measurements restricted to one or two months of the ablation season. The alternative relationship shown in Fig. 7 is based on velocities of selected Alaskan glaciers observed over one-year periods in 1977-78 and in 1983-84 (Pelto, 1987).

Using the empirical relationship the calving flux Q_c can be obtained from

$$Q_c = \overline{W}_T \times \overline{h}_T \times \overline{V}_c \qquad (2)$$

where $\overline{W}_T$ = mean glacier width, $\overline{h}_T$ = mean glacier thickness and $\overline{V}_c$ = mean calving velocity.

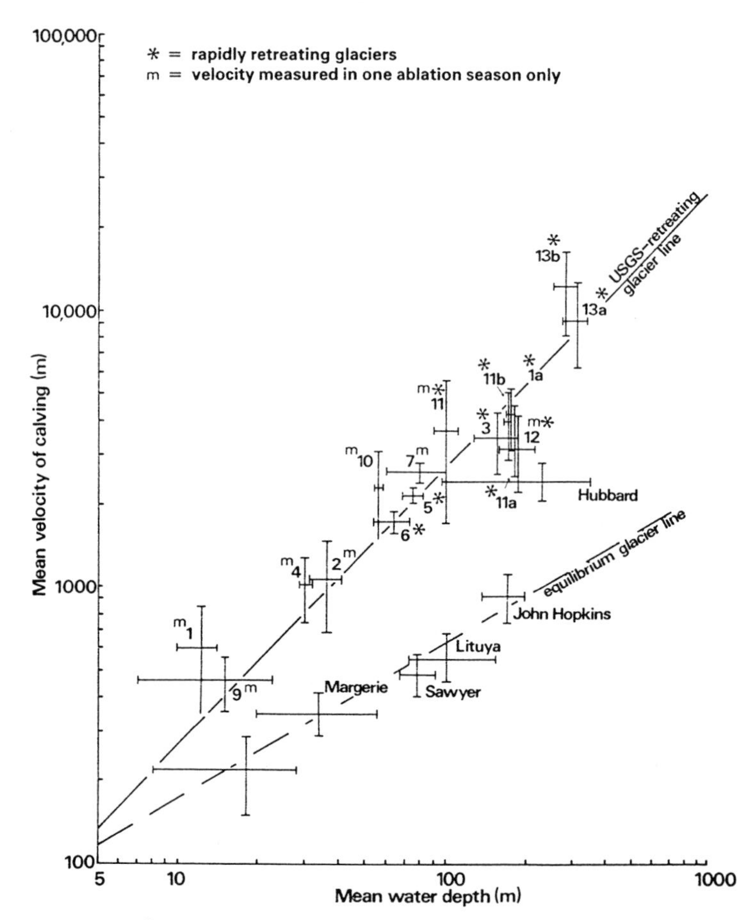

Figure 7. Logarithmic plot of mean calving velocity v_c as a function of mean water depth $\overline{h}_w$, showing Brown et al.'s 1982 relationship and Pelto's alternative relationship.

Table III shows that the calving flux of Moraine Fjord glacier at its deepest point would be 250×10^6 m^3. This is a figure about seven times larger than the available ice discharge (Fig. 8a) and, so big is the discrepancy, that it seems safe to conclude that the glacier could not have advanced down the fjord without a moraine bank. One can go further and derive the depth of water necessary for the glacier to advance, assuming an available ice discharge of 33×10^6 m^3. Taking

Table III. The calving flux necessary for Moraine Fjord glacier to remain in equilibrium at different points in Moraine Fjord, given existing fjord depths. For comparison, the available discharge calculated from mass balance reconstruction is 33×10^6 m^3.

Glacier	Mean water depth $\overline{h}_W$ (m)	Mean glacier width $\overline{W}_T$ (m)	Mean glacier thickn. $\overline{h}_T$ (m)	Calving velocity $\overline{V}_C$ (m)	Calving flux Q_C (m^3)
Harker (present)	35	900	75	240	16.2×10^6
Hamberg (present)	30	800	70	195	10.92×10^6
Moraine Fjord (Little Ice Age)	75	1500	115	500	86.25×10^6
Moraine Fjord Neoglacial/ deepest	110	2800	150	600	250×10^6

values of $\overline{W}_T$ = 1500 m and $\overline{h}$ = 120 m, (based on the glacial geomorphology) then the mean calving velocity ($\overline{V}_C$) is 150 m a^{-1}. Such velocities reflect water depths of < 10 m (Fig. 8b). This implies that Moraine Fjord glacier could have only advanced down-fjord if pinned to a moraine bank rising to within < 10 m of the sea surface. Such a conclusion is reinforced by calculation of the calving flux of the present day and Little Ice Age snouts. The Little Ice Age calving flux would have been 86×10^6 m^3, an amount in excess of what was available, and explains why the glacier could not extend further down the fjord at the time. We calculated the present day calving flux of Hamberg Glacier as 27×10^6 m^3, which is in good agreement with the observations of Smith (1960). This is less than that available during periods of cooling and would explain why the glacier was able to advance c. 500-700 m into deeper water during the Little Ice Age.

The implication of these calculations is that a particular blend of fjord topography, sediment availability and climatic deterioration caused an advance of Moraine Fjord glacier which was anomalous in size and age. The distance of the advance of 6.5 km was determined by the topography of the fjord and it only ceased where the fjord opened out. The large size of the moraines compared to other glaciers is perhaps the result of the incorporation of soft fjord sediments (Fig. 8c). It is also possible that the moraine is composite; a palaeosol with a radiocarbon age of 2000 ± 50 BP lies beneath 3 m of moraine and is developed on an earlier till forming part of the moraine bank. Thus Moraine Fjord glacier may have advanced to the fjord mouth earlier in the Holocene and/or during a Late-glacial stage. In contrast, other fjord glaciers such as the Nordenskjöld advanced only a few hundred metres in front of their present margins and Little Ice Age moraines during the Neoglacial (Fig. 2). Presumably these positions are constrained by water depth and will remain so unless a moraine bank builds up. The small size of moraines reflects a normal sediment supply to the grounded edges of Nordenskjöld glacier.

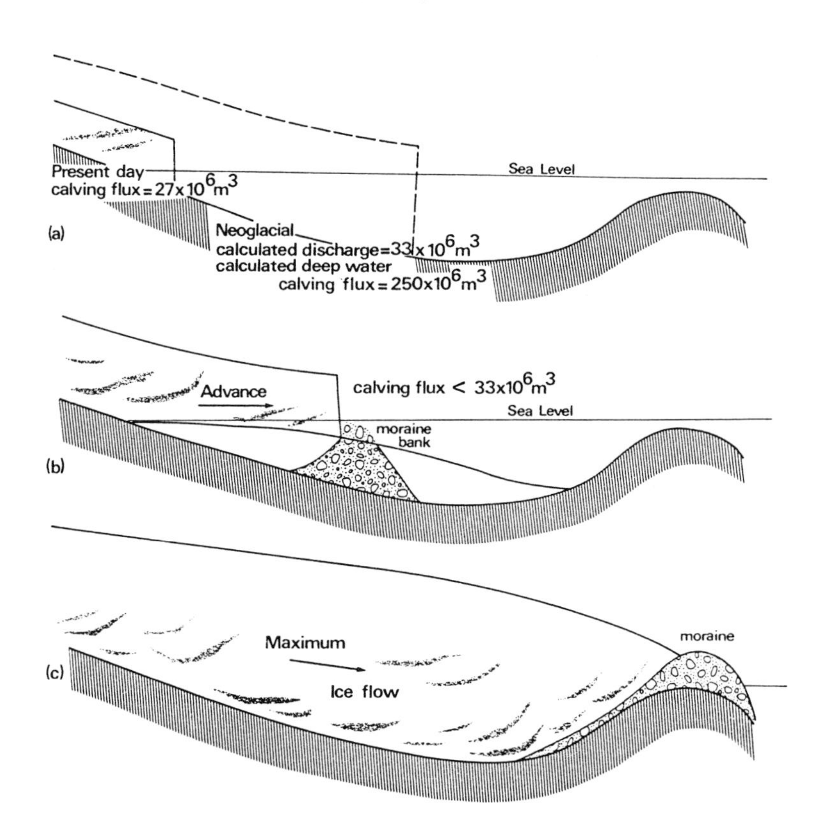

Figure 8. Schematic diagram to show the manner in which Moraine Fjord glacier advanced to from large Neoglacial moraines. The calculations show (a) The apparent impossibility of advance given the existing basin topography, (b) the postulated mechanism of advance and (c) the construction of the moraine on the basin lip.

The final question to be considered is why the Moraine Fjord advance is apparently not synchronous with comparable Neoglacial moraines in South America. One answer is that there may be no reason to expect them to be synchronous. The Moraine Fjord advance could have been triggered by the progressive accumulation of sediment reaching a critical limiting water depth and thus have no climatic implications whatsoever. On the other hand the increased ice discharge associated with cooling is an advantage, so that such advances are most likely to have been confined to the Neoglacial period. It is likely, however, that glacier advances determined partly by sediment flux and moraine bank construction will culminate some time after the initial cooling. For example, Mann (1986) has shown that glaciers in the Lituya Bay area, Alaska, took 200 years to retreat c. 10 km during the Little Ice Age, but require at least 600 years to readvance an equivalent distance. This is because the calving retreat of large fjord glaciers in Alaska may reach typical rates of 500 - 10,000 m a^{-1} (Meier et al., 1980), whereas rates of advance are more commonly 10 - 40 m a^{-1}.

We therefore suggest the following sequence of events in Moraine Fjord as a working hypothesis. Neoglacial cooling caused Moraine Fjord glacier to advance as far as the fjord mouth sometime before 2000 radiocarbon yrs BP; part of the moraine bank was deposited at that time and a shallow soil formed during a subsequent ice-free period lasting at least until c. 1700 radiocarbon yrs BP. It is not known how far the glacier receded during this interval, but a withdrawal of 6.5 km to the fjord head could have been accomplished in less than 15 years if the calving rate averaged c. 500 m a^{-1}. A second readvance buried vegetation growing on the soil soon after 1700 radiocarbon yr BP, but before 1460 radiocarbon yr BP. It is possible that the post 1700 BP readvance was a lagged response to the global cooling that caused some Patagonian glacier to advance at 2000 - 2700 BP (Mercer, 1976). If Moraine Fjord glacier responded to the same climatic reversal and advanced down-fjord at a rate of 10 - 14 m a^{-1}, then a lag of 450-650 years behind land-terminating Patagonian glaciers is to be expected.

This hypothesis now requires testing, for example, by discovering whether other Neoglacial moraines in South Georgia are older or whether other palaeoenvironmental indicators confirm an earlier period of cooling.

8. CONCLUSIONS

The geographical location of South Georgia implies that Holocene climatic fluctuations affecting the island should have been synchronous with and of similar scale to, those in Patagonia and coastal Antarctica. A limited Neoglacial advance of several hundred metres, but in one or two cases by more than 6 km, represents a cooling of 1°C. An as yet poorly dated Little Ice Age advance represents a cooling of 0.5°C. Both estimates agree with global trends.

Overall the scale of the glacier advances in South Georgia are small when compared to Patagonia or Alaska. This may reflect the relative aridity of the northeastern side of the island, where glacier mass balance characteristics are similar (on small cirque glaciers) to those in the more continental parts of Norway (Timmis, 1986). Alternatively, the amount of temperature depression in Sub-Antarctic latitudes may have been slightly less than in temperate latitudes.

The relatively large extent of the Neoglacial advance of Moraine

Fjord glacier and the large size of its terminal moraine complex reflect the behaviour of a moraine-banked fjord glacier. The advance radiocarbon dated at c. 1700-1460 BP lagged behind any Neoglacial fluctuations of land-terminating glaciers because of the time required to move a moraine bank all the way down-fjord. This is an important conclusion because, by virtue of the extent of the advance and the clarity of the depositional features, this kind of morainic assemblage is most commonly studied in detail, date, and directly related to periods of climatic deterioration; yet typically, such advances may lag behind the climatic perturbation by several hundred years.

ACKNOWLEDGEMENTS

Fieldwork in South Georgia was supported by Grant No. GR3/5199A from the U.K. Natural Environmental Research Council. We appreciate the field help of Dr. R.V. Birnie and Dr. J. Birnie, the detachment of H M troops at King Edward Cove, and the naval and air support of H M S Endurance. We are most grateful to Larry Williams for assistance in calculating ELAs. Dr. D.D. Harkness of the NERC Radiocarbon Laboratory, East Kilbride, kindly provided the radiocarbon dates.

REFERENCES

Braithwaite, R.J. 1984. Can the mass balance of a glacier be estimated from its equilibrium line altitude? Journal of Glaciology 30 (104), 364-368.

Brown, C.S., Meier, M.F. and Post, A. 1982. Calving speed of Alaskan tidewater glaciers with applications to the Columbia Glacier, Alaska. U.S. Geological Survey Prof. Paper 1258-C.

Clapperton, C.M. 1971. Geomorphology of the Stomness Bay - Cumberland Bay area, South Georgia. British Antarctic Survey, Scientific Report no. 70.

Clapperton, C.M. 1983. The glaciation of the Andes. Quat. Science Reviews 2, 83-155.

Clapperton, C.M. and Sugden, D.E. 1980. Geomorphology of the St. Andrews Bay - Royal Bay area, South Georgia. British Antarctic Survey, Miscellaneous Map Series, sheet 1.

Clapperton, C.M. and Sugden, D.E. 1987. Holocene glacier fluctuations in South America and Antarctica. XIIth INQUA Congress Symposium (in press).

Clapperton, C.M., Sugden, D.E., Birnie, R.V., Hansom, J. and Thom, G. 1978. Glacier fluctuations in South Georgia and comparison with other island groups in the Scotia Sea. In: Antarctic Glacial History and World Palaeoenvironments, E.M. van Zinderen Bakker (Ed.). SCAR Symposium Proc. Xth INQUA Congress, Birmingham, U.K., 1977. Balkema Rotterdam 98-104.

Gordon, J. 1987. Radiocarbon dates from Nordenskjöld Glacier, South Georgia, and their implications for late Holocene glacier chronology. British Antarctic Survey Bulletin (in press).

Hayward, R.J.C. 1983. Glacier fluctuations in South Georgia, 1883-1974. British Antarctic Survey Bulletin 52, 47-61.
Karowski, L. 1891. Die Hohe der Schneegrenze mit besonderer Berücksichtigung der Finsteraarhorn-Gruppe. Geographische Abhandlungen hrsg. von Albrecht. Penck, Band 5, Heft 1, 119-160.
Liestøl, O. 1967. Storbreen glacier in Jotunheimen, Norway. Norsk Polarinstitutt, Skrifter nr. 141.
Jenne, R.L., Crutcher, H.L., Van Loon, H. and Taljaard, J.J. 1974. A selected climatology of the Southern Hemisphere. NCAR-TW/STR-92, National Centre for Atmosphere Research, Boulder, Colorado.
Manabe, S. and Hahn, D.G. 1977. Simulation of the tropical climate of an Ice Age. Journal of Geophys. Res. 82 (27), 3889-3911.
Mann, D.H. 1986. Reliability of a fjord glacier's fluctuations for palaeoclimatic reconstructions. Quaternary Research 25, 10-24.
Meier, M.F., Rasmussen, L.A., Post, A., Brown, C.S., Sikonia, W.G., Bindschadler, R.A. Mayo, L.R. and Trabant, D.C. 1980. Predicted timing of the disintegration of the lower reach of Columbia Glacier, Alaska. U.S. Geological Survey Open File Report 80-582.
Mercer, J.H. 1961. The response of fjord glaciers to changes in the firn limit. J. of Glaciology 3, 1053-1062.
Mercer, J.H. 1976. Glacial history of southernmost South America. Quaternary Research 6, 125-166.
Mercer, J.H. 1983. Cenozoic glaciation in the Southern Hemisphere. Annual Review of Earth and Planetary Science 11, 99-132.
Pelto, M.S. 1987. Mass balance of southeast Alaska and northwest British Columbia glaciers from 1976-1984, methods and results. Annals of Glaciology 9, 111-117.
Porter, S.C. 1981. Glaciological evidence of Holocene climatic change. In: Climate and History, T.M.L. Wigley, M.J. Ingram and G. Farmer (Eds.), 82-110. Cambridge University Press.
Porter, S.C. 1986. Pattern and forcing of Northern Hemisphere glacier variations during the last millenium. Quaternary Research 26, 27-48.
Schwerdtferger, P. 1970. The climate of the Antarctic. In: Climates of the polar regions, S. Orvig (Ed.), 253-355. Elsevier, Amsterdam. World Survey of Climatology.
Schwerdtferger, P. 1975. The effect of the Antarctic Peninsula on the temperature regime of the Weddell Sea. Monthly Weather Review 103, 45-51.
Sissons, J.B. and Sutherland, D.G. 1976. Climatic inferences from former glaciers in the south-east Grampian Highlands, Scotland. J. of Glaciology 17, 325-346.
Sugden, D.E. and Clapperton, C.M. 1977. The maximum ice extent of island groups in the Scotia Sea, Antarctica. Quaternary Research 7, 268-282.
Timmis, R. 1986. Glacier changes in South Georgia and their relationship to climatic trends. Unpublished PhD thesis, University of East Anglia.
Weller, G. 1980. Spatial and temporal variation in the south Polar surface energy balance. Monthly Weather Review 108, 2006-2014.

GLACIER CHANGES FOLLOWING THE LITTLE ICE AGE - A SURVEY OF THE INTERNATIONAL DATA BASIS AND ITS PERSPECTIVES

W. Haeberli, P. Müller, P. Alean and H. Bösch
Versuchsanstalt für Wasserbau,
Hydrologie und Glaziologie (VAW)
ETH Zürich, Switzerland

ABSTRACT

International monitoring of glacier variations started in 1894. Today, the World Glacier Monitoring Service (WGMS) of ICSI/FAGS and GEMS/UNEP collects standardized glacier information. The data basis is briefly described and illustrated. It includes observations on changes in length and, since World War II, also in mass, as well as recent glacier inventories. By far the highest information density is found for the Alps and Scandinavia.

About 90 years of observations clearly reveal a general shrinkage of mountain glaciers on a global scale. This effect was most pronounced during the first half of the 20th century; more recently, however, glaciers have started to grow again in several regions, especially on the maritime slopes of mountain ranges. For the first time, length changes of mountain glaciers as a reaction to well documented and strong signals in mass balance history can now be measured. Important empirical information has started to become available on the complex relationship between climate and glaciers.

1. INTRODUCTION

Worldwide collection of information about ongoing glacier changes was initiated in 1894 with the foundation of the International Glacier Commission at the 6th International Geological Congress in Zürich, Switzerland. It was hoped that long-term observation of glaciers would provide answers to two fundamental questions in the research of climatology and earth sciences at that time (Forel, 1895), namely:

1. What are the mechanisms of modern variations of climate and glaciers; in particular, are glacier fluctuations globally uniform, synchronous and, hence, caused by some extra-terrestrial influences or are they regionally variable, asynchronous and therefore brought about by terrestrial conditions?

J. Oerlemans (ed.), Glacier Fluctuations and Climatic Change, 77–101.

2. How should one understand the dramatic processes of the geologically most recent past, the Ice Age, the recognition of which having significantly contributed - in the historically most recent past - to the break-through of the theory of evolution?

Since then, the goals of international glacier monitoring have evolved and multiplied. The present contribution aims at summarizing the historical development, at briefly describing the various types of presently available data, and at giving some hints on the possibilities for their interpretation.

2. HISTORICAL DEVELOPMENT OF INTERNATIONAL GLACIER MONITORING

In the years 1895 to 1913, the various presidents of the Commission published annual reports on glacier variations in several regions of the world (Forel and Du Pasquier, 1896, 1897; Richter, 1898, 1899, 1900; Finsterwalder and Muret, 1901, 1902 and 1903; Fielding-Ried and Muret, 1904, 1905, 1906; Brückner and Muret, 1907, 1909, 1910, 1911a, 1911b; Rabot and Muret, 1912, 1913; Rabot and Mercanton, 1913, Hamberg and Mercanton, 1914). The reports were written in French, German, Italian and English, and gave mainly qualitative information (advance/retreat). Results of quantitative measurements were presented for a limited number of cases only. The most detailed data was available for glaciers in Europe (Alps, Scandinavia) and information from the southern hemisphere remained extremely rare. A general tendency for glacier shrinkage was recorded as well as some spectacular changes, such as the rapid advance of Vernagtferner in the Austrian Alps or the catastrophic disappearance of ice in Glacier Bay, Alaska - the largest retreat of glaciers ever directly observed by man.

During and between the two World Wars, the reports now prepared by the secretary of the newly formed Commission of Snow and Ice (later the International Commission on Snow and Ice, ICSI) became thinner and appeared at larger time intervals (Mercanton, 1930, 1933, 1936). They contained numerical data on length changes of glaciers in the Alps and Scandinavia. References to various interesting national reports can be found. Matthes (1934), for instance, gave the whole series of length variation measurements of Nisqually glacier since 1857 in his US-report for 1931-32. Signs of shrinking and glacier retreat clearly predominated with the exception of a short, but marked, advance of glaciers in the Alps around 1920. The reports which followed (Mercanton, 1948, 1952, 1954, 1957, 1960) continued to give numerical values on length changes of glaciers in the Alps, Scandinavia and Iceland. Glacier retreat still clearly predominated there during this time period.

In 1967, the Permanent Service on the Fluctuations of Glaciers (PSFG) was established as one of the services of the Federation of Astronomical and Geophysical Services (FAGS) of the International Council of Scientific Unions (ICSU). This resulted in the publication of the "Fluctuations of Glaciers" at 5-yearly intervals. With the first volume (Kasser, 1967), mass balance data from various countries including USSR, USA and Canada entered the reports for the first time, thus forming the essential link between climate fluctuations and glacier length changes. In the second volume (Kasser, 1973), length variation data - showing signs of intermittent glacier advance - from the USA and the USSR as well as from other countries completed the corresponding records from the Alps, Scandinavia and Iceland, where most glaciers continued to retreat. The third and fourth volumes (Müller, 1977; Haeberli, 1985a) saw a major step towards standardization and computer based processing of data. Length variation data from the southern hemisphere (Peru, Argentina, Kenya, New Zealand and others) were now also regularly included. Glacier advances were reported from various parts of the world, especially from the Alps, where mass balances had been predominantly positive since the mid 60s. For the first time, therefore, empirical information has started to become available about glacier reactions to well documented and strong signals in mass balance history.

The World Glacier Inventory (WGI) was planned to be a snapshot of ice conditions on Earth during the second half of this century. Within the framework of the Global Environmental Monitoring System (GEMS) of the United Nations Environment Programme (UNEP), a Temporary Technical Secretariat (TTS/WGI) started operations in 1976, as another service of ICSI. Detailed and preliminary regional inventories were compiled all over the world to form a modern statistical basis on the geography of glaciers and to update and complete earlier compilations (especially Field, 1975 and Mercer, 1967). From the very beginning, inventory data were formatted to allow computerized data processing.

1986 saw the start of the new World Glacier Monitoring Service (WGMS), combining the former two ICSI-services (PSFG and TTS/WGI). The importance of glacier fluctuation and inventory data as high priority key variables in climate system monitoring (WMO, 1983), as a basis for hydrological modelling with respect to possible CO_2-effects (US Department of Energy, 1985) and as fundamental information in glaciology, glacial geomorphology and quarternary geology had become undisputable. The tasks of the WGMS for the near future are:

- to complete and continuously upgrade a global inventory of perennial surface ice masses;
- to continue collecting and publishing standardized glacier fluctuation data at 5-yearly intervals;
- to publish results of mass balance measurements from selected reference glaciers at about 2-yearly intervals;
- to include satellite observation of remote glaciers in order to reach global coverage;
- to periodically assess ongoing changes.

This work is being carried out at the VAW/ETH Zürich under the auspices of ICSI, FAGS, UNESCO and especially GEMS/UNEP. Data from WGMS flow into the World Data Center for Glaciology (WDC) and the Global Resources Information Database (GRID) of GEMS. At present, Volume V of the "Fluctuations of Glaciers" (1980-1985) and a World Glacier Inventory publication (status, 1985) are in preparation.

COUNTRY/REGION: P A K I S T A N : NANGA P DRAINAGE BASIN: 5 Q 1 3 0 / KABUL

COMPILED BY KICK, W., REGENSBURG, F.R.G.

IDENT	GLACIER NAME	-LATITUDE- DEGR	MIN	-LONGITUDE- DEGR	MIN	--COORDINATES--		ASPECT AC	AB	CLASS	TONGUE ACT	MOR	TOP MAP SC	YR	PHOTO T	YR	NR OF STATES	SURF IN PK [KM2]	DRAIN BASINS	NR OF CARDS
02 701	DIAMIR GLACIER	N 35	16.66	E 74	27.74	3905160	451150	SW	NW	510523	34 61	44	50	34			1	39.41	1	4
02 702		N 35	15.22	E 74	28.27	3902470	451850	N	N	630110			100	34			1	.73	1	3
02 703		N 35	15.18	E 74	28.06	3902400	451530	N	N	630110			100	34			1	.43	1	3
02 704		N 35	15.28	E 74	27.24	3902620	450300	N	N	620110			100	34			1	2.22	1	3
02 705	LOIBA GLACIER	N 35	13.94	E 74	25.76	3900150	448100	NW	NW	520110			100	34			1	11.90	1	3
02 706		N 35	13.62	E 74	25.33	3899550	447400	NW	NW	540120			100	34			1	4.07	1	3
02 707	AIRL GLACIER	N 35	12.76	E 74	24.73	3898000	446500	N	N	540110			100	34			1	10.57	1	3
02 801		N 35	18.68	E 74	28.90	3908860	452900	NW	NW	640110			50	34			1	.38	1	3
02 802		N 35	18.11	E 74	29.72	3907780	454150	N	N	650120		99	50	34			1	.21	1	3
02 803		N 35	17.97	E 74	30.10	3907550	454740	N	N	640210		99	50	34			1	.62	1	3
02 804	W OF PATRO GL	N 35	17.87	E 74	30.37	3907360	455120	N	N	650220		99	50	34			1	.13	1	3
02 805	PATRO GLACIER	N 35	18.46	E 74	30.67	3908490	455600	NW	N	520020		22	50	34			1	7.55	1	3
02 806	N OF S JILIPER P	N 35	19.65	E 74	32.45	3910680	458300	N	N	630110		22	50	34			1	1.13	1	3
02 901		N 35	21.31	E 74	31.12	3913750	456300	N	NW	640110		99	50	34			1	.28	1	3
02 902	HAENGEGL. IN MAP	N 35	21.67	E 74	31.71	3914410	457230	N	N	650220		22	50	34			1	.45	1	3
02 903		N 35	21.19	E 74	32.72	3913500	458710	N	N	640113	34	42	50	34			1	1.32	1	3
03 101	GANALO GLACIER	N 35	19.54	E 74	34.62	3910450	461610	N	NE	520320		22	50	34			1	4.84	1	3
03 102	OLD BRNCH O RAKH	N 35	19.42	E 74	34.90	3910230	462010	N	N	520320		22	50	34			1	5.86	1	3
03 103	RAKHIOT GLACIER	N 35	22.00	E 74	35.12	3915015	462408	NW	N	510311	34 54	42	50	34	D	34	1	38.72	1	6
03 104	W OF GR CHONGRA	N 35	19.49	E 74	38.68	3910370	467750	SW	SW	630210		88	50	34			1	.34	1	3
03 105		N 35	20.18	E 74	38.34	3911650	467260	SW	SW	640210		88	50	34			1	.20	1	3
03 106		N 35	20.39	E 74	38.13	3912010	466930	S	S	630210		00	50	34			1	.11	1	3
03 107	S OF BULDAR PEAK	N 35	20.71	E 74	37.97	3912600	466710	W	W	630210		00	50	34			1	.17	1	3
03 108	S OF BULDAR PEAK	N 35	20.90	E 74	37.86	3912940	466540	W	W	650210		22	50	34			1	.15	1	3
03 109	W OF BULDAR PK	N 35	21.01	E 74	37.57	3913150	466160	W	SW	630310		22	50	34			1	.54	1	3
03 110	NW OF BULDAR PK	N 35	21.56	E 74	37.34	3914180	465800	W	W	630210		28	50	34			1	.57	1	3
03 111		N 35	21.86	E 74	37.80	3914720	466500	W	W	630210		88	50	34			1	.22	1	3
03 201	N.KAPPENGL.INMAP	N 35	23.00	E 74	38.69	3916860	467770	NE	NE	600110		22	50	34			1	1.92	1	3
03 202	S.KAPPENGL INMAP	N 35	22.62	E 74	39.43	3916140	468900	NE	NE	630110		22	50	34			1	1.25	1	3
03 203	S OF BULDAR PEAK	N 35	21.00	E 74	38.70	3913170	467780	NE	NE	630320		88	50	34			1	.26	1	3
03 204		N 35	20.59	E 74	38.85	3912380	468000	NE	NE	650220		88	50	34			1	.12	1	3
03 205	UPPER BULDAR GL	N 35	22.22	E 74	41.00	3915420	471310	N	NE	510320		29	50	34			1	7.91	1	3
03 206		N 35	20.82	E 74	40.16	3912820	470010	NW	NW	630320		88	50	34			1	.53	1	3
03 207	BULDAR GLACIER	N 35	23.83	E 74	41.17	3918400	471520	N	N	510320		22	50	34			1	10.86	1	3
03 208	IN MAP BU R GL U	N 35	22.53	E 74	42.68	3915970	473900	NW	NW	680120		42	50	34			1	2.28	1	3
03 209	IN MAP BU R GL 2	N 35	23.16	E 74	43.05	3917150	474480	W	W	630010		24	50	34			1	1.16	1	3
03 210	GL 3 IN MAP	N 35	23.68	E 74	43.24	3918050	474760	W	W	630110		40	50	34			1	.54	1	3
03 211	GL 1 IN MAP	N 35	24.48	E 74	42.71	3919550	473940	NW	W	640100		20	50	34			1	1.10	1	3
03 212	MUTHAT GLACIER	N 35	25.62	E 74	42.44	3921650	473520	NW	W	630110		42	50	34			1	1.08	1	3
03 301		N 35	27.00	E 74	40.89	3924190	471140	N	N	650120		00	50	34			1	.08	1	3
03 302	W LICHAR GL	N 35	26.46	E 74	42.03	3923200	472850	NW	NW	630110		49	50	34			1	.34	1	3
03 303	LICHAR GLACIER	N 35	26.57	E 74	42.69	3923400	473880	N	NE	630110		44	50	34			1	1.14	1	3

NUMBER OF GLACIERS IN BASIN 5Q130 0: 42

COUNTRY/REGION: P A K I S T A N : NANGA P DRAINAGE BASIN: 5 Q 1 3 0 / KABUL

COMPILED BY KICK, W., REGENSBURG, F.R.G.

IDENT	--SURFACE AREA-- TOTAL [KM2]	A	EXPOSED [KM2]	MEAN WIDTH [KM]	MEAN LENGTH [KM]	-MAX LENGTH- TOTAL [KM]	EXPOS [KM]	MAX ELEV [M]	MED ELEV [M]	-LOW ELEV- TOT [M]	EXP [M]	-MED ELEV- ACC [M]	ABL [M]	-----SNOWLINE---- ELEV [M]	A	--DATE-- DA/MO/YR	-ABLAT.AREA- SURFACE [KM2]	LENG [KM]	MEAN DEPTH [M]	A	IDENT
02 701	39.41	1	37.74	.6	11.9	14.3	10.3	8000	5500	3615	4000	6000	4100	4650	3	61	9.15	7.9	130	5	701
02 702	.73	3	.00	.2	1.8	1.9	-	5570	5050	4180	-	5100	4500	4700	3	34	.18	.8	25	5	702
02 703	.43	3	.00	.2	1.2	1.4	-	5450	4900	4300	-	5000	4550	4750	3	34	.13	.6	20	5	703
02 704	2.22	3	.00	-	2.0	2.1	-	5570	4700	4300	-	4900	4470	4600	3	34	1.00	1.1	30	5	704
02 705	11.90	2	10.40	.7	5.2	5.5	3.3	5815	4900	4150	4650	5200	4650	4850	4	34	2.85	2.6	60	5	705
02 706	4.07	2	2.60	.5	4.7	4.7	3.6	5815	4700	4000	4200	4800	4300	4600	4	34	1.57	2.8	40	5	706
02 707	10.57	3	10.00	.7	6.0	6.2	5.8	5713	4750	3950	4050	4950	4270	4600	4	34	3.17	3.7	75	5	707
02 801	.38	2	.00	.3	1.2	1.2	-	4700	4370	4060	-	4450	4250	4300	3	9 34	.17	.7	20	5	801
02 802	.21	2	.00	.2	1.1	1.1	-	5200	4900	4470	-	5020	4600	4700	3	34	.07	.4	15	5	802
02 803	.62	2	.00	.2	1.4	1.5	-	5376	5000	4280	-	5100	4500	4650	3	34	.11	.6	18	5	803
02 804	.13	2	.11	.2	.8	.8	.7	4950	4550	4300	4350	4650	4450	4550	3	34	.06	.3	15	5	804
02 805	7.55	2	7.05	.8	4.2	5.1	4.5	6600	4900	3930	4040	5130	4200	4500	3	34	2.21	2.5	60	5	805
02 806	1.13	1	1.11	.3	1.9	1.9	1.8	5206	4850	4480	4500	4920	4680	4775	2	34	.39	.9	25	5	806
02 901	.28	2	.28	.2	.8	.9	.9	5100	4840	4560	4560	4900	4700	4775	2	34	.11	.4	15	5	901
02 902	.45	2	.40	.3	1.5	1.5	1.4	5180	4650	4360	4400	4850	4500	4550	3	34	.16	.6	17	5	902
02 903	1.32	2	1.32	.7	1.4	1.6	1.6	5043	4700	4500	4500	4750	4600	4650	2	34	.42	.6	25	5	903
03 101	4.84	2	2.78	.6	5.1	6.0	3.4	6606	4500	3900	4210	5450	4160	4650	3	54	2.62	3.9	40	5	101
03 102	5.86	2	4.75	.6	5.5	6.1	4.0	6500	4600	3780	4100	5400	4170	4550	3	54	2.77	3.9	50	5	102
03 103	38.72	1	34.88	.7	15.2	17.2	14.2	7840	5335	3210	3400	5900	3980	4650	3	54	11.23	12.8	100	5	103
03 104	.34	2	.34	.2	.9	.9	.9	5700	5450	5150	5150	5500	5270	5350	3	34	.09	.4	10	5	104
03 105	.20	2	.20	.3	.5	.5	.5	5400	5100	4900	4900	5170	5000	5050	3	34	.05	.2	10	5	105
03 106	.11	2	.11	.2	.6	.6	.6	5520	5320	5100	5100	5360	5170	5250	3	34	.02	.2	10	5	106
03 107	.17	2	.17	.2	.5	.6	.6	5587	5350	5050	5050	5380	5130	5200	3	34	.04	.2	10	5	107
03 108	.15	2	.15	.2	.7	.7	.7	5500	5280	4950	4980	5330	5060	5150	3	34	.04	.3	10	5	108
03 109	.54	2	.50	.3	1.3	1.4	1.3	5602	5250	4700	4780	5280	4900	5000	3	34	.12	.7	15	5	109
03 110	.57	2	.57	.3	1.5	1.6	1.6	5450	5050	4400	4400	5150	4680	4800	3	34	.15	.7	15	5	110
03 111	.22	2	.22	-	.5	.5	.5	5423	5220	5000	5000	5250	5050	5100	3	34	.05	.1	10	5	111
03 201	1.92	3	1.88	.8	2.3	2.5	2.4	5200	4940	4400	4430	4980	4670	4800	3	34	.51	1.1	30	5	201
03 202	1.25	2	1.20	.3	2.8	2.9	2.8	5423	4900	4130	4150	5000	4450	4700	3	34	.35	1.3	25	5	202
03 203	.26	2	.24	.2	1.1	1.1	1.0	5600	5300	4770	4820	5350	4900	5050	3	34	.06	.5	10	5	203
03 204	.12	2	.12	.2	.9	.9	.9	5240	4850	4630	4660	5030	4750	4850	3	34	.05	.4	10	5	204
03 205	7.91	2	6.79	.4	6.7	7.1	4.1	6830	5050	4000	4320	5250	4340	4500	3	34	2.29	4.5	50	5	205
03 206	.53	2	.53	.2	2.5	2.5	2.5	6565	5300	4420	4420	5500	4570	4800	3	34	.12	.8	20	5	206
03 207	10.86	2	8.09	.4	8.8	9.2	4.2	6565	4580	3245	4100	5450	4130	4550	3	34	5.16	6.5	60	5	207
03 208	2.28	2	2.20	2.0	1.4	1.6	1.5	5100	4690	4320	4350	4770	4550	4650	3	34	.92	.8	35	5	208
03 209	1.16	2	1.12	.6	1.6	1.6	1.5	5220	4870	4520	4540	4950	4650	4750	3	34	.33	.7	25	5	209
03 210	.54	1	.53	-	.8	.9	.8	5310	5000	4720	4750	5040	4830	4900	2	34	.19	.4	15	5	210
03 211	1.10	1	1.10	-	1.4	1.6	1.6	5250	4770	4460	4460	4900	4600	4700	2	34	.41	.8	25	5	211
03 212	1.08	2	.98	-	1.5	1.7	1.2	5040	4800	4480	4670	4850	4690	4750	2	34	.30	1.0	25	5	212
03 301	.08	3	.08	-	.5	.5	.5	4650	4550	4445	4445	4575	4530	4560	2	34	.05	.3	15	5	301
03 302	.34	2	.34	-	.7	.9	.8	5030	4825	4715	4720	4850	4760	4780	2	34	.11	.5	15	5	302
03 303	1.14	1	1.14	-	1.4	1.6	1.6	5030	4800	4580	4590	4850	4680	4750	2	34	.44	.6	20	5	303
TOTALS	163.69		142.02	MEAN ELEVATIONS				5630	4927	4367	4493			4763							
				STANDARD DEVIATIONS				738	279	442	367			223							
				NUMBERS OF GLACIERS				42	42	42	36			42							

3. MODERN DATA BASIS AND OBSERVATIONAL NETWORK

The international data basis on glaciers of the 20th century contains two different kinds of information: (1) glacier inventory data describing the spatial variability and (2) glacier fluctuation data documenting changes in time of perennial surface ice masses.

Instructions and guidelines for the compilation of standardized glacier inventory data have been developed by UNESCO (1970), Müller et al. (1977), Müller (1978) and Scherler (1983). Reports on the status of the World Glacier Inventory were published by Müller and Scherler (1979a, 1979b) and Scherler (1980).

Detailed glacier inventories include information on surface area, width, length, orientation, (highest, mean, median and lowest) elevation, morphology, tongue activity, moraine characteristics and snow line position. Ice volumes are usually estimated using simple and empirical (nonlinear) relations between surface area and mean ice depth (Müller et al., 1976; Driedger and Kennard, 1986). Table I is an example of a printout from a detailed glacier inventory, whereas Figs. 1 to 3 present information from a detailed glacier inventory in graphic form.

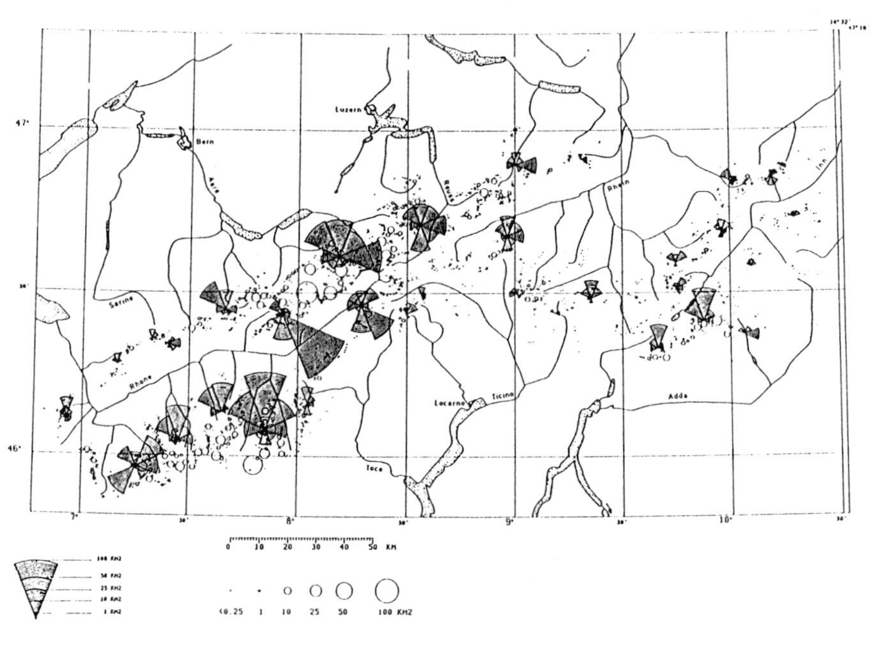

Figure 1. Surface areas and aspects of glaciers in the Swiss Alps. Data basis from F. Müller et al. (1976).

Surface areas are basic to modelling energy balance and runoff. With an anticipated CO_2-induced rise of the equilibrium line by 200 to 300 m in the coming 100 years (cf. Kuhn, 1985), a great number of small and thin glaciers in mountain ranges such as the Alps could disappear

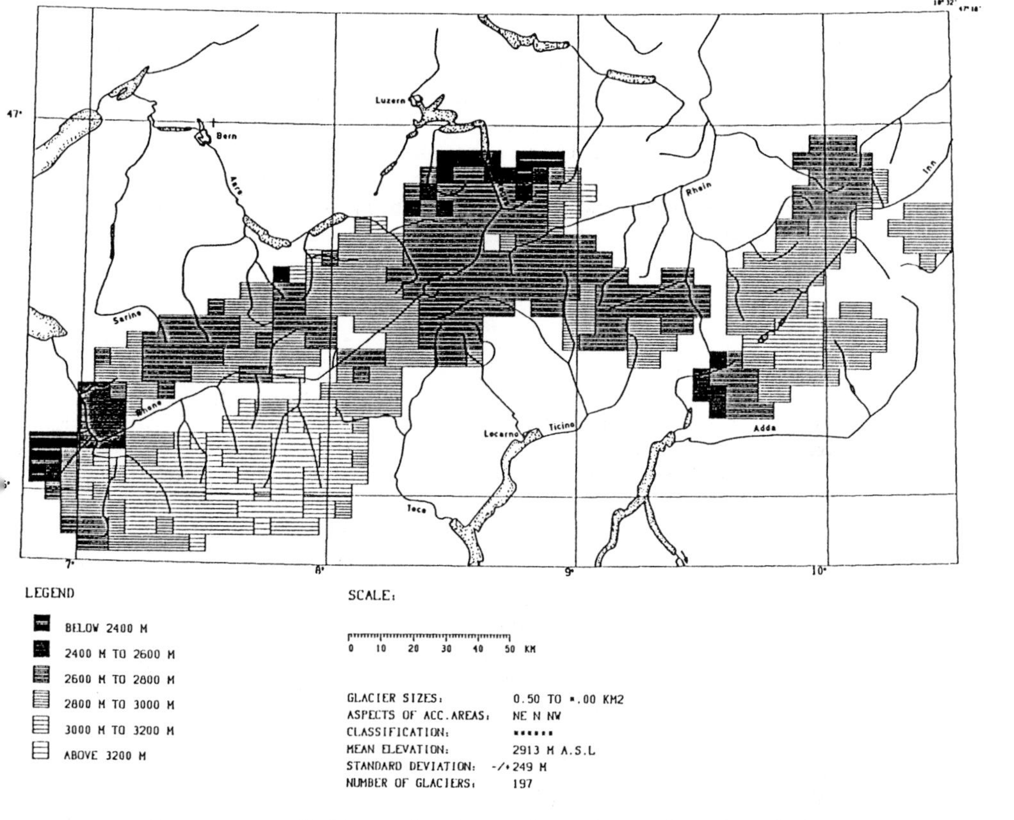

Figure 2. Median glacier elevations in the Swiss Alps. Data basis from F. Müller et al. (1976).

more or less completely, while the most drastic temporary changes in runoff would take place in drainage basins containing large glaciers. Glacier inventory information such as depicted in Figs. 1 and 3 helps to estimate such effects.

Mean glacier elevation is a rough approximation to equilibrium line altitude and, as such, is connected with continentality and, hence, with precipitation, mass balance gradients, activity (mass turnover), englacial temperature and glacier-permafrost relations (Haeberli, 1983). The Swiss example shown in Fig. 2, for instance, clearly reveals the contrast between the marginal zones (Mont Blanc, Central Alps) with heavy precipitation and low snowlines (mostly active, temperate glaciers ending below the permafrost belt) and the interior regions (Wallis, Engadin) with reduced precipitation and elevated snowlines (less active, partially cold glaciers often surrounded by permafrost).

Four levels exist for preliminary inventories. On level 1, satellite imagery only is used and the total glacierized area of a country is given. Levels 2 and 3 are also based on satellite imagery but, in addition, give names and co-ordinates of individual ice areas and drainage basins. Finally, level 4 combines information from satellite imagery and topographic maps. This level mainly deviates from detailed inventories in that aerial photographs are not used and resolution is correspondingly lower. In many countries such as Argentina, Canada or China, the national inventory is made up of a combination of both detailed and preliminary inventories. It is planned to continuously upgrade preliminary inventories and to periodically repeat detailed inventories at longer intervals.

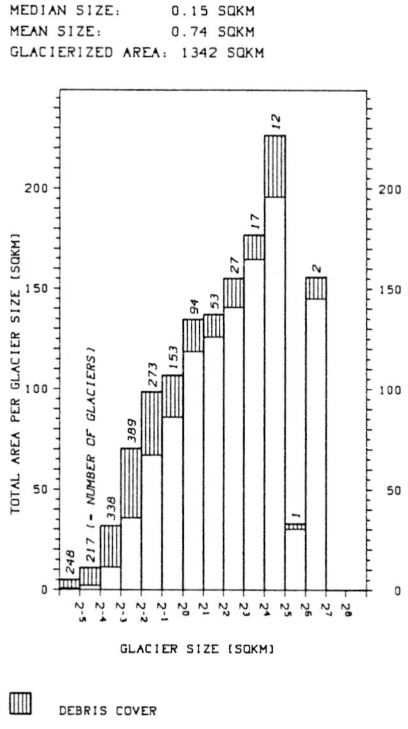

Figure 3. Distribution of glacier sizes in the Swiss Alps. Data basis from F. Müller et al. (1976).

Collection of standardized glacier fluctuation data follows recommendations published by UNESCO (1969) and regularly updated instructions for submission of data for the "Fluctuations of Glaciers" (cf. Haeberli, 1985a: p. 95–116). Due to the historical development of international glacier monitoring, these instructions first of all regard conditions usually encountered at latitudes poleward of about 40°. Methodological problems which should be solved as soon as possible still exist for tropical and subtropical zones. They mainly concern penitentes (cf. Leiva et al., 1986), seasonality of mass balance (cf. Ageta and Higuchi, 1984) and contacts between debris covered glaciers and perennially frozen rock glaciers (cf. Corte and Espizua, 1981; Haeberli, 1985c).

Length variation data are now being regularly collected for several hundred glaciers by using a variety of geodetic and photogrammetric techniques. Even though they are not interpretable in a straightforward way, these simple measurements still represent the essential link to the past and provide the most complete information about the spatial variability of glacier changes. Thereby, glaciers of different size indicate climate and balance changes of different frequencies, the strength of the obtained signal being approximately inversely proportional to the frequency of the recorded variation (Haeberli, 1985b). Figures 4 to 6 give examples from the well-documented Swiss Alps. Pizol Glacier (Fig. 4) is a small cirque glacier or "glacier réservoir" (Lliboutry, unpublished) which flows slowly under low basal shear stresses. Its altitudinal range is smaller than the annual variations of the equilibrium line altitude and its length changes are thus directly related to the mass balance in the corresponding year. The curve of length variations of Pizol Glacier at least qualitatively reflects the periods of distinctly positive mass balances during the late 60's and again during the late 70's - a clear, strong and fairly representative double signal in the mass balance history of Alpine glaciers (cf. Patzelt, 1987; Reynaud et at., 1984). Stein Glacier (Figs. 5 and 7) is an active mountain glacier or "glacier évacuateur" flowing under high basal shear stresses and reacting in a dynamic way to changes in mass balance. The length variation curve demonstrates a pronounced advance after the late 60's. Two maxima of advance velocity thereby occurred, each one delayed by a few years with respect to the corresponding maxima of cumulative mass balance and of related advances of cirque glaciers such as Pizol Glacier. Gorner Glacier, on the other hand (Figs. 6 and 8), one of the largest valley glaciers in the Alps, does not yet show any marked change in the length variation curve. Its reaction time is assumed to be longer than about 25 years.

Summary results of mass balance measurements - mainly gained by applying the "direct glaciological method" (stakes and pits) - are now being regularly reported for about 75 glaciers all over the Northern Hemisphere (cf. Collins, 1984). Selected time series of cumulative mass balances for 7 glaciers are presented in Fig. 9. The development is obviously not uniform, but a recent trend exists for some glaciers in maritime environments to increase in mass, whereas mass losses continue consistently in zones with dry, continental climates. Mass balance versus altitude, i.e. information on "balance gradients" is available in detail for about 20 glaciers. This information is fundamental to a deeper understanding of direct climate-glacier relationships (Kuhn, 1981), for modelling glacier flow and behaviour (Kruss, 1983) and for interpreting paleoglaciological reconstructions (e.g. Sugden, 1977; Haeberli and Penz, 1985). Changes in area, volume and thickness over time periods of a few years to a few decades document overall mass changes for another 30 to 40 glaciers. These results of mass balance measurements using the "geodetic method" can be used to check glaciological or hydrological mass balance determinations, but can also

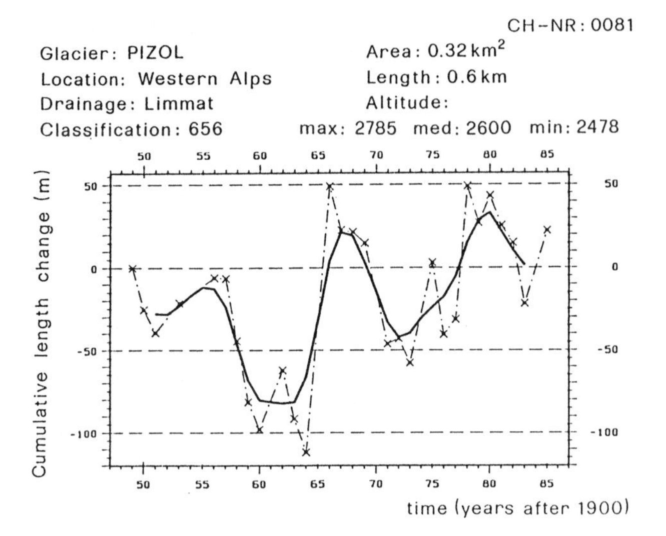

Figure 4. Recent cumulative length changes of Pizol Glacier, a small cirque glacier in the Swiss Alps. From P. Müller (1988). Solid line is a five-year centre-weighted running average.

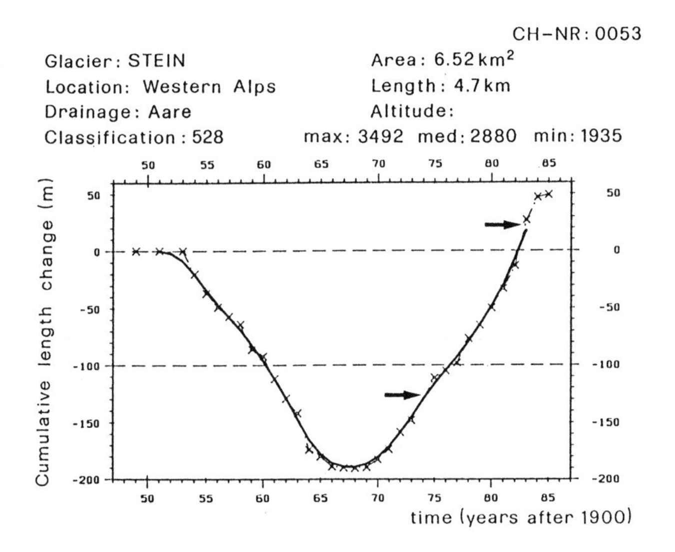

Figure 5. Recent cumulative length changes of Stein Glacier, an active mountain glacier in the Swiss Alps. Arrows point to maxima of advance velocity. From P. Müller (1988). Solid line is a five-year centre-weighted running average.

have an important predictive value. Based on data from Finsterwalder and Rentsch published in 1981 (cf. also Haeberli, 1985a), Patzelt (1985) demonstrates that an increase in ice thickness had been observed at high altitudes on eight Alpine glaciers as early as 1950 - 1959 (cf. the slight readvance of Pizol Glacier during the same period, Fig. 4). The main zone of increased thickness was then shifted towards lower altitudes in 1959 - 1969 and reached the glacier tongues in 1969 - 1979. The fact that surface elevations at high altitudes remained stationary during 1969 - 1979 is interpreted by Patzelt to indicate that the Alpine advance period is nearing its final stage.

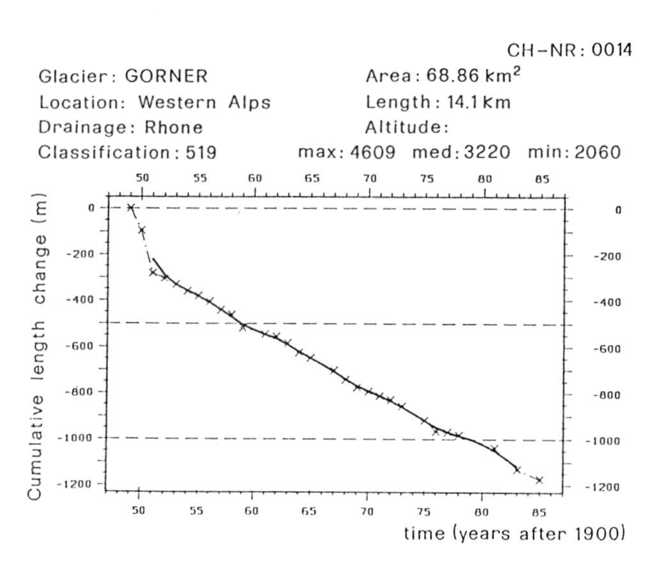

Figure 6. Recent cumulative length changes of Gorner Glacier, a relatively large valley glacier in the Swiss Alps. From P. Müller (1988). Solid line is a five-year centre-weighted running average.

Besides the data on glacier length changes and mass balance, information about the availability of hydrometeorological data is also given to facilitate climatic interpretation and modelling of observed glacier changes. In addition, an effort has now been made to internationally collect and publish short abstracts on special events such as glacier surges (Fig. 10), ice avalanches and glacier floods (Figs. 11 and 12), drastic retreats of tidal glaciers (Fig. 13), glacier - volcano interactions and others.

For all glaciers included in the regular fluctuations reports, general topographic and morphological characteristics, based on glacier inventory data, are given. This information helps to correctly interpret fluctuation data for individual glaciers and at the same time allows a description of the present-day observational network to be given. Figure 14 is a collection of four particularly instructive histograms which

Figure 7. The advancing snout of Stein Glacier. The ice front is now in contact with the proglacial lake, a situtation which could significantly alter the advance characteristics of the glacier in a manner independent of climatic evolution. Foto. P. Müller 1986.

afford a general overview. The sample of presently monitored glaciers is clearly dominated by the Alps (roughly 50% of all observed glaciers). This explains the peak percentage of glaciers around 10° longitude and 47° latitude. The high percentage of missing values is due to Russian glaciers for which no coordinates are reported. The Alps aside, the observed glaciers are quite well distributed in longitude (Eurasia), but glaciers at very high latitudes are under-represented. The same is valid for glaciers at very low latitudes and within the southern hemisphere. From the frequencies of altitudinal intervals (= difference between maximum and minimum elevation, which determines mass turnover, flow activity and basal shear stress of a glacier, cf. Haeberli (1985b)), roughly one third of the monitored glaciers have to be considered as being inactive cirque glaciers with more or less direct, non-dynamic response characteristics ("glaciers réservoirs"). Two thirds of the monitored glaciers are short (less than 6 km long); they can be assumed to have reaction times of a few years only (Müller, 1987) and to give clear and only moderately smoothed length reactions to mass balance changes, whereas 10 to 20% are long valley glaciers (12 km and more) probably giving strong, albeit heavily delayed and drastically smoothed, long-term signals.

4. DISCUSSION AND PERSPECTIVES

Worldwide glacier fluctuation data are being compared and interpreted in an increasing number of publications. Kislov and Koryakin (1986), for example, confirm that glaciers around the North Atlantic generally shrank after the end of the Little Ice Age. Makarevich and Rototayeva (1986) give a detailed compilation of the fluctuations of mountain glaciers in the Northern Hemisphere since 1950. They demonstrate that a retardation of the earlier retreat tendency is clearly documented, and that the development has become rather heterogeneous over the past decades with almost equally large numbers of advancing and retreating glaciers in various mountain ranges. Patzelt (1987) summarizes that the advance period of glaciers in the Alps is a consequence of a general mass gain of nearly 3 m between 1964 and 1980, which in turn is mainly due to a decrease in mean summer temperature by almost 1°C. The recent growth of Alpine glaciers is not an isolated phenomenon. Signs of mass gain and glacier advance are quite frequent on the maritime slopes of mountain ranges (North America, Iceland, Scandinavia, Himalayas), whereas mass losses and glacier retreat continue to predominate in dry continental areas (especially central Eurasia). Patzelt (1987) also emphasizes that

Figure 8. Gorner Glacier. The strikingly white (and polythermal) ice of the main tongue (Grenz Glacier) is a result of the fact that this air-bubble rich ice forms in the infiltration recrystallization zone at high altitudes (dry, cold firn metamorphosis). Photo: W. Haeberli 1985.

Lewis Glacier on Mt. Kenya lost about 50% of its volume between 1963 and 1983 and, hence, no longer follows the development of glaciers at higher latitudes in a distinct way. The heterogeneity of the development in the northern hemisphere may, in fact, be the most clearly pronounced difference with respect to the first half of the 20th century which was characterized by general atmospheric warming and glacier retreat (Wallén, 1984).

Statistical analysis of mass balance data (Letréguilly, 1984; Reynaud, 1980; Reynaud et al., 1984) from different regions of the world and based on a simplified version of the linear balance model by Lliboutry (1974) resulted in the important conclusion that the temporal

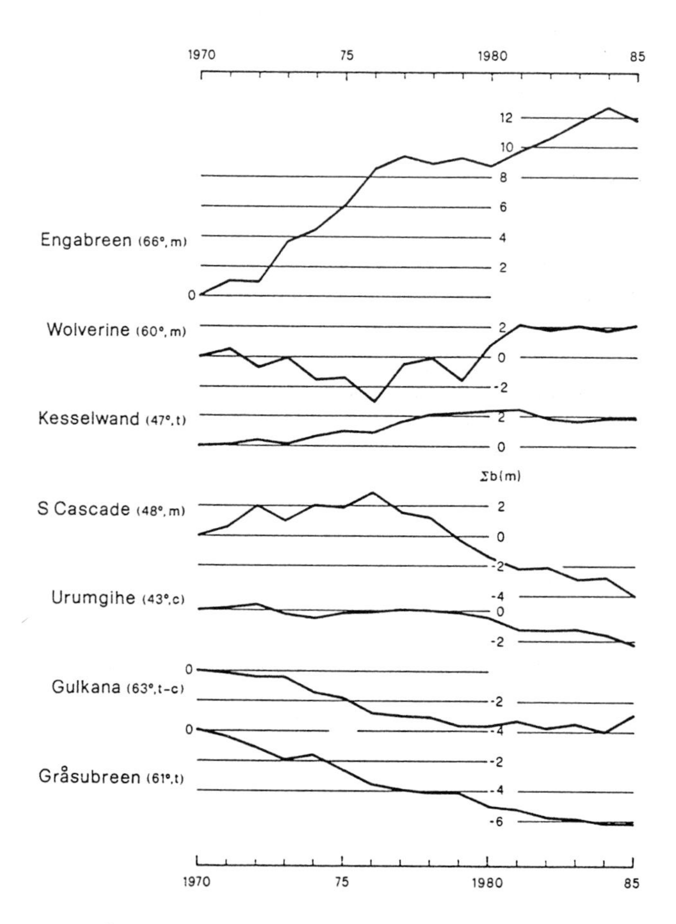

Figure 9. Cumulative mass balances of seven glaciers in the Northern Hemisphere. Data from Müller (1977), Haeberli (1985a) and from "Fluctuations of Glaciers 1980-1985" (in preparation), m = maritime, t = transitional, c = continental-type climatic conditions.

Figure 10. Glaciar grande del Nevado del Plomo in the Andes of Argentina immediately after its surge in 1985. The sudden advances of the glacier in the tributary valley occur with a return period of roughly 50 years, block the main valley and lead to increased flood hazards. Foto: W. Haeberli 1985.

variation of mass balance within a mountain range such as the Alps is strikingly predominant over the spatial variation. This means that the interpretation of glacier fluctuation data is facilitated by the homogeneous climatic signal. The homogeneity, however, reduces when a comparison between different mountain ranges or even different continents is made, and it reappears with the consideration of decadal trends only. Because mass balance measurements only started after World War II, reconstructed mass balance series based on statistically calibrated relations between mass balance and meteorological parameters have to be used for a global comparison of long-term trends. Vallon et al. (1986) indeed show that very similar trends in mass balance history since 1896 prevail in various regions of Eurasia, but that strongly contrasting developments may exist in North America. The assumption that local/regional mass balance gradients and, hence, relations between meteorological parameters and glacier mass balance remain constant in time is crucial to both the linear balance model and the reconstructed mass balance time series. It may be justified as a first order approximation for rough estimates and comparisons, but is not true in general. Mass balance gradients can undergo distinct temporal changes

(Kuhn, 1984, 1986) and the relative importance of parameters for the calibration of climate/balance models, such as air temperature and precipitation, can change with time (Müller, 1988). To further investigate these aspects, it is fundamentally important to continue the few already existing long time series of mass balance measurements at a high level of methodological sophistication and quality.

Figure 11. Ghiacciaio del Belvedere, Macugnaga, Italian Alps. The glacier is heavily covered with debris from the roughly 2000m-high east face of Monte Rosa and steadily builds up an elevated sediment bed. During the recent advance, ice from the orographic right tongue started to override lateral moraines dating from the Little Ice Age. In addition, Lago delle Locce at the foot of Monte Rosa was dammed up until 1979, when it produced a sudden outburst. The flood wave triggered a mudflow after having cut the orographic right lateral moraine near Alpe Pedriola. Foto: J. Alean 1985.

Figure 12. Advancing snout of Belvedere Glacier (orographic right tongue). The ice front (right) is completely covered with debris and small (now deformed) trees. It overrides a lateral moraine from the Little Ice Age (left) - a development which is probably not due to special climatic conditions, but to reduced melting during the 20th century (because of the insulating effect of the debris cover) and to the ongoing build up of a thick sediment bed. Foto: W. Haeberli 1984.

This leads directly to the limitations of the existing data basis and observational network; mass balance observations are sparse, cover short time periods only, are not well distributed over the continents and are therefore not representative for large areas. Length variation data are more multitudinous, but unevenly distributed over the continents. In addition, they are inhomogeneous with respect to accuracy and investigated time periods, and not easily comparable or interpretable. To design a completely new and better network would obviously be an academic exercise due to practical reasons. However, it should be feasible to complete the existing network and to make the best possible use of it.

Detailed mass balance measurements using the direct glaciological method periodically checked by the geodetic method can probably continue on a small number of glaciers only (about 10 to 20). The glaciers to be selected should already have a long time series of detailed observations and should represent maritime, transitional and continental type climates at high, middle and low latitudes. Tropical and subtropical glaciers should soon be included. The aim of these observations is to document long-term developments as a basis for modelling climate/glacier relationships in detail and for applying simpler methods to other glaciers. A much larger number of mass balance measurements (around 100) can be carried out using index methods (Körner, 1986; Reynaud et al., 1986) and repeated mapping at longer intervals (Finsterwalder and Rentsch, 1981) to investigate and document spatio-temporal variation patterns all over the world. The most important gap again has to be filled in the southern hemisphere and at very low latitudes. On flat ice

Figure 13. Columbia Glacier, Alaska, before its recent fast retreat. Catastrophic length reductions of tidal glaciers are the counterparts of glacier surges in that they are due to special mechanical conditions and not linked directly to climatic changes. Foto: W. Haeberli 1978.

caps without a too complicated topography, a combination of ice core drilling and the linear balance model could be applied to reconstruct mass balance time series for the past decades (cf. Reeh et al., 1978; Thompson et al., 1979 and 1988). The difficult problem of defining annual layers in temperate firn could possibly be solved by the application of pollen analysis (Vareschi, 1942) - a method which has not yet been fully exploited in glaciology. The proposed differentiation of mass balance measurements seems to be feasible and appropriate for documenting - at a reasonable expense - climate/glacier relations in general as well as regional trends.

A corresponding differentiation of length variation observations is also necessary. The high density of annual measurements carried out in the Alps cannot easily be matched in other parts of the world. The unique observational material being collected in the Alps, however, is an ideal basis for determining glacier reactions to mass balance changes. In analogy to the above-mentioned detailed mass balance

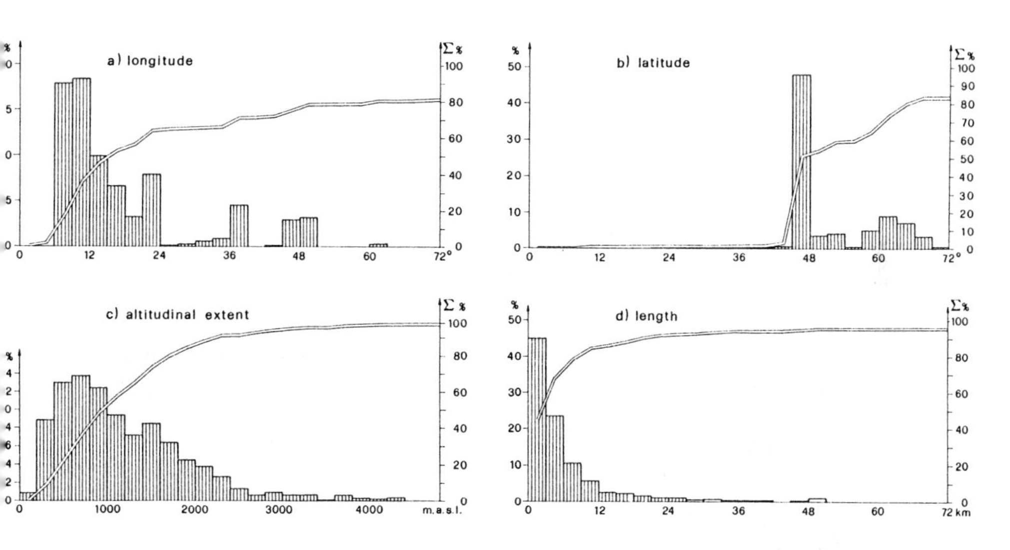

Figure 14. Histograms describing characteristics of glaciers which are being periodically monitored. Data basis: Haeberli (1985a).

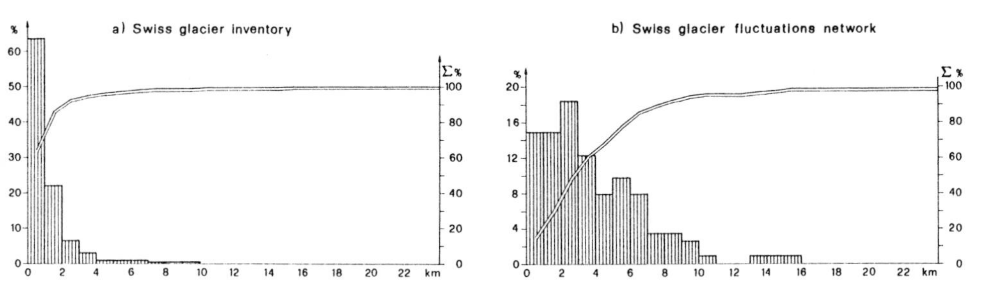

Figure 15. Histograms of glacier lengths from al Swiss glaciers (a) and the Swiss fluctuations network (b). Data basis: Haeberli (1985a) and F. Müller et al. (1976).

measurements, the well-documented length changes of Alpine glaciers should be used as a basis for developing models of climate/glacier relations and for developing standard curves for a worldwide intercomparison of less frequent measurements. From the contrasting examples shown in Figs. 7, 8 and 10 to 13, it is obvious that only groups of glaciers with similar characteristics can be used for an intercomparison of length variation data. As a first step, a subdivision into:

1. small, non-dynamically reacting cirque glaciers ("glacier réservoirs") with reaction times close to 0 years,
2. short and active mountain glaciers ("glaciers évacuateurs") with reaction times of a few years and
3. long valley glaciers with reaction times of decades

may be helpful (cf. Lliboutry, unpublished, Müller, 1988). Surging and calving glaciers should be treated separately as well as heavily debris covered glaciers with steadily upbuilding sediment beds (Figs. 11 and 12). Observations at longer intervals are useful for long-term comparison and can, in the case of large and remote glaciers, now be based on earlier maps and airphotography in combination with modern images from high-resolution satellites (e.g. RBV, Thematic Mapper, SPOT, Large-Format Camera etc.). Glacier inventory data are best suited for selecting appropriate glaciers and for assessing their representativity. As an example, Fig. 15 compares the size distribution of glaciers within the Swiss Alps as a whole and within the regular observational network of length variation measurements. It is obvious that the two samples have different frequency characteristics. Especially, the tongues of long valley glaciers with their delayed low-frequency signal have all been included into the fluctuations network, simply because they reach low altitudes and are easily accessible. On the other hand, small glaciers which experience nearly undelayed, high frequency changes are underrepresented. Simple averages of length changes calculated on the basis of the historically grown fluctuations network therefore give a systematically wrong impression of the overall development. Together with statistically derived parameters such as reaction time or velocity of advance/retreat (Müller, 1988), inventory data can be used as weighing functions to correctly estimate and test regionally representative trends. Repetition of detailed glacier inventories in the future will furnish a wealth of information about long-term glacier changes.

When international glacier monitoring started almost a century ago, it was clear that this activity would bear fruit only after decades of patient work. This is now the case, just in time to face the severe challenges of global environmental planning. The results of efforts undertaken by a great number of local observers make it worthwhile to continuously analyse the steadily accumulating information and to further contribute to a wisely planned and fascinating long-term experiment on glacier fluctuations and climate.

ACKNOWLEDGEMENTS

Thanks are due to the parent and funding organizations ICSI, FAGS, GEMS/UNEP and UNESCO, as well as to the national correspondents and scientific consultants of the World Glacier Monitoring Service. The constant encouragement and support by the ETH Zürich and especially by the Laboratory of Hydraulics, Hydrology and Glaciology (VAW) under the directorship of Prof. Dr. D. Vischer is gratefully acknowledged.

REFERENCES

Ageta, Y. and Higuchi, K. 1984. Estimation of mass balance components of a summer-accumulation type glacier in Nepal, Himalaya. Geografiska Annaler 66A, (3), 249-255.

Brückner, E. and Muret, E. 1907. Les variations périodiques des glaciers. XIIme rapport, 1906. Zeitschrift für Gletscherkunde II, 161-198.

Brückner, E. and Muret, E. 1909. Les variations périodiques des glaciers. XIIIme rapport, 1907. Zeitschrift für Gletscherkunde III, 161-185.

Brückner, E. and Muret, E. 1910. Les variations périodiques des glaciers. XIVme rapport, 1908. Zeitschrift für Gletscherkunde IV, 161-176.

Brückner, E. and Muret, E. 1911a. Les variations périodiques des glaciers. XVme rapport, 1909. Zeitschrift für Gletscherkunde V, 177-202.

Brückner, E. and Muret, E. 1911b. Les variations périodiques des glaciers. XVIme rapport, 1910. Zeitschrift für Gletscherkunde VI, 81-103.

Collins, D.N. 1984. Water and mass balance measurements in glacierised drainage basins. Geografiska Annaler 66A, (3), 197-214.

Corte, A.E. and Espizua, L.E. 1981. Inventario de glaciares de la cuenca del Rio Mendoza. CONICET/IANIGLA, Mendoza/Argentina, 64 pp.

Driedger, C.L. and Kennard, P.M. 1986. Glacier volume estimation on Cascade volcanoes: an analysis and comparison with other methods. Annals of Glaciology 8, 59-64.

Field, W.O. (Ed) 1975. Mountain glaciers of the Northern Hemisphere. CRREL, Hanover, Vol. 1, 698 pp., Vol. 2, 932 pp., and Atlas, 49 plates.

Fielding Reid, H. and Muret, E. 1904. Les variations périodiques des glaciers. IXme rapport, 1903. Archives des Sciences physiques et naturelles, Genève, Année 109, Période 4, t. 18, 160-195.

Fielding Reid, H. and Muret, E. 1905. Les variations périodiques des glaciers. Xme rapport, 1904. Archives des Sciences physiques et naturelles, Genève, Année 110, Période 4, t. 20, 62-74, 169-190.

Fielding Reid, H. and Muret, E. 1906. Les variations périodiques des glaciers. XIme rapport, 1905. Zeitschrift für Gletscherkunde I, 1-21.

Finsterwalder, R. and Rentsch, H. 1980. Zur Höhenänderung von Ostalpengletschern im Zeitraum 1969-1979. Zeitschrift für Gletscherkunde und Glazialgeologie 16, (1), 110-115.

Finsterwalder, S. and Muret, E. 1901. Les variations périodiques des glaciers. VIme rapport, 1900. Archives des Sciences physiques et naturelles, Genève, Année 106, Période 4, t. 12, 56-69, 118-131.

Finsterwalder, S. and Muret, E. 1902. Les variations périodiques des glaciers. VIIme rapport, 1901. Archives des Sciences physiques et naturelles, Genève, Année 107, Période 4, t. 14, 282-302.

Finsterwalder, S. and Muret, E. 1903. Les variations périodiques des glaciers. VIIIme rapport, 1902. Archives des Sciences physiques et naturelles, Genève, Année 108, Période 4, t. 15, 661-677.

Forel, F.-A. 1895. Les variations périodiques des glaciers. Discours préliminaire. Archives des Sciences physiques et naturelles, Genève, t. XXXIV, 209-229.

Forel, F.-A. and Du Pasquier 1896. Les variations périodiques des glaciers. Ier rapport, 1895. Archives des Sciences physiques et naturelles, Genève, Année 101, Période 4, t. 2, 129-147.

Forel, F.-A. and Du Pasquier 1897. Les variations périodiqes des glaciers. IIme rapport, 1896. Archives des Sciences physiques et naturelles, Genève, Année 102, Période 4, t. 4, 218-245.

Haeberli, W. 1983. Permafrost-glacier relationships in the Swiss Alps - today and in the past. Permafrost Fourth International Conference, Proceedings, NAP/Washington D.C., 415-420.

Haeberli, W. 1985a. Fluctuations of glaciers 1975-1980. IAHS and UNESCO, Paris.

Haeberli, W. 1985b. Global land ice monitoring: present status and future perspectives. US Department of Energy, DOE/EV/60235-1, 216-231.

Haeberli, W. 1985c. Creep of mountain permafrost: internal structures and flow of alpine rock glaciers. Mitteilungen der Versuchsanstalt für Wasserbau, Hydrologie und Glaziologie, ETH Zürich, no.77, 142 pp.

Haeberli, W. and Penz, U. 1985. An attempt to reconstruct glaciological and climatological characteristics of 18ka BP Ice Age glaciers in and around the Swiss Alps. Zeitschrif für Gletscherkunde und Glazialgeologie 21, 351-361.

Hamberg, A. and Mercanton, P.-L. 1913. Les variations périodiques des glaciers. XIXme rapport, 1913. Zeitschrift für Gletscherkunde IX, 42-65.

IAHS 1980. World Glacier Inventory - Proceedings of the workshop at Riederalp, Switzerland, 17-22 September 1978. IAHS Publication no. 126, 351 pp.

Kasser, P. 1967. Fluctuations of glaciers 1959-1965. IAHS and UNESCO, Paris.

Kasser, P. 1973. Fluctuations of glaciers 1965-1970. IAHS and UNESCO, Paris.

Kislov, A.V. and Koryakin, V.S. 1986. Spatial and temporal regulations of glacier fluctuations in Eurasian Arctic. Academy of Sciences of the U.S.S.R., Soviet Geophysical Committee, Data of Glaciological Studies, Publication no. 57., 236-241.

Koerner, R.M. 1986. A new method for using glaciers as monitors of climate. Academy of Sciences of the U.S.S.R., Soviet Geophysical Committee, Data of Glaciological Studies, Publication no. 57, 175-179.

Kruss, P. 1983. Climate change in East Africa: a numerical simulation from the 100 years of terminus record at Lewis glacier, Mount Kenya. Zeitschrift für Gletscherkunde und Glazialgeologie 19 (1), 43-60.

Kuhn, M. 1981. Climate and glaciers. IAHS Publication no. 131, 3-20.

Kuhn, M. 1984. Mass budget imbalances as criterion for a climatic classification of glaciers. Geografiska Annaler 66A, (3), 229-238.

Kuhn, M. 1985. Reactions of mid-latitude glacier mass balance to predicted climatic changes. US Department of Energy, DOE/EV/60235-1, 248-254.

Kuhn, M. 1986. Meteorological conditions of mass balance extremes. Academy of Sciences of the U.S.S.R., Soviet Geophysical Committee, Data of Glaciological Studies, Publication no. 57, 149-153.

Leiva, J.C., Cabrera, G. and Lenzano, L.E. 1986. Glacier mass balances in the Cajon del Rubio, Andes Centrales Argentinos. Cold Regions Science and Technology 13, 83-90.

Letréguilly, A. 1984. Bilans de masse des glaciers Alpins: méthodes de mesure et répartition spatio-temporelle. CNRS, Laboratoire de Glaciologie et de Géophysique de l'Environnement, Grenoble, Publication no. 439, 274 pp.

Lliboutry, L. 1974. Multivariate statistical analysis of glacier annual balances. Journal of Glaciology 13 (69), 371-392.

Lliboutry, L. (unpublished). Note sur le "Raport du Sous-Comité sur les variations des glaciers existants" à la Commission des Neiges et des Glaces, 3pp.

Makarevich, K.G. and Rototayeva, O.V. 1986. Present-day fluctuations of mountain glaciers in the Northern Hemisphere. Academy of Sciences of the U.S.S.R., Soviet Geophysical Committee, Data of Glaciological Studies, Publication no. 57, 157-163.

Matthes, M.F.E. 1934. Committee on glaciers of the American Geophysical Union - report for 1931/32. IAHS Bulletin no. 20, 251-264.

Mercanton, P.-L. 1930. Rapport sur les variations de longueur des glaciers de 1913 à 1928 (Chaine des Alpes; Scandinavie). Bulletin no. 14 of the IAHS, 53 pp.

Mercanton, P.-L. 1934. Rapport sur les variations de longueur des glaciers de 1928 à 1931 (1932) <Chaine des Alpes; Scandinavie). Bulletin no. 20 of the IAHS, 229-250.

Mercanton, P.-L. 1936. Rapport sur les variations des glaciers de 1931 à 1935 (Chaine des Alpes; Scandinavie et Islande). IAHS publication no. 22, 430-456.

Mercanton, P.-L. 1948. Rapport sur les variations des glaciers de 1935 à 1946 (1947) (Alpes françaises, suisses, italiennes et autrichiennes; variations des glaciers en Suède, Islande et Norvège). IAHS publication no. 30, 233-261.

Mercanton, P.-L. 1952. Rapport sur les variations de longueur des glaciers européens de 1947 à 1950. Alpes françaises, suisses, italiennes et autrichiennes; Pyrénées, Appennins; Norvège, Suède et Islande. IAHS publication no. 32, 107-119.

Mercanton, P.-L. 1954. Rapport sur les variation de longueur des glaciers d'Europe en 1950-51; 1951-52 et 1952-53. IAHS publication no. 39, 478-490.

Mercanton, P.-L. 1958. Rapport sur les variations de longueur des glaciers européens en 1953-54; 1954-55 et 1955-56. IAHS publication no. 46, 358-371.

Mercanton, P.-L. 1961. Rapport sur les variations de longueur des glaciers européens en 1956-57; 1957-58 et 1958-59. IAHS publication no. 54, 366-378.

Mercer, J.H. 1967. Southern Hemisphere glacier atlas. US Army Natick Laboratories Technical Report 67-76-ES, 325 pp.

Müller, F. 1977. Fluctuations of glaciers 1970-1975. IAHS and UNESCO, Paris.

Müller, F. 1978. Instructions for compilation and assemblage of data for a World Glacier Inventory. Supplement: Identification/Glacier Number. TTS/WGI, Dept. of Geography, ETH Zürich, 7 pp.

Müller, F., Caflisch, T. and Müller, G. 1976. Firn und Eis der Schweizer Alpen. Gletscherinventar. ETH Zürich, Geographisches Institut, publ. no. 57, 174 pp.

Müller, F. Caflisch, T. and Müller, G. 1977. Instructions for compilation and assemblage of data for a World Glacier Inventory. TTS/WGI, Dept. of Geography, ETH Zürich, 19 pp.

Müller, F. and Scherler, K.E. 1979a. Report on World Glacier Inventory - status December 1978. TTS/WGI, Dept. of Geography, ETH Zürich, 67 pp.

Müller, F. and Scherler, K.E. 1979b. Report on World Glacier Inventory - status October 1979. TTS/WGI, Dept. of Geography, ETH Zürich, 21 pp.

Müller, P. 1988. Parametrisierung der Gletscher-Klima-Beziehung für die Praxis: Grundlagen und Beispiele. Mitteilungen der Versuchsanstalt für Wasserbau, Hydrologie und Glaziologie. ETH Zürich, nr. 95, 228 pp.

Patzelt, G. 1985. The period of glacier advances in the Alps, 1965 to 1980. Zeitschrift für Gletscherkunde und Glazialgeologie 21, 403-407.

Patzelt, G. 1987. Gegenwärtige Veränderungen an Gebirgsgletschern im weltweiten Vergleich. Verhandlungen des Deutschen Geographentages Bd. 45, Stuttgart, 259-264.

Rabot, C. and Muret, E. 1912. Les variations périodiques des glaciers. XVIIme rapport, 1911. Zeitschrift für Gletscherkunde VII, 37-47.

Rabot, C. and Muret, E. 1912. Supplément au VII rapport sur les variations périodiques des glaciers. Zeitschrift für Gletscherkunde VII, 191-202.

Rabot, C. and Mercanton, P.-L. 1913. Les variations périodiques des glaciers. XVIIIme rapport, 1912. Zeitschrift für Gletscherkunde VIII, 42-62.

Reeh, N., Clausen, H.B., Dansgaard, W., Gundestrup, N., Hammer, C.U. and Johnson, S.J. 1978. Secular trends of accumulation rates at three Greenland stations. Journal of Glaciology 20 (82), 27-30.

Reynaud, L. 1980. Can the linear balance model be extended to the whole Alps? IAHS publication no. 126, 273-282.

Reynaud, L., Vallon, M., Martin, S. and Letréguilly, A. 1984. Spatio temporal distribution of the glacial mass balance in the Alpine, Scandinavian and Tien Shan areas. Geografiska Annaler 66A (3), 239-247.

Reynaud, L., Vallon, M. and Letréguilly, A. 1986. Mass balance measurements: problems and two new methods of determining variations. Journal of Glaciology 32 (112), 446-454.

Richter 1898. Les variations périodiques des glaciers. IIIme rapport, 1897. Archives des Sciences physiques et naturelles, Genève, Année 103, Période 4, t. 6, 22-55.

Richter 1899. Les variations périodiques des glaciers. IVme rapport, 1898. Archives des Sciences physiques et naturelles, Genève, Année 104, Période 4, t. 8, 31-61.

Richter 1900. Les variations périodiques des glaciers. Vme rapport, 1899. Archives des Sciences physiques et naturelles, Genève, Année 105, Période 4, t. 10, 26-45.

Scherler, K.E. 1980. Report on World Glacier Inventory - status December 1980. TTS/WGI, Dept. of Geography, ETH Zürich, 34 pp.

Scherler, K.E. 1983. Guidelines for preliminary glacier inventories. TTS/WGI, Dept. of Geography, ETH Zürich, 16 pp.

Sugden, D.E. 1977. Reconstruction of the morphology, dynamics and thermal characteristics of the Laurentide Ice Sheet at its maximum. Arctic and Alpine Research 9 (1), 21-47.

Thompson, L., Hastenrath, S. and Morales, B. 1979. Climatic ice core records from the tropical Quelccaya Ice Cap. Science 203, 1240-1243.

Thompson, L.G., Wu Xiaoling, Mosley-Thompson, E. and Xie Zichu 1988. Climatic ice core records from the Dunde Ice Cap, China. Annals of Glaciology, 178-182.

UNESCO 1969. Variations of existing glaciers. UNESCO/IAHS Technical Papers in Hydrology no. 3, 19 pp.

UNESCO 1970. Perennial ice and snow masses. A guide for compilation and assemblage of data for a World Glacier Inventory. UNESCO/IAHS Technical Papers in Hydrology no. 1, 59 pp.

US Dept. of Energy 1985. Glaciers, ice sheets, and sea level: effect of a CO_2-induced climatic change. Report of a Workshop held in Seattle, Washington, September 13-15, 1984. DOE/EV/60235-1, 330 pp.

Vallon, M., Letréguilly, A. and Reynaud, L. 1986. Glacier mass balance reconstructions for the Northern Hemisphere covering the past century and their climatic significance. Academy of Sciences of the U.S.S.R., Soviet Geophysical Committee, Data of Glaciological Studies, Publication no. 57, 153-157.

Vareschi, V. 1942. Die pollenanalytische Untersuchung der Gletscherbewegung. Veröffentlichung des Geobotanischen Institutes Rübel in Zürich, 19. Heft, 142 pp.

Wallén, C.C. 1984. Present century climate fluctuations in the Northern Hemisphere and examples of their impact. WMO/TD-no. 9, 85 pp.

WMO 1983. Report of the meeting on climate system monitoring, Geneva, 5-9 December 1983. WMO/WCP-64.

A DESCRIPTION OF THE UNITED STATES' CONTRIBUTION TO THE WORLD GLACIER INVENTORY

C. Suzanne Brown
U.S. Geological Survey
1201 Pacific Avenue
Tacoma. Wa 98446, U.S.A.

ABSTRACT

Although the immensity of the glacier-covered area and its relative inaccessibility makes a complete inventory of glaciers in the United States currently impracticable, the present listing covers about 95 per cent of the total number of glaciers in detail, is the most comprehensive to date, and has been submitted as the United States' contribution to the Temporary Technical Secretariat for a World Glacier Inventory.

1. INTRODUCTION

The present inventory includes information from Mountain Glaciers of the Northern Hemisphere (Field, W.O., 1975), which has been updated for large tidewater glaciers that have undergone substantial change in the past decade; regional inventories by several authors covering 266 glaciers in the Olympic Mountains, 742 glaciers in the North Cascade Mountains, 497 glaciers in the Sierra Nevada, and 1001 glaciers in the Brooks Range; and data gathered for more than half the glaciers in Alaska not previously inventoried. Each glacier has been assigned an unique number on the basis of hydrologic basins; this is the first systematic numbering of glaciers applied to the entire United States.

There are more than 8000 glaciers in the United States that cover at least 66,000 km^2 , over 6000 of which are in Alaska covering an area of more than 65,000 km^2. The data from this inventory are primarily for use in the world water assessment by UNESCO, but they will also provide a basis for studies on glacier fluctuations and climatic change.

2. EARLY HISTORY

The glacier inventory submitted in 1985-86 to the Temporal Technical Secretariat (TTS) for a World Glacier Inventory as the United States' contribution represents about 95 per cent of the total number of glaciers and about 93 per cent of the total area covered by glaciers in

J. Oerlemans (ed.), Glacier Fluctuations and Climatic Change, 103–108.

the United States and is the most comprehensive to date. Because of the immensity of the glacier-covered area in the United States and the relative inaccessibility of much of it, a complete inventory of glaciers in the United States is presently impracticable.

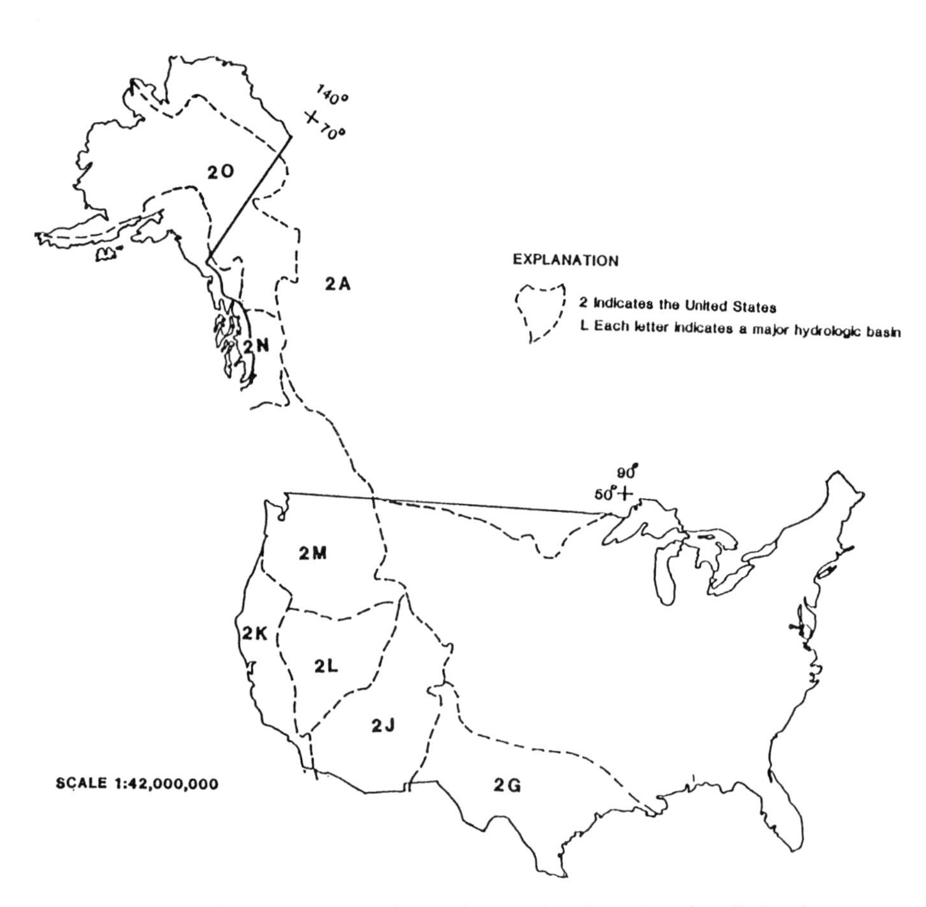

Figure 1. The major hydrologic basins in the United States and the numbers assigned by the Temporary Technical Secretariat for the World Glacier Inventory.

Russell (1885) was the first to attempt a census of glaciers in the conterminous United States This was followed by a brief survey of glaciers, again in the conterminous United States, by Wentworth and Dole (1931). Field and others (1958) recorded the general state of knowledge of and many of the geographical aspects of glaciers in the Northern Hemisphere prior to the International Geophysical Year (1957-58); for Alaska, glaciers 8 km or more in length were included, and most of them were being inventoried for the first time. Parts 2a and 3 of that publication constitute the first substantial inventory ever compiled of glaciers in the entire United States.

Stimulated by the program of the International Hydrological Decade (1965-74) to assemble inventories of glaciers, Field proposed an update of the 1958 report. Because much glaciological activity had taken place in the interim, particularly in Alaska, it was necessary to compile a whole new text, revise many of the old maps, and draft several new ones. Volume II of the finished work (Field et al., 1975) remains the classic work of decriptive and tabulated inventory data for glaciers in the entire United States

The inventory submitted for the World Glacier Inventory includes information from: (1) the inventory of glacier by W.O. Field in 'Mountain Glaciers of the Northern Hemisphere' (1975), which has been updated for about 25 of the large tidewater glaciers in Alaska that have undergone a substantial change, most commonly a retreat, in the past decade; (2) some more recent and more detailed regionalized inventories by several authors covering 266 glaciers in the Olympic Mountains (Spicer, in prep.), 742 glaciers in the North Cascade Mountains of Washington (Post et al., 1971), 497 glaciers in the Sierra Nevada of California (Raub et al., in prep.), and 1001 glaciers in the Brooks Range of Alaska; and (3) data gathered for those Alaska glaciers less than 8 km in length, which amounted to more than half the glaciers in Alaska, not previously inventoried.
Figure 1 shows the major hydrologic basins in the United States and the numbers assigned to them by the TTS for systematic numbering of the glaciers. One of the most important aspects of the present inventory, if not the most important, is that each glacier in the United States, whether previously inventoried or not, was assigned a unique number in this international system of hydrologic basins. This is the first time a systematic identification of the glaciers has been applied to the entire United States, with every glacier having a unique number.

3. GLACIERS IN THE CONTERMINOUS UNITED STATES

In the conterminous United States (Fig. 2) glaciers exist only in the mountains of the western states, usually on an isolated peak such as the volcanic peaks Mt. Rainier in Washington and Mt. Hood in Oregon. Making a detailed glacier inventory in these states is not difficult; the glaciers are not large and they are easy to diffentiate from one another. In the most glacierized region of the conterminous United States, the North Cascade Mountains of Washington, the glaciers average 0.35 km^2 in area. Because the mountainous regions over which the glaciers are situated are quite compact, aerial photographs needed to map and inventory the glaciers in a region can be taken in one day. The two largest regions that contain glaciers - the North Cascades of Washington and the Sierra Nevada in California - cover approximately 31,000 km^2 and 25,000 km^2, respectively. Many glaciers in the conterminous United States also are accessible by road or trail for on-site verification of characteristics inferred from maps or aerial photographs.

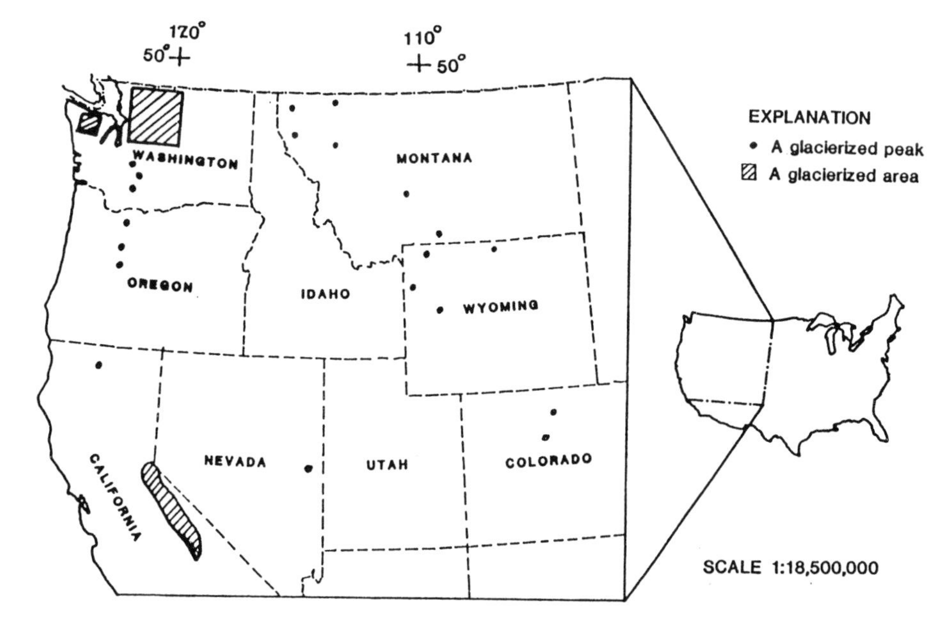

Figure 2. The location of glaciers in the conterminous United States.

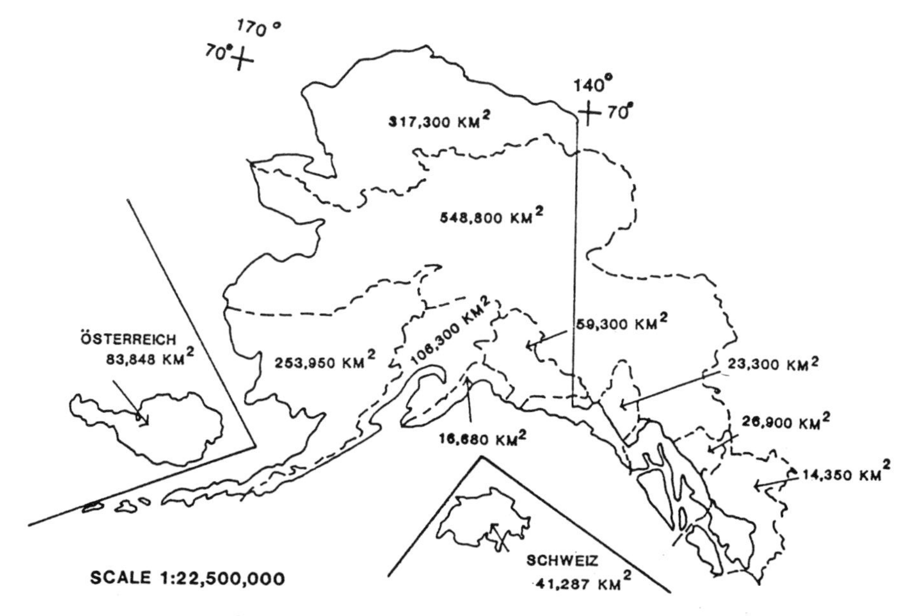

Figure 3. The primary hydrologic basins in Alaska and their areas, not including offshore islands or that part of the basins that lies in Canada. Switzerland and Austria are shown for comparison.

4. GLACIERS IN ALASKA

More than 75 per cent of the glaciers in the United States are located in Alaska, and making a detailed inventory is difficult. The state covers more than 1.5 million km^2; the distance between the two largest cities, Anchorage and Fairbanks, is around 380 km. Figure 3 shows Alaska with an inset of Switzerland and Austria to illustrate the relative sizes of those highly glacierized regions.

There are nine primary hydrologic basins in Alaska, subdivisions of those shown in Fig. 1. The area of each primary basin is given (not including offshore islands or that part of the basins that lies in Canada); the areas of Switzerland and Austria are also given for comparison. The smallest primary basin in Alaska has more than 6000 km^2 of ice.

Owing to the large glacierized area, it was most practicable to use the 1:250,000 scale U.S.Geological Survey topographic quadrangle maps, with reference being made to the 1:63,360 scale quadrangles when more detailed information was needed. On the small scale maps, the large glaciers are easy to locate and inventory, but the small glaciers - less than 3 km in length in Alaska - are easily lost. Another problem encountered in dealing with such an extensive glacierized region is the difficulty of obtaining good aerial photography and complete photographic coverage, both necessary for inventory work. Because of the remoteness of most of the glacierized area of Alaska, the possibility of on-site verification that may be necessary in an inventory, let alone of doing any scientific work, is greatly reduced.

Determining where the glacier boundaries are is a problem common to all glacier inventories, and Alaska is no exception. Many glaciers there are so large that a main glacier drainage system may have several named component parts, each of which could be (and frequently is) considered to be a separate glacier. The Bering Glacier for example, with five component parts, extends into two countries, two major drainage basins, and two mountain ranges (Post and Meier, 1980).

Satellite imagery was not used for this inventory because of insufficient resolution to obtain the necessary detail, the presently incomplete coverage of Alaska, and of the prohibitive expense.

The inventory of glaciers in te conterminous United States is complete. Some of the smallest glaciers reported 20 years ago need to be updated; otherwise, the inventory is accurate as of 1975. The inventory of glaciers in Alaska is not complete. The individual lengths and areas of most of the glaciers less than 8 kilometers in length (more than half the total number) still are unmeasured. An estimate of their combined area was made and added to the known glacier area to obtain the number given in this report (65,800 km^2). The measured total glacierized area in Alaska probably will fall between that number and the 74,700 km^2 estimated by Post and Meier (1980). The distribution of glaciers by state is shown in Table I.

Table I. Distribution of Glaciers in the United States

State	No. of Glaciers	Total area (km^2)
Alaska	6200	65,800
Washington	1085	420
Oregon	36	20
California	507	55
Montana	106	27
Wyoming	81	49
Colorado	10	2
Nevada	1	> 1

REFERENCES

Field, W.O. 1958. Geographic study of mountain glaciation in the Northern Hemisphere. American Geographical Society, mimeographed report, Parts 1, 2a, 2b and 3.

Field, W.O. 1975. Mountain glaciers of the Northern Hemisphere, volume 2. Cold Regions Research and Engineering Laboratory, Hanover, N.H.

Post, A., Richardson, D., Tangborn, W.V. and Rosselot, F.L. 1971. Inventory of glaciers in the North Cascades, Washington. U.S. Geological Survey Professional Paper 705-A, 26 pp.

Post, A. and Meier, M.F. 1980. World Glacier Inventory. Proceedings of the Riederalp Workshop, September 1978. IAHS-AISH Publ. no. 126, 45-47.

Raub, W., Brown, C.S. and Post, A. (in prep.). Inventory of glaciers in the Sierra Nevada, California. U.S. Geological Survey, PP 705-B.

Russel, I.C. 1885. Existing glaciers of the United States. U.S. Geological Survey 5th Annual Report, 1883-84, 344-346.

Spicer, R. (in prep.). Inventory of glaciers in the Olympic Mountains, Washington. United States Geological Survey PP 705-C.

Wentworth, C.K. and Dole, D.M. 1931. Dinwoody Glacier, Wind River Mountains, Wyoming: with a brief survey of existing glaciers in the United States. Geological Society Bulletin 42, 605-620.

HISTORIC GLACIER VARIATIONS IN SCANDINAVIA

J. Bogen, B. Wold and G. Østrem
Norwegian Water Resources and Energy Administration
Glaciology Section, P.O. Box 5091 mj, Oslo, Norway

ABSTRACT

It is generally believed that most glaciers in Scandinavia melted away during the climatic optimum 7000-3000 B.P. A majority of the glaciers reached their maximum extension during the climatic deterioration in the 18th century, but some of the maritime glaciers in western Norway reached their maximum or near maximum as late as in the first part of the 20th century. Glacier advances in the early part of this century are associated with periods of below normal summer temperatures. A number of glaciers did respond simultaneously to these events. After the extensive glacier melt during the following decades, it seems that various glaciers reacted differently. Some dit not respond to the increase in net balance in the 1960s until a long build-up period had passed. The paper gives a summary of present knowledge on historic glacier variations in Scandinavia.

1. INTRODUCTION

Most of the glaciers probably disappeared from Scandinavia during the climatic optimum from 5000 to 1000 B.C. The equilibrium line was then more than 400 m higher than today, and glaciers could probably only survive in the highest parts of Jotunheimen, the Jostedalen and in the Svartisen area. Most of our present glaciers reappeared approx. 2500 years ago in southern Norway and in the Svartisen and Sarek/Kebnekaise areas in northern Scandinavia (Rekstad, 1903, 1904). Liestøl (1960) has constructed the diagram in Fig. 1 of the relative variation of the equilibrium line altitude in South Norway.

The oldest written information on the behaviour of glaciers in the Nordic countries is found in Iceland. From the beginning of the Norwegian colonization about 870 A.D. up to the middle of the 13th century the glaciers of Iceland were considerably smaller than they are today. Later the glaciers began to increase, and at the end of the 17th and the beginning of the 18th century several farms located close to the glaciers had to be abandoned. Ahlmann (1953) says that according to the Icelandic glaciologist Thorarinsson it is most probable that the

J. Oerlemans (ed.), Glacier Fluctuations and Climatic Change, 109–128.

climatic deterioration had begun in the 13th century, strengthened in the next century and culminated during the 17th and 19th century (Ahlmann, 1953).

Until the last decades of the 17th century there were no written records of any glaciers in Scandinavia. The first written information about glacier variations in Norway is found in documents made by officials sent from the king because farmers asked for reduced taxes due to loss of arable land. Later tourists and scientists have given various kinds of documentation. In this paper we emphasize well documented historic sources and data from direct observations of front variations.

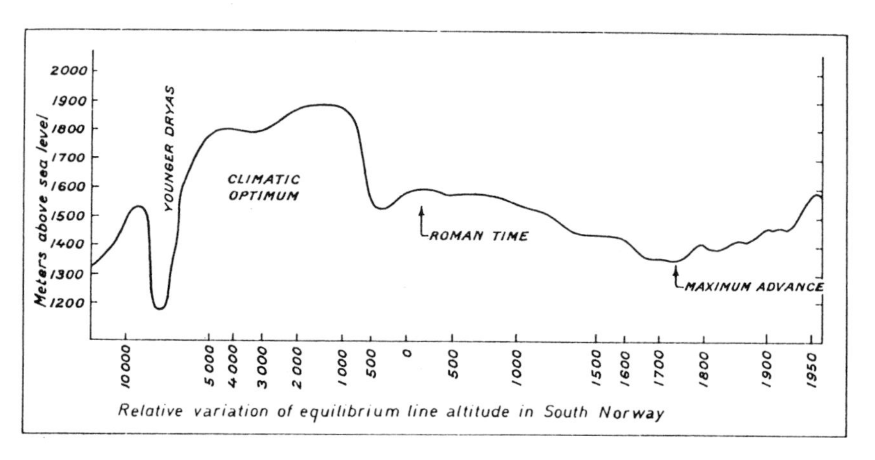

Figure 1. Relative variations of the equilibrium line altitude in Southern Norway. From Liestøl (1960).

2. MAIN GLACIER AREAS IN SCANDINAVIA

Today a total number of 1591 glaciers covering an area of 2745 km^2 are found in Norway (Østrem, Haakensen and Melander, 1973), and 308 glaciers totalling 314 km^2 in Sweden. There are no glaciers in Finland.

The most important glacierized areas in Scandinavia are shown in Fig. 2. All the largest glaciers are icecaps with outlet glaciers draining into neighbouring valleys. Jostedalsbreen (no. 1 in Fig. 2) is the largest glacier on the European mainland. Its present area is 486 km^2 and the elevations range from 1980 m a.s.l. to approx. 300 m a.s.l.

Folgefonni (no. 2 in Fig. 2) is divided in three icecaps, of which the southern is the second largest ice mass in southern Norway, with an area of 172 km^2. The height distribution goes from 1660 m a.s.l. down to approx. 490 m a.s.l. This glacier is presently the southernmost glacier in Norway today, except for some very small remnants. Hardangerjo/kulen (no. 3 in Fig. 2)has an area of 78 km^2, and its height ranges from 1850 m a.s.l. to 1000 m a.s.l. Jotunheimen (nos. 4-6 in Fig. 2) is the highest mountain area in Norway, situated in south central Norway. Many valley and cirque glaciers are situated there.

In northern Norway the largest glacier complex is the Svartisen area (no. 8 in Fig. 2). It consists of two large icecaps and several smaller glaciers. The western icecap covers 221 km^2 with an elevation range from 1580 m a.s.l. to approx. 20 m a.s.l.

Several relatively small glaciers are situated along the Swedish-Norwegian border (nos. 9-10 in Fig. 2). In this area there is a very high gradient from maritime to continental climate. In the Lyngen area (no. 12 in Fig. 2), there are mostly valley glaciers and these are thought to be more or less of the continental type. On the other hand, the Øksfjord - Seiland area (no. 11 in Fig. 2) has mainly icecaps which are thought to be of the maritime type.

The Swedish glaciers are generally valley glaciers of a continental (almost sub-polar) type. They are mainly located in the highest mountain massifs in northern Sweden, Kebnekaise (no. 13 in Fig. 2), and Sarek (no. 14 in Fig. 2).

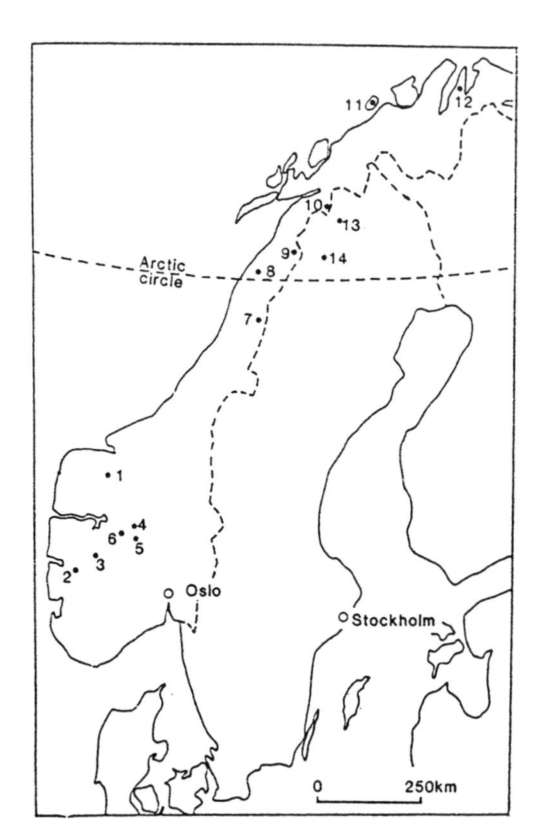

Figure 2. The main glacier areas in Scandinavia. Numbers are referred to in the text.

3. VARIATIONS IN HISTORIC TIME

3.1 Folgefonni

It has been assumed that Norwegian glaciers in general had their greatest extent about 1750 and that no greater advance has occurred since then. However, for parts of the Folgefonni Icecap this does not seem to be the case. In several places signs are found, indicating that the largest extent was reached in the last part of the 19th century or even later. This advance is well documented by a series of terrestrial photographs.

The southern part of the icecap has an outlet named Blomsterskardsbreen. This glacier advanced 200-250 meters between 1904 and 1971, and it is probably the only glacier in Norway where a net advance has been so clearly documented during this period. Tvede and Liestøl (1977) found that the major advance took place between 1920 and 1940, and that the glacier reached its maximum historic extension as late as about 1940. Since then the retreat has been approx. 50 meters (Tvede, pers. comm.).

Regarding Bondhusbreen, another outlet from Folgefonni, Rekstad (1905) says that the outermost moraine was a result of a glacier advance between the years 1865 and 1875. Thus, this glacier seems to have had its maximum historic extension in 1875. This is supported by an old local inhabitant in the area who told Tvede (1972) that his great-grandfather could rememeber that trees growing on the plain in front of the glacier were pushed forward and covered by the ice. Rekstad (1905) made some measurements of the ice front and he says that it had retreated about 150 m from the 1875-moraine in 1904. Since then routine observations have been made of the front position and the data are plotted in Fig. 12. This shows that the glacier advanced markedly also at two later occasions, having its maximum size about 1911 and 1930. These two advances nearly reached the historical maximum. A very rapid retreat took place between 1930 and the late 1950s. From then the glacier front has been more or less stable, compare Fig. 13.

An analysis of the climatic conditions in this part of Norway has indicated that a significant increase in the glacier mass must have taken place between 1918 and 1925 (Tvede and Liestøl, 1977).

3.2 The Jostedalsbreen ice cap

Large valley glaciers extend from the ice plateau at 1900 m a.s.l. down to 400-500 m a.s.l. The oscillations of the outlet glaciers on the western side of the ice cap have shown to be slightly different from that of the outlets on the eastern side.

Nigardsbreen is one of the largest outlets from Jostedalsbreen, flowing south-east from the middle part of the icecap. It extends from 1950 m a.s.l. down to approx. 350 m a.s.l. (Fig. 3).

The first known written information about Nigardsbreen can be traced back to 1735 when the glacier advanced so far that it started to damage valuable grasslands in the valley. The farmers asked the king to reduce taxed due to this loss of land areas. Most of the present knowledge of Nigardsbreens advance originates from reports written by officials sent by the king to check the complaints. A minister in the valley later wrote that the glacier destroyed the houses of the farm Nigard in 1743. Later the glacier was named after this farm. Another farm, Bjørkehaug, lost all its arable land during the following few years (Foss, 1802-1803). According to Foss, the glacier advance culminated in 1748. Since the the glacier has retreated more or less continuously, only with intermittent advances. During these advances small end-moraines were formed and their age has later been dated by various scientists (Faegri, 1933; Andersen and Sollid, 1971). No direct measurements of the front variations were made until Rekstad started in 1899 (Rekstad, 1902). Many single observations were, however, made by various visitors to the area earlier in the 19th century (Smith, 1817; Bohr, 1819; Naumann, 1824; Lindblom, 1839; Durocher, 1847; Forbes, 1853; Blytt, 1869; Larsen, 1874). From 1907 annual measurements have been made by different researchers.

Figure 3. Oblique air photograph showing the Nigardsbreen outlet glacier. The lake Nigardsvatn was still partly glacier covered in 1947 when this photo was taken (by Widerøes Flyveselskap).

Based upon these various sources it has been possible to construct a table showing front variations from 1710 upto the present (Østrem, Liestøl and Wold, 1977). The results are plotted in Fig. 5. From 1931 to 1972 the glacier had a calving front in the newly formed Lake Nigardsvatn (Fig. 6).

Probably the lake (presently about 30 m deep) was more or less completely sediment-filled when the glacier advanced during the 17th and 18th century. Since the "1750-moraine" was formed the glacier has receeded almost 5 km and its thickness has been reduced by 200-300 m (Fig. 7).

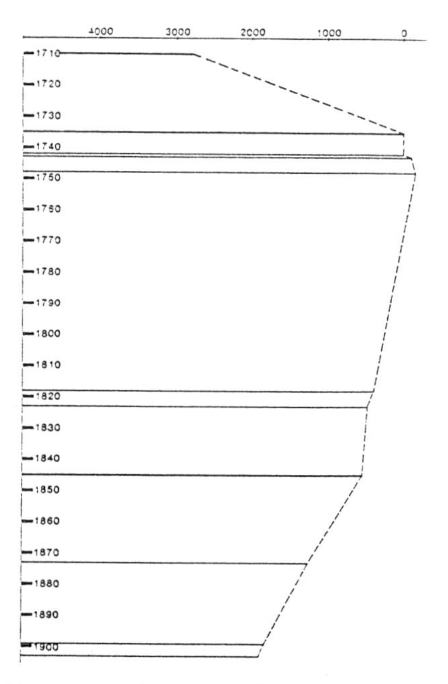

Figure 4. Front positions of Nigardsbreen 1710-1900 (data from Østrem, Liestøl and Wold, 1977).

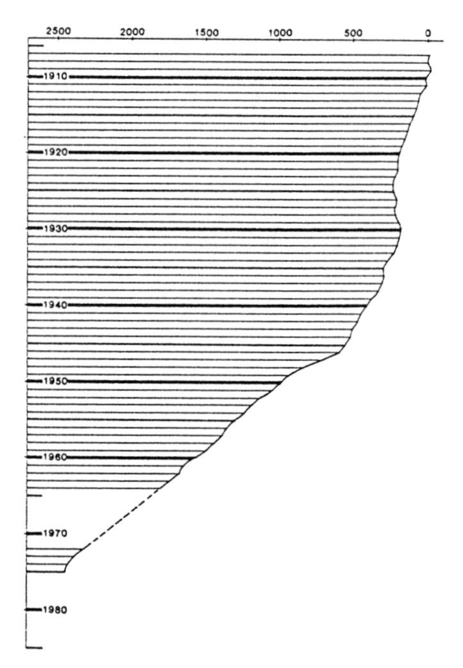

Figure 5. Front positions of Nigardsbreen 1907-1975 (data from Østrem, Liestøl and Wold, 1977).

The other eastern outlet glaciers from Jostedalsbreen also advanced during the 1800s and 1900s. Their subsequent retreat was, however, not simultaneaous everywhere. The outlet glacier Lodalsbreen kept its advanced position near the 1750 moraines much longer than most of the others. However, during the 20th century it has retreated more than 2 km.

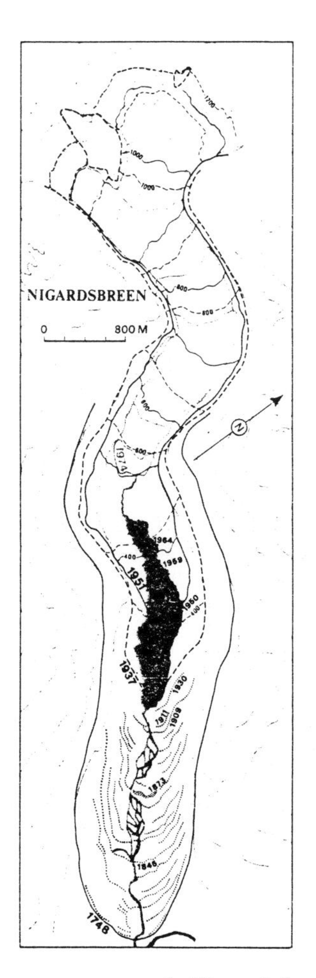

Figure 6. The retreat of Nigardsbreen. The map is based upon moraine ridges, dated by various scientists, and existing maps. Contour lines are indicated for the years 1937, 1951 and 1974. (From Østrem, Liestøl and Wold, 1977).

Observations of front variations on the eastern side of Jostedalsbreen are shown in Figs. 8 and 9. Some significant advances were recorded in the periods 1905-12 and 1927-32. Later, a general

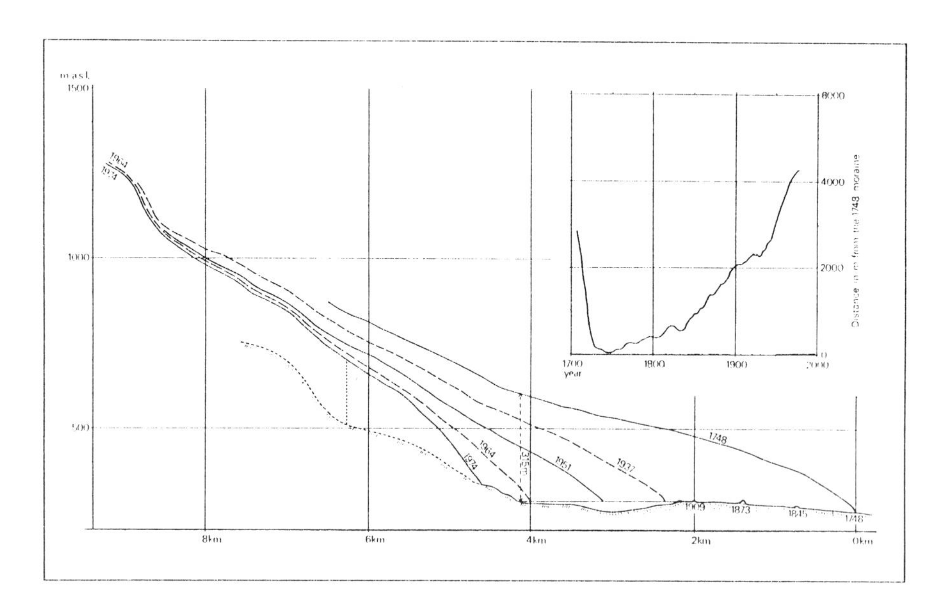

Figure 7. The decay of the tongue of Nigardsbreen is shown in this longitudinal profile. Note that the ice thickness has been reduced by more than 300 m in areas where the icefront is situated today. (From Østrem, Liestøl and Wold, 1977)

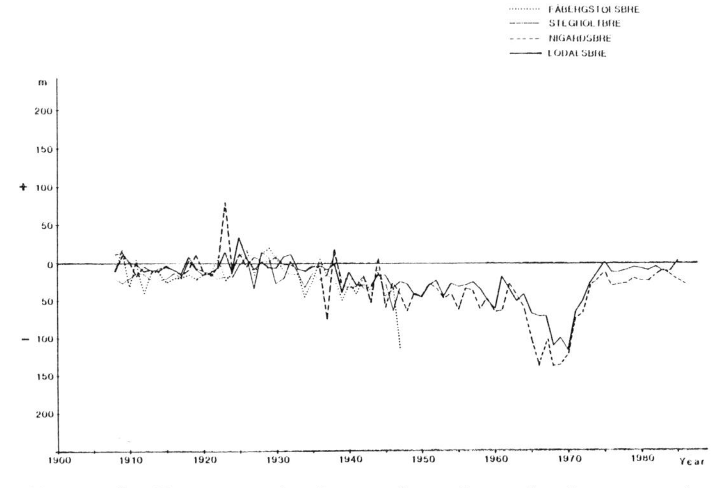

Figure 8. Front variations of outlet glaciers on the eastern side of the Jostedalsbreen icecap: Stegholtbreen, Lodalsbreen, Fåbergstølsbreen and Nigardsbreen. Compare also Figs. 9 and 11.

retreat occurred up to about 1975. Small advances were observed at Fåbergstølsbreen and Austerdalsbreen in the 1970s, whereas Tunsbergdalsbreen, the longest in Norway (18 km), has shown a continuous retreat in the 20th century. Only a small advance in 1911 has been reported.

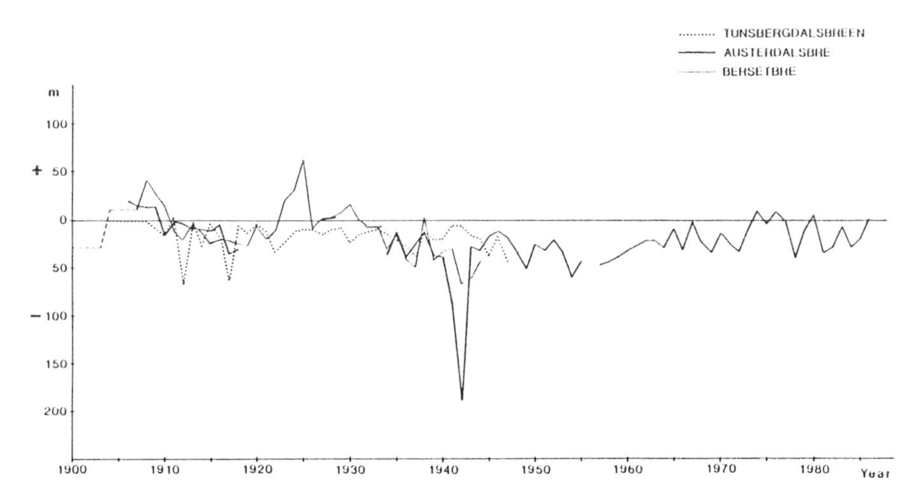

Figure 9. Front variations of outlet glaciers on the eastern side of the Jostedalsbreen icecap: Austerdalsbreen, Bergsetbreen and Tunsbergdalsbreen. Compare Figs. 8 and 11.

Grove (1972) and Grove and Battagel (1983) have made a frequency diagram of tax reductions allowed for inhabitants in the area due to inundations associated with climatic deterioration in the 17th, 18th and 19th century (Fig. 10).

Archeologic findings near the mountain Cecilikruna indicate that the glacier was smaller before the said period (Eide, 1955). At Briksdalsbreen, a very steep outlet glacier, Pedersen (1976) found that historic documents indicate that the outermost moraine must have been formed before 1600. Abrekkebreen which is now a re-generated glacier, advanced about 4 km and damaged the farm Tungøyane between 1734 and 1743 (Eide, 1955; Rekstad, 1902; Husebye, 1983).

The 20th century glacier front variations west of the Jostedalsbreen icecap are very similar to those observed on its eastern side (Fig. 11).

Rogstad (1941) studied hydrological data from the Nordfjord area to calculate glacier variations for adjacent parts of the Jostedalsbreen ice cap. Comparison with front variations of the steep western outlet glaciers Briksdalsbreen and Mjølkevollsbreen indicate that their tongues show similar variations, but with a delay of approximately four years. Thus, a series of years of positive mass balance produced a front advance four years later.

Pedersen (1976) dated moraine ridges in the Briksdalen valley from historic documents by lichenometry and calculations of the glacier front variations based on a method developed by Nye (1960) using calculations of the net balance from hydrological data. The results indicate glacier advances in 1840, 1870, 1910 and 1929. It is also believed that older moraines were formed in the 17th and 18th century in Briksdalen.

A significant retreat occurred in the period 1932-42 when, for example the Lake Briksdalsvatn was uncovered, Lake Nigardsvatn started to appear almost simultaneously. Calving in the lake accelerated the retreat in both cases. A small lake, Saetrevatnet, was also formed in front of Bødalsbre.

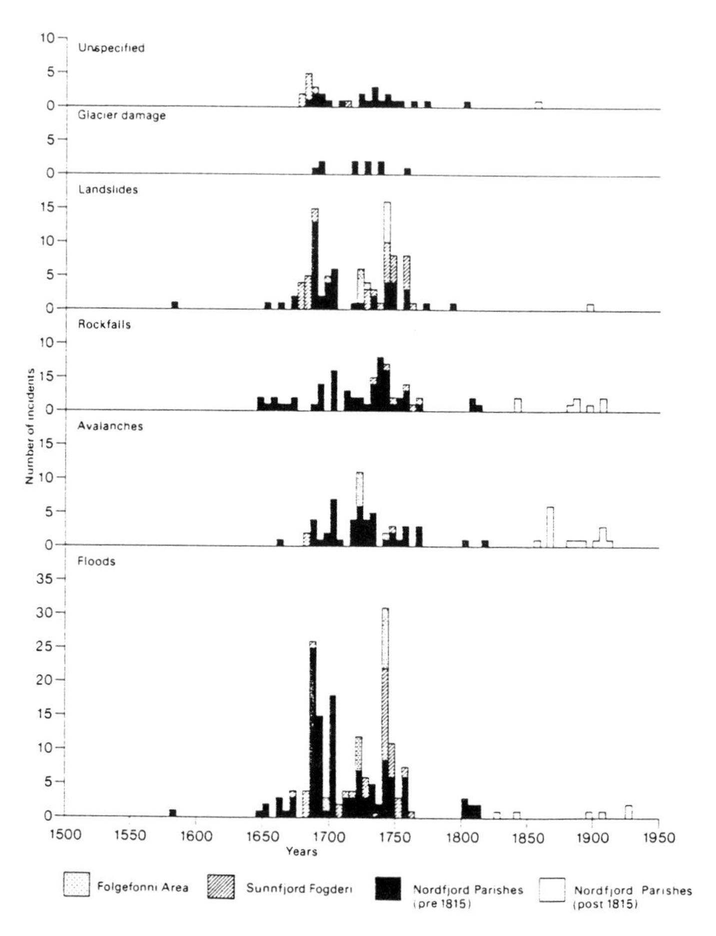

Figure 10. Frequency diagram of tax reductions due to glacier inundation on farm land on the western side of the Jostedalsbreen icecap. (From Grove and Battagel, 1983)

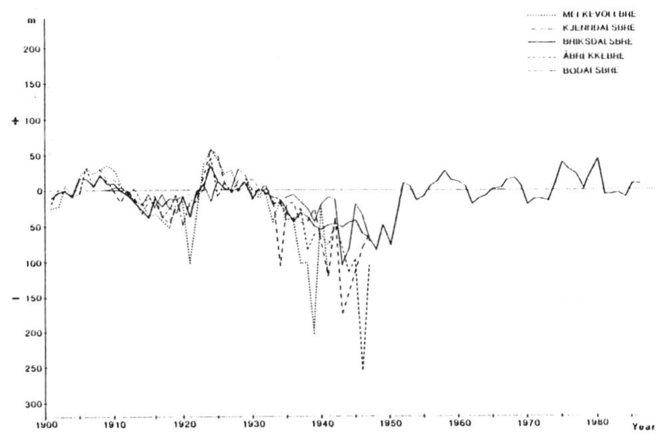

Figure 11. Front variations of outlet glaciers on the western side of the Jostedalsbreen icecap: Briksdalsbreen, Abrekkebreen, Melkevollsbreen, Kjenndalsbreen and Bødalsbreen. Compare Figs. 8 and 9.

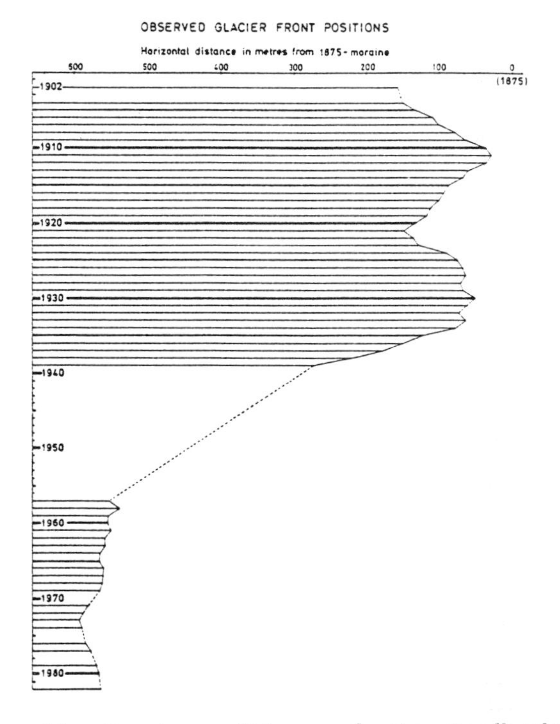

Figure 12. Front positions of the Bondhusbre, a west-facing outlet glacier from the Folgefonni icecap. Compare Fig. 13.

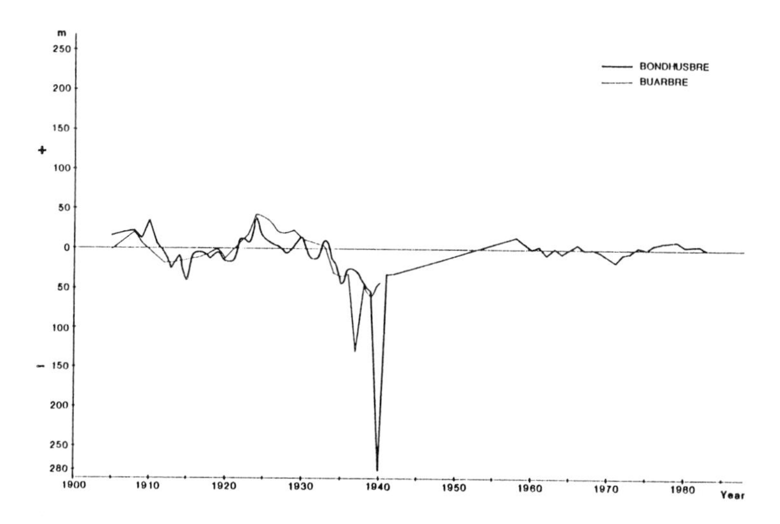

Figure 13. Comparison between front variations at Bondhusbreen and Buarbreen, and east-facing outlet glacier from the Folgefonni icecap. Compare Fig. 12.

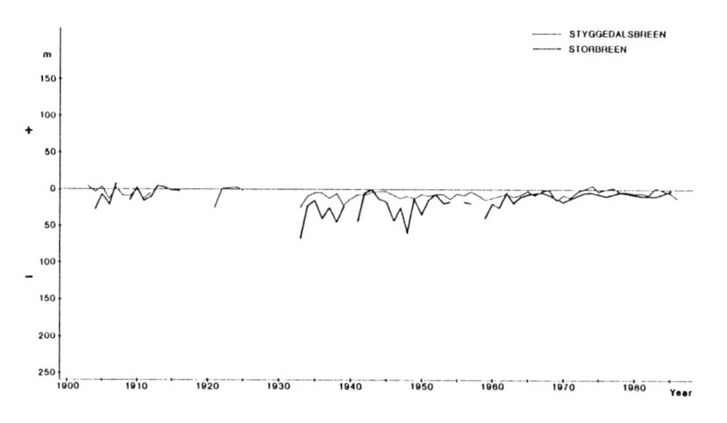

Figure 14. Front variations of two valley glaciers in the Jotunheimen mountains: Styggebreen and Storbreen.

3.3 Jotunheimen

The formation of the outermost moraines in front of glaciers in Jotunheimen has not been dated. Øyen (1893) indicates a formation after 1800, whereas Matthews (1974) suggest that the outermost moraines were formed about 1750. Østrem (1965) dated ice under the numerous ice-cored moraine ridges in front of Gråsubreen and found about 1300 years old ice under the outermost, 2600 years old ice under the next and much older ice under several other moraine ridges. The glacier had apparently overridden older moraines to form the younger ones.

P.A. Øyen, a geologist at the Geological Museum at the University of Oslo, started his glacier front observations in Jotunheimen in 1901 (Øyen, 1901, 1910). They were abandoned in 1912 due to lack of funds, but were resumed in 1933 by W. Werenskiold and were later continued by The Norwegian Polar Research Institute. The measurements were made once a year, ordinarily towards the end of the summer and in total 17 glaciers were measured. All these glaciers are of the valley type and not outlets from an icecap. Two of the longest observation series, from Storbreen and Styggebreen, are shown in Fig. 14.

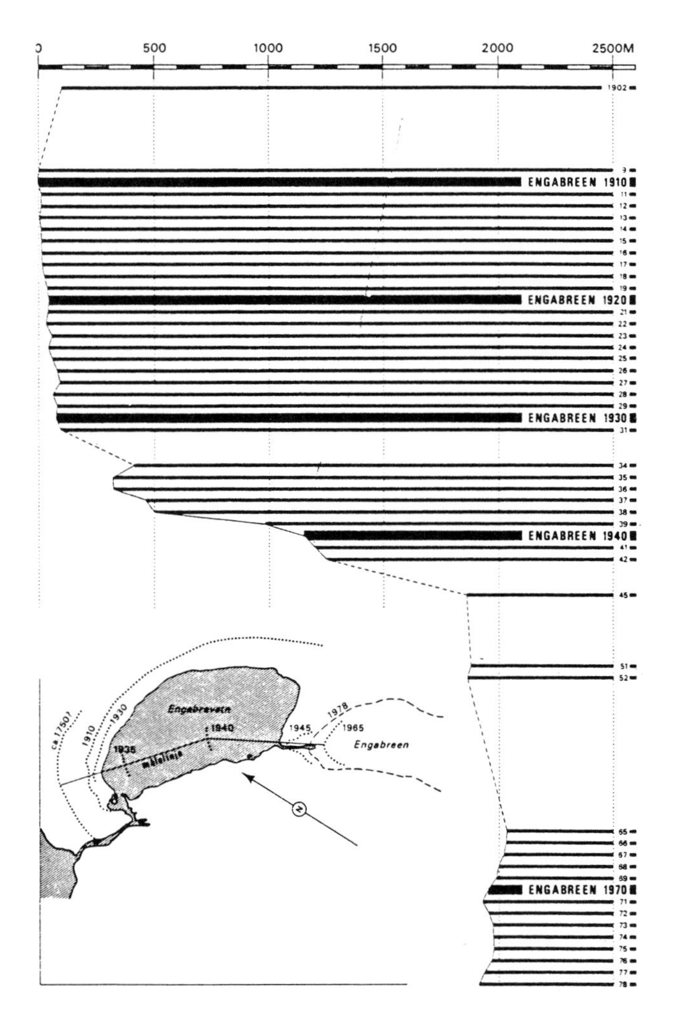

Figure 15. Front positions of Engabreen, an outlet glacier from the Svartisen icecap (northern Norway). (From Liestøl, 1979). Compare Fig. 16.

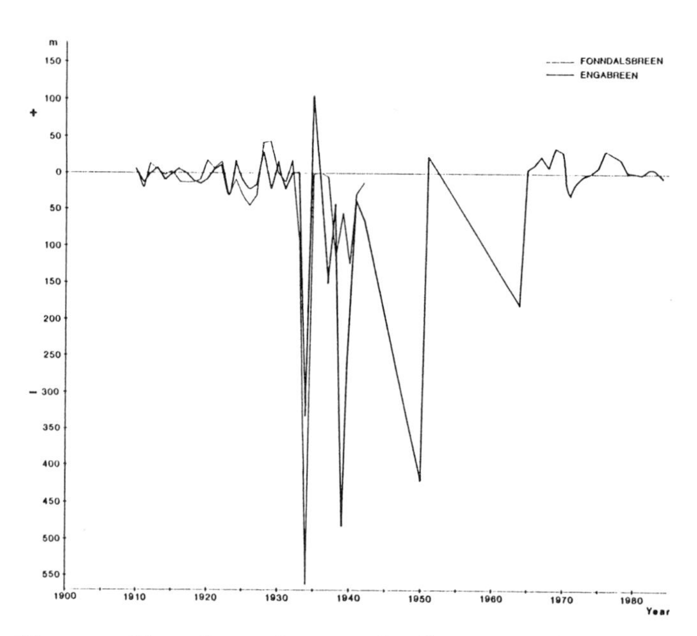

Figure 16. Comparison of front variations of Engabreen and the adjacent Fonndalsbreen. Both are west-facing outlets from the Svartisen icecap. Compare Fig. 15.

3.4 Svartisen

The largest outlet glacier from the Svartisen icecaps is the west-facing Engabreen (36 km^2), which almost reaches sea level. A glacier lake was formed in front of Engabreen in the period between 1931 and 1944. Front variations have been measured since 1902. Fig. 15 is based upon Liestøls (1979) observations and our own recent data. The earliest observations were made by Rekstad (1892).

A tree stump was found in front of Engabreen in 1953, in a newly deglaciated location. The C^{14} dating indicated that this was a tree which was overridden by ice about A.D. 1600, and this ice is supposed to have grown 180 m in thickness during the following century (Bergersen, 1953; Liestøl, 1962, 1979).

Engabreen had most probably its greatest extension since the last glaciation in the early part of the 18th century. From official documents it is known that the glacier destroyed one farm and damaged another in 1723 (Liestøl, 1962). Observed front variations at Engabreen and Fonndalsbreen are compared in Fig. 16.

3.5 The nothernmost Scandinavia

In the beginning of the 19th century there was increasing interest among geographers and geologists to start detailed observations of glacier variations (Wahlenberg, 1808). The French geographer Charles Rabot made valuable observations during his extensive travels in North Scandinavia around 1880 (Rabot, 1882). Also several Swedish researchers started glacier studies at this time (Svenonius, 1884, 1899; Hamberg, 1892; Westman, 1899; and others). In recent years several Swedish geographers have made extensive investigations, (Enquist, 1918; Wallén, 1949; Schytt, 1968 and others).

Rekstad (1892, 1893) was probably the first Norwegian researcher to start systematic studies in the Svartisen area. From 1909 he performed annual or bi-annual front observations at a great number of glaciers in North Norway. Similarly Hoel started measurements at other glaciers, e.g. in the Okstindan area (Hoel and Werenskiold, 1962). Theakstone (1964) summarized the observations made by earlier workers in the Svartisen area.

Table I. Retreat of Swedish glaciers 1965-1985 (according to Holmlund, 1986).

Glacier	area (km^2)	length (km)	retreat (m)	remarks
Salajekna	24.5	10.0	209	to 1984
Mikkajekna	7.6	4.6	325	
Ruotesjekna	5.4	4.6	544	to 1984
Rabots glaciär	4.2	4.1	225	
Riukojietna	5.5	3.3	224	from 1963
Suottasjekna	8.1	4.4	174	to 1984
Vartasjekna	3.6	3.0	72	to 1984
Storglaciären	3.1	3.7	98	
Ruopsojekna	3.6	3.9	137	
Ö. Påssusjietna	1.8	1.9	117	from 1968
U. Reitaglac.	2.0	2.1	80	
Stuor Reitglac.	2.0	2.6	147	
Hyllglaciären	1.5	2.2	80	to 1984
Isfallsglac.	1.4	2.1	163	
S.Ö. Kaskasagl.	0.6	1.4	150	
Kårsojietna	1.6	2.2	131	from 1968

Note: More detailed information on these glaciers (location, glacier type, orientation, morphology, etc.) can be found in Østrem et al. (1973).

Bergström (1955, 1973) has described old end moraines which must have been formed about 2500 years ago. They are situated far from present glacier tongues, e.g. in the Sulitjelma-, Kebnekaise-, and Abisko-areas. Some glaciers have, however, advanced up to or even beyond these very old moraines.

Several glacier advances have been assumed or directly observed about 1900-1915. For example, the Reintindbreen in Skjomen advanced 53 m between 1906 and 1911. Later this glacier has retreated again. Similar short-term, but very significant advances are known from northern Sweden, where F. Enquist took several good photographs of glaciers in the Kebnekaise area. Many of the glaciers are thought to have culminated about 1911-1917; later an almost continuous retreat has occurred (Schytt, 1959). Bergström (1955, 1973) has shown that many glaciers in North Scandinavia had their maximum postglacial extent as late as during the first decades of the 20th century.

Holmlund (1986) has compiled data from various front observations made on Swedish glaciers during the period 1965-1985 (Table I). All of the observed glaciers are of the continental type (with possibly one or two exceptions) and their fronts have been steadily retreating. This is in contrast to maritime glaciers in northern Norway, where advancing fronts have been observed during the time period (Fig. 16). On the other hand, similar conditions are seen in southern Norway, where maritime glaciers have shown signs of advance, whereas continental glaciers (e.g. in Jotunheimen) are still retreating, although slower than in the 1930s and 1940s.

4. DISCUSSION

Graphs showing front variations with time for 17 glaciers in southern Norway are combined in Fig. 17. In this diagram, front observations are plotted for all outlet glaciers at the ice caps Jostedalsbreen and Folgefonni. A common trend can easily be seen. In the periods 1900 to 1911, 1923-1933 and 1974-1986 there are frequent glacier advances.

From the plot of temperature anomalies it is clear that the glacier advances in the early part of this century are associated with periods of below normal temperatures. A general retreat of a number of glaciers starts around 1933, due to a series of excessively warm summers. These relations have also been pointed out by Faegri (1948) and Liestøl (1967).

From around 1960 there was a change in this trend. The summers then became colder, when compared with the preceding period. The summer temperatures were, however, above the mean for the period 1900-1930. Annual mass balance measurements in Norway started in 1948 on Storbreen, but from 1963 a greater number of glaciers were measured. Therefore, it can be shown that the continental glaciers in Jotunheimen have experienced a mass deficit whereas a surplus has been observed on the more maritime ones like Ålfotbreen and Nigardsbreen, without any clear reaction to the tongue positions.

Whereas the glaciers responded simultaneously to climatic fluctuations during the three first decades of this century, the various glaciers reacted differently to the extensive glacier melt during the following decades. Some did not respond to the increase in net balance until a long build-up period had passed. Several smaller outlet glaciers had a shorter response time and advanced several times during this period.

Whereas regional summer temperature variations more or less uniformly cause similar variations in glacier melt for most glaciers within an area, accumulation on the same glaciers may show individual variations. For glaciers in South Norway this is obvious: Many glaciers in maritime areas are no more retreating - some are advancing, whereas glaciers in continental areas (e.g. Jotunheimen) are still slightly retreating. This must be a result of regional accumulation pattern. Such 'irregularities' in snow accumulating on the ground have been observed several times during the last decades - southwestern Norway may receive significantly different amounts of solid winter precipitation than southeastern Norway.

The pattern of front variations in northern Scandinavia seems to be different from that of southern Norway. However, the presence of lakes front of glacier snouts and calving of glaciers in lakes caused a rapid recession of the Fonndalsbre and Engabre during the 1930s and 40s. These glaciers are outlets from the Svartisen Ice Cap in northern Norway.

Also during the 'Little ice Age' there seem to have been individual variations between glaciers. Karlén (1979) says that lichenomentric data suggest periods of glacial expansion at the end of the 1500's and in the mid and late 1600's. He has also dated several later advances and concludes that, although periods of general glacial advances occurred, the lichenometric results indicate that no single climatic event caused all glaciers to advance to maximum positions in northern Norway. Thus, although some glaciers were at advanced positions in the mid-18th century, the results do not support the assumption that the outermost 'Little Ice Age' moraines were all formed at the same time, about 1750.

Similarly, in southern Norway many outlet glaciers from the Folgefonni Ice Cap reached their most recent maximum during the late

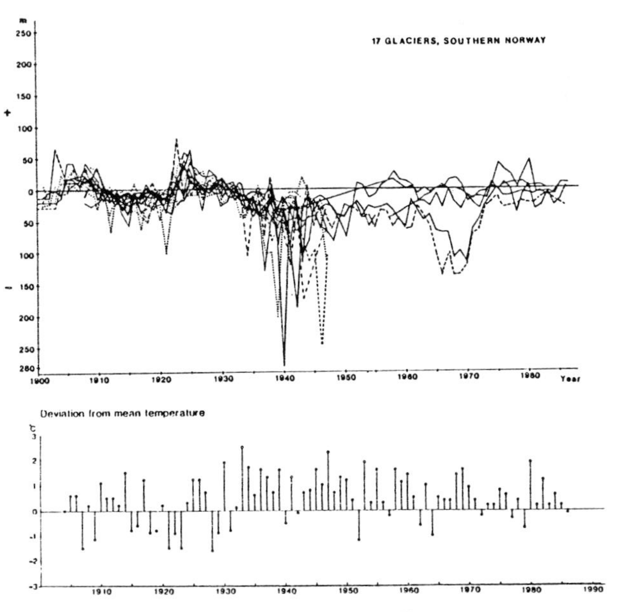

Figure 17. Front variations of 17 glaciers in South Norway plotted against the mean temperature of the summer months June-September in Bergen

1800s and in the case of Blomterskardsbreen, as late as in the 1940s (Tvede and Liestøl, 1977).

In a lichenometric study, Erikstad and Sollid (1986) found no evidence which could challenge the assumption of a regional advance around 1750. Mattews (1984) suggests large differences between Neoglacial glacier variations in southern compared with northern Scandinavia. In the south the 'Little ice Age' events appeared to have been larger, relative to any other Neoglacial glacier advance.

ACKNOWLEDGEMENTS

The present paper is based upon various data collected and processed by such a great number of people that it is impossible to mention them all. However, the names of authors in the reference list reflects the variety of scientists interested in the problem of glacier variations. We are grateful to all of them for the valuable work they have done, making it possible for us to prepare this paper.

REFERENCES

Ahlmann, H.W. 1948. Glaciological research on the North Atlantic Coasts. Royal Geogr. Soc. Research Series No. 1 (London) 83 pp.

Ahlmann, H.W. 1953. Glaciär och klimat i Norden under de senaste tusentalen år. Norsk Geografisk Tidsskrift XIII, 56-75.

Andersen, J.L. and Sollid, J.L. 1971. Glacial chronolgy and glacial geomorphology in the marginal zones of the glaciers Midtdalsbreen and Nigardsbreen, South Norway. Norsk Geografisk Tidsskrift 25, 1-38.

Bergersen, A. 1953. En undersøkelse av Svartisen ved Holandsfjorden 1950-51. Unpublished thesis, Dept. of Geography, University of Oslo, 1953. 55 pp. + ill.

Bergström, E. 1955. Studies of the variations in size of Swedish glaciers in recent centuries. UGGI, Ass. Gen. de Rome 1954 tome IV, publ. no. 39, 356-366.

Bergström, E. 1973. Den prerecenta lokalglaciationens utbredningshistoria inom Skanderna (The history of the prerecent local glaciation in the Scandinavian mountains). Research report STOU-NG 16 from Dept. of Physical Geography, University of Stockholm, 214 pp.

Blytt, A. 1869. Botaniske Observationer fra Sogn. Nyt Mag. Naturvid 16, 81-226.

Bohr, Chr. 1820. Om Iisbraeerne i Justedalen og om Lodalskaabe. Blandinger 2, 289-317. Christiania.

Durocher, M.J. 1847. Etudes sur les glaciers du Nord et du centre de l'Europe. Annales des Mines, Ser. 4, tome XII, Paris, 104.

Eide, T.O. 1955. Breden og bygda. Norveg. Tidsskrift for folkelivsgransking 5, 1-40.

Enquist, F. 1918. Die glaziale Entwicklungsgeschichte Nordwest-Skandinaviens. Sv. Geol. Unders. 251, 1-143.

Erikstad, L. and Sollid, J.L. 1986. Neoglaciation in South Norway using lichenometric methods. Norsk Geografisk Tidsskrift 40, 85-105.

Forbes, J.D. 1853. Norway and its glaciers visited in 1851. 349 pp.

Chapter VII: Justedal - The Fillefield. Edinburg.

Foss, M. 1802-1803. Justedalens kortelige beskivelse. Mag. Dan. og Nor. topogr. oecon. og stat. beskr. 2, 3-44.

Faegri, K. 1933. Ueber die Längenvariationen eniger Gletscher der Jostedalsbre und die dadurch bedingten Pflantzen-Sukzessionen. Bergen Museum Arbok 1933 2, 1-255.

Faegri, K. 1948. On the variations of western Norwegian glaciers during the last 200 years. IGGU, General Assembly Oslo, 293-303.

Grove, J.M. 1972. The incidence of landslides, avalanches and floods in western Norway during the Little Ice Age. Arctic and Alpine Research 4 (2), 131-138.

Grove, J.M. and Battagel, A. 1983. Tax recods from western Norway as an index of Little Ice Age environmental and economic deterioration. Climate Change 5, 265-282.

Hamberg, A. 1892. Hafsis, glaciäris och glaciärrörelse. Geol. Fören. Förh. 14, 558-599.

Hoel, A. and Werenskiold, W. 1962. Glaciers and snowfields in Norway. Norsk Polarinstitutts Skrifter nr. 114, 291 pp.

Holmlund, P. 1987. Mass balance of Storglaciären during the 20th century. Geografiska Annaler 69A (3-4), 439-447 and map.

Hysebye, S. 1983. Oldevatn i Nordfjord, Sediment av minerogent materiale. Unpl. thesis. Geography dept. University of Oslo.

Karlén, W. 1979. Glacier variations in the Svartisen area, northern Norway. Geografiska Annaler 61A (1-2), 11-28.

Larsen, Joh. 1874. To Vestlandsruter. Den Nor. Turistfor. Aarb. 1874, 7-14.

Liestøl, O. 1960. Glaciers of the present day. In: Geology of Norway. Norges Geol. Undersøkelse no. 208, 482-489.

Liestøl, O. 1962. Discovery of a tree stump in front of Engabreen, Svartisen. Norsk Polarinstitutt årbok 1960, 64-65.

Liestøl, O. 1967. Storbreen glacier in Jotunheimen, Norway. Norsk Polarinst. Skr. 141, 63 pp.

Liestøl, O. 1979. Svartisen. Den Norske Turistforening Årbok 1979, 137-143.

Lindblom, A.E. 1839. Vandring i Norrige, sommaren år 1839. Kgl Svenska Vetensk. Akad. Handl. 242-299.

Mattews, J.A. 1984. Limitations of ^{14}C dates from buried soils in reconstructing glacier variations and Holocene climate. In: N.A. Mörner and W. Karlén (eds.) "Climate changes on a yearly to millennial basis", Reidel, Dordrecht, 281-290.

Naumann, C.F. 1824. Beyträge zur Kenntniss Norwegens. Leipzig (2 volumes), 407 pp.

Nye, J.F. 1960. The response of glaciers and ice-sheets to seasonal and climatic changes. Proc. Roy. Soc. 256, 559-584.

Pedersen, K. 1976. Briksdalsbreen, Vest-Norge. Glaciologiske og glacialgeologiske undersøkelser. Unpubl. thesis. Dept. of Geology, University of Bergen.

Rabot, C. 1882. Reisen in Lappland 1880 and 1881. Petermans Mitt. (Gotha) no. 28, 339-342.

Rekstad, J. 1892. Om Svartisen og dens gletschere. Det Norske geografiske selskap Årbok 1891-92, 71-90.

Rekstad, J. 1893. Beretning om Undersøgelse af Svartisen, foretagen i sommeren 1890 og 1891. Archiv for Matematik og Naturvidenskab 16, 266-321.

Rekstad, J. 1902. Iagttagelser fra braeer i Sogn og Nordfjord. Norg. geol. unders. Aarb. 3, 42.

Rekstad, J. 1903. Skoggraensens og snelinjens større høire i det sydlige Norge. Norges Geologiske Undersøkelse nr. 36., 1-18 (with English summary).

Rekstad, J. 1904. Fra Jostedalsbreen. Bergen Museum Aarbok 1904, 1, 1-95.

Rekstad, J. 1905. Iagttagelser fra Folgefonnens braeer. Norges geologiske undersøkelse nr. 43, 1-17.

Rogstad, O. 1941. Jostedalsbreens tilbakegang. Norsk Geografisk Tidsskrift 8, 273-298.

Schytt, V. 1959. The glaciers of the Kebnekaise-Massif. Geografiska Annaler 41 (4), 213-227.

Schytt, V. 1968. Notes on glaciological activities in Kebnekaise, Sweden, during 1966 and 1967. Geografiska Annaler 50A (2), 111-120.

Smith, C. 1817. Nogle iagttagelser, isaer over Isfjeldene (Gletscher) paa en fjeldreise i Norge 1812. Topografisk-statistiske samlinger Del 2, Bd. 2. Kgl. Selsk. f. Norges Vel., 1-62.

Svenonius, F. 1884. Studier vid svenska jöklar. S. Geol. Unders. C.61, 1-37.

Svenonius, F. 1899. Öfersikt af Stora Sjöfallets och angränsnade fjällrakters geologi. Geol. Fören. Förhandl. 21, 540-570.

Theakstone, W.H. 1964. Recent changes in the glaciers of Svartisen. Journ. of Glac. 5, 411-431.

Tvede, A.M. 1972. En glasio-klimatisk undersøkelse av Folgefonni. Unpubl. thesis, University of Oslo.

Tvede, A.M. and Liestøl, O. 1977. Blomsterskardbreen, Folgefonni, mass balance and recent fluctuations. Norsk Polarinstitutt Årbok 1976, 225-234.

Wahlenberg, G. 1808. Berättelse om mätningar och observationer för att bestämma lappska fjällens höjd och temperatur vid 67 graders polhöjd, förrättade år 1807. Stockholm 1808. 58 pp.

Wallén, C.C. 1949. Glacial-meteorological investigations on the Kårsa glacier in Swedish Lappland 1942-1948. Geogr. Ann. 30 (3-4), 451-672.

Westman, J. 1899. Beobachtungen über die Gletscher von Sulitjelma und Ålmajäkna. Bull. geol. inst. Uppsala, no. 4, 45-78.

Østrem, G. 1965. Problems of dating ice-cored moraiens. Geogr. Ann. 47A (1), 1-58.

Østrem, G., Haakensen, N and Melander, O. 1973. Atlas over breer i Nord-Skandinavia. Meddelelse nr. 22 fra Hydrologisk avd., 315 pp.

Østrem, G., Liestøl, O. and Wold, B. 1977. Glaciological investigations at Nigardsbreen, Norway. Norsk Geogr. Tidsskr. 30, 187-209.

Øyen, P.A. 1901. Variation of Norwegian glaciers. Nytt Mag. Nat. 39, 73-116 Kristiania.

Øyen, P.A. 1910. A brief summary of the evidence furnished by glacial phenomena and fossiliferous deposits in Norway as to late Quaternary climate. In: Die Veränderungen des Klimas seit dem Maximum der letzten Eiszeit, Stockholm, 339-343.

THE DECLINE OF THE LAST LITTLE ICE AGE IN HIGH ASIA COMPARED WITH THAT IN THE ALPS

W. Kick
Regensburg
Federal Republic of Germany

ABSTRACT

Mayewski and Jeschke and other authors of compilations relating to the glacier history of High Asia conclude that there has been a "maximum position about 1850 and a general state of retreat since that time". Such a historical profile would be parallel to the pattern of occurrence in the European Alps. But an analysis of all sources quoted for this claim leads to the result that in not a single case can such a parallelization to the Alps be justified. A secular decline of the last Little Ice Age has been obviously proceeding also in Asia. But the point of time when this general recession began, cannot be confined to such a clear short period as is the case in the Alps. The starting point for the individual glaciers may be anywhere between the beginning of the 19th century and the beginning of the 20th. Even within a group of neighbouring glaciers, the fluctuations and the time data vary considerably. This greater "standard variation" remains valid also after an exclusion of those ice streams which are larger than the biggest glaciers in the Alps and of the surging glaciers which are much more numerous in the Asiatic mountains. Only since about the 1920's can a general decline of the Little Ice Age be confirmed also in Asia.

At last the different debris-cover as a possible reason for these deviations is discussed in a short outline.

1. THE SUBJECT OF COMPARISON

Firstly the behaviour of the glaciers in the European Alps during the middle of the 19th century and thereafter - the decline of the last LIA - may easily be characterized in a brief outline.

For the approximately 4000 glaciers of the Alps a clear tendency from the 1850's until the present-day is definitely proved by reliable historical records. Each of these glaciers has decreased in size. It is true that there had already been a maximum stage in 1820 or somewhat later, in some cases greater than that of the 1850's, and that during the last 130 years of secular decrease there have been short periods -

J. Oerlemans (ed.), Glacier Fluctuations and Climatic Change, 129–142.

about a decade each - of stagnation or re-advancement in the 1890's (until 1900), the 1920's (in the E-Alps about 1913-1924), and the 1970's (beginning 1965). But during these secondary episodes not one of the several thousand glaciers reached again the length or volume which it had in the decade after 1850. In this decade all the glaciers of the Alps obtained for the last time a great extension, similar to that of their first advancement - after centuries of a lower extent - in about 1600 and then several times between 1600 and 1850. The beginning of the succeeding decline fell in a clearly defined time period between 1855 and 1870. Not one exception is known. The fluctuations of the glaciers in the Alps have occurred with a distinct uniformity.

2. PRESENT VIEW ABOUT THE 19TH C. GLACIER HISTORY OF HIGH ASIA WITH AN ANALYSIS OF ITS FOUNDING

Almost everywhere in the Himalayas, the Karakorum, the Hindukush, the Tien Shan and the Pamirs, high lateral moraines, in their appearance quite similar to the "1850 moraines" in the Alps, are most striking. The question is whether this Great Lateral Moraine (GLM) was piled up and overshed by ice and debris for the last time at a date in the middle of the 19th century similar to that in the Alps. Looking at these striking ridges one is inclined to call them the "1850 moraines" as one is accustomed and justified to do in the Alps. In fact several authors support the view that the 33-times greater glacier areas of High Asia underwent a similar ice maximum in the middle of the 19th century.

2.1

Hermann von Wissmann (1960) wrote in his excellent work on the present glaciation and the snow limit of High Asia: "Also in High Asia the moraines, which have their origin in the high stand of the glaciation in the middle of the last century, are widespread." Wissmann cited no sources for this statement because the recent glacier variations were not subject of his book. But his judgement was based on a broad knowledge of the records referring to the glaciers of High Asia (670 cited publications). He had considered the literature up until 1958/59. Two more recent compilations are those of Mercer (1963) and Mayewski and Jeschke (1979, 1980).

2.2 John H. Mercer - Karakorum glaciers

In 1963 Mercer published a review about the records on the glacier variations in the Karakorum. In the introduction he summarizes (p.20): "The general picture that emerges (most data consist of isolated scientific observations or sporadic observations by travellers) is of a maximum position about 1850 or 1860, and re-advances in the present century."

Only a very small part of the 70 cited publications about 50 glaciers of the Karakorum deals with the 19th century, the time mainly in question in this paper. The oldest records are those of the Aktash, the Kishik and the Chong Kumdan Glaciers of the Upper Shyok Drainage. They mention several advanced states occuring in a number of years between 1770 and 1958, not always synchronous for the three glaciers. They do not provide any proof for a maximum peculiar to the middle of the 19th century. In strong contrast to the lake disasters, which were caused by Alpine glacier advances and which ceased definitely in the middle of the 19th century, those of the Shyok glacier lakes continued at least into the middle of the 20th century.

For Siachen (73 km, 1180 km^2), the greatest of the Karakorum glaciers, Mercer remarks that in Longstaff's (1910) photograph "certainly no shrinkage from the lateral moraine is apparent". This means that in 1909 the volume of the glacier tongue was - in any case at the spots seen in the photograph - the same as at the time when the GLM was piled up. Therefore the GLM of the Siachen in no way marks such an outstanding volume maximum right in the middle of the 19th century which was never reached again. Longstaff (1910) observed the same at the Bilafond Glacier: "The glacier appears to be overtopping its lateral moraine."

In the north of K2 the Gasherbrum Glacier was visited by the Italian expedition of 1929. "A photograph (Desio, 1930) shows the glacier against and overtopping a large terminal moraine". Biapo (68 km, 635 km^2) and Panmah (44km, 515 km^2) indeed seem to have been at their maximum length when Godwin-Austen sketched them in 1861 (Godwin-Austen, 1864). Unfortunately a very natural painting by A. Schlagintweit which shows the terminus of the Panmah in 1856 has not yet been used to reconstruct the location of the snout and the situation of the lower tongue as it was in the middle of the 19th century. The painting is published in a black and white reproduction (Kick, 1958 and 1969).

The number and reliability of records of the 45 km long Chogo Lungma Glacier is higher than those of the other great Karakorum glaciers. The former demonstrates a last volume maximum (last overtopping of ice over the GLM) not in the middle but in the last third of the 19th century (Kick, 1956). Even more different from the situation in the Alps is the time date of the last length maximum which occurred as late as 1913, not during the 19th century at all. The last overtopping of the GLM - in the lowest section of the tongue! - of the Tippuri Glacier nearby did not happen until the 20th century, at least half a century later than in the Alps (Kick, 1956).

The records which Mercer (1963) cited in his compilation do not prove his statement of a general maximum position "about 1850 or 1860", especially not one last outstanding maximum like that in the Alps.

Mercer mentions only one case for which the date of the origin of the present GLM **seems** to have been proved: for the Mani Glacier (north side of Haramosh) "the ages of trees indicated a recent maximum round about 1850" (Mercer, 1963). For this statement he refers to Wiche (1958). But the latter wrote a rather different text which was probably misunderstood because it is in German:"... An important advance ... almost near to the extension of the last pleistocene ice age ... (marked by) a 30 to 40 m high lateral moraine ... According to the trees (Baumbewuchses) growing at the moraine, the latter may have originated from a recent advance (1850?)." Wiche did no dendrochronological research as Mercer's interpretation may suggest. From the evidence of the tree vegetation, Wiche thought the moraine's age to be of at least several decades (far from being pleistocene) and therefore - in a widespread alpino-centric way of thinking - suggested a comparison with the 1850 moraine of the Alps. But he added a question mark! He was not thinking of a genuine dating when he mentioned the idea of a parallelization with the 1850-maximum in the Alps.

All in all there is no real proof for an outstanding 1850 maximum and a succeeding general decline of the LIA in the Karakorum only since 1850/70 as in the Alps.

Mercer takes the Nanga Parbat area as a part of the Karakorum, as other authors do, but it undoubtedly is part of the Himalayas which will be treated in the next chapter.

In the last lines of Mercer's compilation one should twice read Chong Kumdan instead of Chogo Lungma Glacier. In the conclusions at the end of his extensive account, Mercer did not repeat the statement of his introduction of "a maximum position about 1850 or 1860".

2.3 Mayewski and Jeschke - Himalayan and Karakorum glaciers

There is a more recent and still more extensive "reevaluation of the existing literature" of Himalayan and "Trans-Himalayan" (= Karakorum) glaciers since 1812 by Mayewski and Jeschke 1979; 1980). These two authors used 138 publications which provide data on the fluctuation history of 112 glaciers, double the number taken into account by Mercer. The summarized result is (1979, abstract): "In a gross regional sense Himalayan and Trans-Himalayan glaciers have been in a general state of retreat since A.D. 1850." In more detail: "Most glaciers in the Himalayas have been in a general state of retreat since 1850. Those in the Trans-Himalayan grouping were either in retreat or advance from 1850 to 1880, reflected near-equivalent influences of retreat, standstill and advance regimes from 1880 to 1940 ..." In this formulation, again a general last maximum in about 1850 and a prevailing retreat since this date is suggested. In the following those sources of Mayewski and Jeschke are analysed which are cited for the statement: "...consistent retreat of the Himalayan glaciers throughout 1850-1960, best exemplified in Lahaul-Spiti, Kolahoi, Nanga Parbat and Garhwal ..." (Mayewski and Jeschke, 1980).

First of all the number of existing (cited) historical records which give evidence for the middle of the 19th century is small. The time date "since 1850" is based on only six out of a sum total of more than 100,000 glaciers in High Asia. These six are: The Pindari and in its neighbourhood the Milam Glacier in the Gargwak- or Kumaun-Himalaya, the Kolahoi, Rakhiot and Chungpar Glaciers in the Nanga Parbat massif and the Baltoro Glacier in the Karakorum.

2.3.1. Pindari and Milam Glaciers

The source for the Pindari was a description and sketch of Richard Strachey (1847). The Pindari was the first glacier Strachey had seen. He knew the phenomenon of glaciers only from the literature of Forbes (1843). His concern was to prove the existence of glaciers in the

Himalayas. In order to demonstrate the movement of the ice - a necessary condition for a glacier - he measured the ice velocity. Thus the first measurement of a glacier velocity in High Asia had taken place. The statement of Mayewski and Jeschke that the Pindari began to recede since Strachey's investigations is based only on Strachey's estimation that the distance of the glacier snout from the lower end of the left lateral moraine (which is situated higher up between two glaciers, see Fig. 1) was "about two miles"; whilst this distance was only about one mile in 1906 (Tewari, 1973). But the stands at these two points of time - 1847

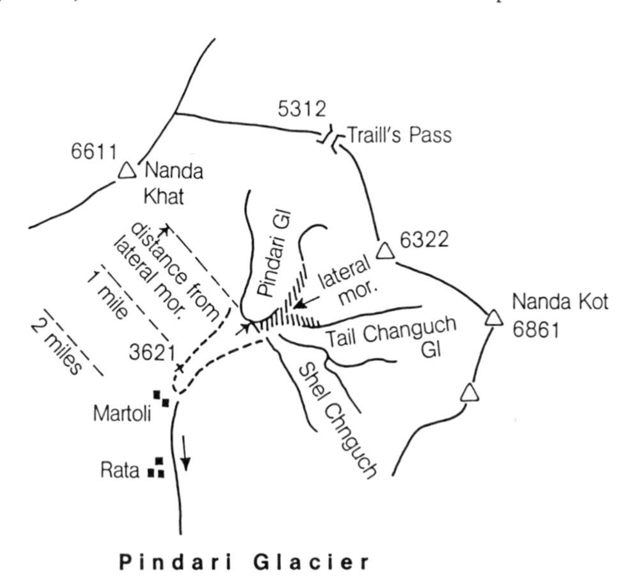

Figure 1.

and 1906 - can in no way prove the beginning of the retreat just in about 1850. At best, this can only be a supposition. Now there is another report which even contradicts such a supposition. In 1846 the inhabitants of the village of Milam told Eduard Madden (1847) that the Milam Glacier - in the northern neighbourhood of the Pindari - had already undergone a recession of 2 to 3 miles. The Pindari too, he was told, had undergone a recession before 1846, though to a smaller extent.

If one believes such reports at all, both glaciers, the Pindari and the Milam, had begun their retreat not in about 1850, but much earlier. The striking left lateral moraine of the Pindari, which is at the same time the right moraine of the Changuch Glacier, was already in 1847 much higher than the surface of the Pindari. A high stand of the volume, perhaps the last maximum in question - much greater than that of 1850 - must therefore have occurred a long time before 1846.

Apart from the report of Madden there is a record referring to the Milam Glacier, by the two geologists Cotter and Brown (1907): "According to Rai Kishen Singh Bahadur of Milam, well known to science as A.K., the explorer of Tibet, the ice-cave 57 years ago was about 800 yards in advance of its present position." But the time date 1849 ("57 years ago") of a certain advanced position does not necessarily define a maximum stage. It is not said whether these 57 years mark a certain time date in the life of the reporter Bahadur which he remembered, or if it really is a turning point in the variation history of Milam Glacier.

Another source has not yet been evaluated. On May 5th, 1855, Adolf Schlagintweit painted the snout of the Pindari. The painting depicts the shape and position of the glacier end in relation to the surrounding landscape so successfully, that one could probably reconstruct the 1855 teminus in the field. I published the picture in Kick (1960) and sent a reproduction to the geologists in Lucknow. But neither the painting nor Schlagintweit's descriptions, measurements and further sketches of the Pindari and the Milam have as yet been used. Through an evaluation of this unpublished research together with a sketch map by R. Strachey and the description of the spot of Strachey's theodolite-measurement on an old moraine, a reconstruction of the 1850 glacier state should be possible.

2.3.2. The Kolahoi Glacier

For the third glacier, quoted for a general 1850 high stand, the Kolahoi, an account of E.F. Neve (1910) was used as the source (Odell, 1969). Neve wrote:"...The trigonometrical survey map was completed in 1857. Reference to this, and ... shows that since that date, in fifty two years (1857-1909) the glacier must have retreated above a mile." Probably the map of 1857 is not reliable and precise enough and above all, the amount of recession since that time does again in no way prove a last maximum just at the mapping time date. The map shows only that the Kolahoi Glacier had a much greater extension in about 1850 than in 1909. It could have been greater still in earlier or later decades.

2.3.3 Rakhiot and Chungpar Glaciers

For these glaciers in the Nanga Parbat massif, Mayewski and Jeschke and Mercer (1963) refer to Finsterwalder (1938) and Kick (1962). The latter again used as sources for the Chungpar a painting, a sketchy drawing, measurements and descriptions of Schlagintweit of the year 1856. The informations originated in a year nearer to the time in question than what F. Drew heard in 1870 from natives of Tarshing about the Rupal lake and the ice level of 1850 (Drew, 1875). But Schlagintweit described the ice surface of 1856 to be some 20 m below a recent moraine (Fig. 2). According to the stage as he saw it, the Chungpar must have had its last greatest stand several years before his visit, probably earlier than 1850. Though the Chungpar's damming of the Rupal lake is known by the descriptions of Drew and Schlagintweit for the years 1850 and 1856 only, this periodic damming may have begun already earlier.

On the other side the Sachĕn at the east side of Nanga Parbat was in 1856 in a much smaller state than even in 1900. This glacier must have reached its last maximum towards the end of the 19th century only.

The evaluation of much unpublished research of Schlagintweit (Kick, 1967) resulted in a rather different behaviour pattern for the glaciers at the east and south side of Nanga Parbat. There is by no means a proof for a last maximum stage in the middle of the 19th century for all of the Nanga Parbat glaciers as Finsterwalder deduced from the situation of only one glacier. It is true that for one, the Chungpar- or Tashing Glacier, do a painting and sketch map document the maximum length, namely down to the Chiche river (Fig. 2), but this extension can have existed for a longer period of time before and after 1856. The result of an analysis and comparison in the field by the author (1958) was: The time period in which the last maximum stages of Toshain-Rupal, Shaigiri, Tap, Bazhin, Chungpar, and Sachĕn Glaciers (south and east side of Nanga Parbat) occurred, ranges from a time one or more decades before 1856 until about 1900 or even later. The last maxima of several glacier individuals are not confined to such a small period of time as in the Alps.

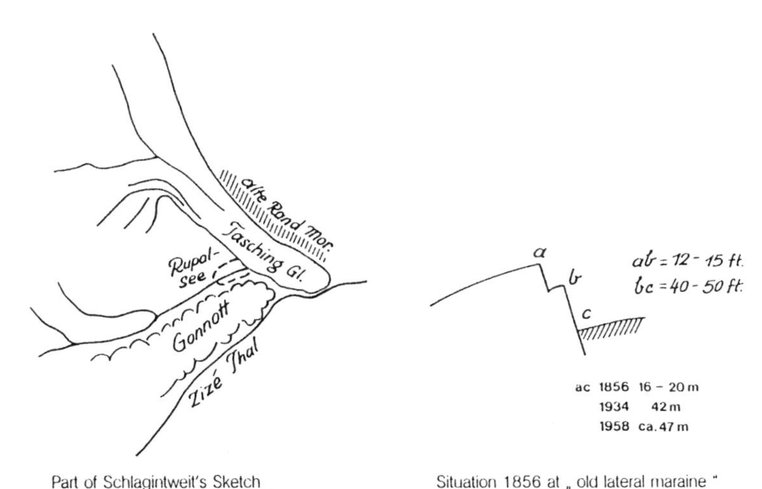

Figure 2. Part of Schlagintweit's sketch "Gletscher der Diamergruppe" September 1856. In 1934 Glacier end 600 m from Chichi river.
Drawing r.h.s.: Situation 1856 at "old lateral moraine"

Subsequent authors will refer to the publication of Mayewski and Jeschke as the most recent and competent compilation. E.G. Goudie (1984) wrote:"The most comprehensive study to date is by Mayewski and Jeschke, who produced regional syntheses for all the Himalayan-Karakorum environments." But, indeed, the compilations by Mercer and Mayewski and Jeschke are examples for the possibility of wrong conclusions, to which sporadic historical records can lead. Mayewski and Jeschke (1979) wrote:"Glaciers of Nanga Parbat have experienced a general but rather minor retreat (average 6 m yr^{-1}) since 1850 AD. The only significant departures fromt this low retreat rate exist for Rakhiot and Chungpar glaciers which between 1930 and 1950 AD had retreat rates as high as 30 m yr^{-1}) ..." The sources for these statements were misunderstood. Only for one amoung 69 glaciers of the Nanga Parbat, namely just for the Chungpar (for which Mayewski and Jeschke maintain a departure!) the position of the terminus in 1850 is known from a sketch map of Schlagintweit (1856) and from Drew (1875). For all other 68 glaciers the extension of 1850 is not known and therefore a "general average retreat of 6 m yr^{-1}" cannot be known. Details are published in Kick (1967). Only in the case of the Chungpar could the difference in its length between 1856 and 1934 be measured. If the recession during these 78 years had occurred continuously, the rate would have been 7.7 m/yr.

For the cited Rakhiot Glacier the stage of 1850 is not known at all, since Schlagintweit had not visited this nothern side of Nanga Parbat. W. Pillewizer measured a recession rate from 1934 (Finsterwalder) to 1954 not of 30, but of 22.5 m/yr, and J.S. Gardner an advance rate from 1954 to 1985 of 5 m/yr (Gardner, 1986). At the Chungpar the photogrammetric resurvey by the author in 1958 showed a recession rate from 1934 to 1958 of an amount between 0 and 4 m/yr, depending on the location along the ice margin at the terminus, and not at all "retreat rates of 30 m/yr^{-1}".

2.3.4 Baltoro

Mercer (1963) mentions - according to Godwin-Austen (1864) and Ferber (1907) - that by the middle of the 19th century a great increase in ice and snow rendered the (old) Mustag Pass to Yarkand impassable. This record relates to a tributary glacier of the Baltoro which leads to the pass. There are some more passes in High Asia which were abandoned during the first half of the 19th century (Kick, 1975). But the dates for the last use are as uncertain as the turning point of the last high glacier stage. The Hispar La was used for the last time in 1837. "From Conway's photograph (1892) the snout of the Baltoro (62 km; 755 km^2) appears to be in the same position as in 1861 ... At least until 1909 the snout remained in the same position" (Mason, 1930). The termini of the Baltoro and the other large debris-covered ice streams of the Karakorum seem to have begun their recession more than half a century later than the largest glaciers of the Alps. But they must - as particular cases - be excluded from a comparison with the Alps. At any rate they cannot be used to sustain the thesis of a general last maximum in about 1850, as it has been done.

3. EXAMPLES FROM THE CAUCACUS, NANGA PARBAT, KARAKORUM, PAMIR

Glaciologists generally interpret the striking lateral moraines also of the Caucasian glaciers as corresponding to a last maximum in the 1850's. This is justified for e.g. the Great Asau Glacier at the Elbrus region, since a historical record supports this view; though even for this Bolshoi Asau a recent more detailed exploration revealed that its stage of 1850 had been overestimated hitherto by 300 m in length (Zolotarev, 1982). But it is not justified for e.g. the Shkhelda, for which it had also often been maintained, most recently by (Dokuking, 1985). A photograph by Von Déchy of 1886 (D., 1905) obliges one to renounce this view (Kick, 1972). Shkhelda reached its last greatest maximum in length only about 1911 or later. In 1979 Kotlyakow wrote (Kick, 1979), that Shkhelda "started to retreat only in 1925, 50-70 years after the other glaciers". But it had reached its maximum also half a century "after the other glaciers", for which it is assumed, though very few records exist. In 1886, vegetation existed on an area between the terminus of the Shkhelda and the location of the later end-moraine of the 20th century, more down-valley, which hitherto had been thought to be the 1850 moraine. The GLM "of 1850" did not yet exist on the photograph of 1886. This is another example of an unproved transfer of Alpine glacier history.

As already mentioned Schlagintweit's fieldbooks of 1856 together with later records show a rather different history for six neighbouring glaciers of the Nanga Parbat massif. In detail: The Rupal (alpine firnfield type; 17 km, 55 km^2) and the Chungpar (Kessel glacier type; 12 km, 22 km^2) had their last maximum before 1856, Bazhin (12 km, 17 km^2) after this date, Shaigiri (7 km, 4 km^2), Tap (5 km, 3 km^2) and Sachĕn (11 km, 11 km^2) near 1900 only, at any rate long after 1856.

Another case with a behaviour pattern quite different from that in the Alps is that of the Batura (59 km, 285 km^2) in the Karakorum. Its terminus remained at its maximum extent in the Hunza river at least until 1925, probably even longer (Shi-Yafeng, 1984). Its recession began only in the 1940's (Goudie, 1984), almost a century later than the glaciers of the Alps. But the variations of Batura had been hidden for at least 1885 because its snout had been eroded by the water of the Hunza (Mason, 1930).

The Fedtshenko in the Pamirs (77 km, 806 km^2) reached its maximum length in 1914, its maximum volume in 1870 (Drygalski-Machatschek, 1942). The Minapin in the Karakorum (18 km, 60 km^2) was at its greatest length in 1913, and retreat seems to have begun in this year or soon afterwards (Mason, 1930).

Y. Ono (in press) designated the year 1815 as a historical high point of glaciation in the Langtang valley (Nepal-Himalaya). "At present (1975) some glaciers of the Wahkan Pamir have still their extension of a modern (last Little Ice Age) maximum stage" (Patzelt, 1978).

4. SUMMARY

The very few historical records which have been brought forward to prove a Himalayan last maximum similar to that of 1850 in the Alps have turned out to be of no use for such a proof. One might suppose that the time date of 1850 has been maintained owing to an alpinocentric prejudice, or because some very few Europeans (the brothers Strachey, the brothers Schlagintweit, Godwin-Austen) who provided these records had visited the glaciers by chance just round about that time.

More recent records demonstrate that the majority of the glaciers in High Asia, but not all of them, have been decreasing for the last few decades. However, the point of time when this general recession started cannot be confined to such a well defined and short period of about one decade after 1855 as is the case in the European Alps. Some glaciers left their last great maximum as early as in the beginning of the 19th century, others in the 2nd half, in about the 1880's, and others in the beginning of the 20th century only. Even neighbouring glaciers vary much in their fluctuations and the referring time data. This greater "standard variation" remains also after an exclusion of a) those ice streams which are larger than the biggest glaciers in the Alps and therefore must naturally experience a longer and more different response lag for the starting of their recession, and b) the surging glaciers which are more numerous in the Karakorum mountains at least. But also the remaining "normal" Asiatic glaciers have fluctuated in a much less uniform way then the glaciers of the Alps.

Some more information about the stages of several glaciers in various regions of the Himalayas and the Karakorum at about 1850 may be gained by a further evaluation of the unpublished research of Schlagintweit (1855-57). But all in all the historical records for the 19th century are not sufficient to provide a reliable profile. Other methods such as dendrochronology, lichenometry, may improve our knowledge. F. Röthlisberger has already applied such biological methods to moraines in various regions of the Himalayas and the Karakorum (Röthlisberger, 1987). But the great differences between the single glaciers will remain a difficulty for the modelling of a general history of the decline of the last LIA. In future it may be overcome by evaluating satellite images to determine the variations of the glacier areas. This has already begun by Institutes of the USSR (see e.g. Ice No. 81, 1986).

Since the glacier areas of High Asia are more than 30-times larger than those of the Alps, we ought to consider them as the "normal" or most common morphologie type of mountain glaciation and therefore should deduce e.g. the course of the Little Ice Age (or most recent holocene period of big glacier stages from the 16th/17th to the 20th century) in its Eurasian extent from the occurrence in Asia and not of Europe, and should rather ask why the fluctuations of the Alpine glaciers in Europe are of such an astonishing uniformity. The Alps must be seen as a particular case within Eurasia, with a rise in tectonic steps, therefore with level land for the wide firnfields which serve as the accumulation areas.

Probably, it may be taken as a symptom of more complexity in the Alps too, and of more similarity to Asia, that before the 1920's the accumulation areas had been filled up so considerably that the high stand of 1850 would have been reached again in the following years, had not the succeeding dry and warm summers of 1928-30 melted off the surplus, and had not the starting position been much "lower" than before 1850. In the case of some rather few glaciers, the Sulden and the Grünau Ferner in the Eastern Alps (Patzelt, 1970), the Tsidjiore Nouve and the Macugnaga (Kinzl, 1932) in the Western Alps, the ice had already overflown the 1850-moraine, mostly at some upper sections of the tongues.

5. A SHORT OUTLINE OF ONE REASON FOR THE EXTREMELY VARYING REACTIONS OF ASIATIC GLACIERS TO CLIMATIC CHANGES

Whereas the Alps are subject to a relatively uniform climate, the Asiatic mountains are influenced by a great range of climates from humid to dry, and beyond that, some sections get their main precipitation (and ablation) during the summer monsoon. From there the glaciers in the various regions must naturally be expected not to fluctuate in the same uniform way as in the Alps. But there are rather different reactions even within the mountain groups, even between neighbouring glaciers with the same climatic conditions. One reason for this local variability may be seen in the younger tectonic history, and thus in a steeper relief, a more intensive erosion, more debris-cover, all different from the Alps. Often the equilibrium line is located in the steep flancs above the glacier and therefore a climatic variation of the snowline does not change the accumulation area ratio much. The lack of elevated flat land produces the "Kessel glacier type", the predominant glacier type in High Asia. It was named and described by A. Schlagintweit in his unpublished fieldbooks. He had studied many glaciers in the eastern and western Alps (1847-48), and afterwards became acquainted with several glacier areas in various parts of the Himalayas and the Karakorum. Because the experience he had accumulated he several times noticed as e.g. in Vol. 21 (Schlagintweit, 1856):"Kessel glaciers in the Himalayas ... their type, contrast to the Alps ... great masses of firn and also rock sometimes break down and cause a predominance of one side of the glacier for many years ..." Schlagintweit saw the importance of avalanches and the different debris-cover for the behaviour of the single glaciers, even between various sides of the same glacier. Avalanches do not furnish the amounts of accumulation in such a continuous homogeneous succession as the flat firnfields of the "Alpine or firn basin type". The more or less thick debris-cover on many, probably on the majority of the glacier surfaces, rather irregular and different for the single glaciers and between various locations of one and the same glacier, causes rather different amounts of ablation. The latter varies in time and space much more than at blank glaciers ("selective ablation").

Independent from Schlagintweit the Dutch Karakorum explorer Ph.C. Visser has made up just the same term "(Firn)Kessel glacier type", when he searched for a name which should not be bound to certain mountain regions, as the "Mustag-", or the "Turkestan-" type of Oestreich, respectively Klebelsberg (Kick, 1964). With a time distance of more than 70 years, both authors, Schlagintweit and Visser, found a common tern and view of what is the characteristic and important morphologic feature of these glaciers which among other effects determines the response to climatic changes.

There have been several efforts to estimate and to measure the effects of debris-cover on heat balance and ablation. But the amount of ablation depends not only on the thickness, but on several parameters which determine the heat transmission coefficient of the covering material (Kraus, 1967). Another difficulty is the rough relief of such glacier surfaces, with single locations of blank ice amidst deeply covered surroundings.

6. CONCLUSION

The changing thickness and quality of debris-cover causes variances in the response time of single glaciers to the same climatic changes by some decades, and therefore the fluctuations of the Asiatic glacier volumes cannot be described by a scale of years, as in the Alps, but only by quarter-, or half-centuries. Therefore the general decline of the last Little Ice Age in Asia has not begun "in the 1850's", but during the 19th century up until the beginning of the 20th century.

REFERENCES

Cotter, G.P. de, and Brown, J.C. 1907. Notes on certain glaciers in Kumaon. Rec. Geol. Survey India 35, Part 4, 148-157.

Dechy, Moritz von, 1905. Kaukasus. Reisen und Forschungen im kaukasischen Hochgebirge. 3 Volumes, Berlin 1905-07.

Desio, A. 1930. Geological work of the Italian Expedition to the Karakorum. Geogr. J. 75, 402-411.

Dokukin, M.D. 1985. Formation of glacial mudflow sites in course of glacier degradation in the Elbrus area (Russ.). Data of Glaciol. Studies; Publ. no. 53, Moskva, 62-71.

Drew, F. 1875. The Jummoo and Kashmir Territories. London.

Drygalski and Machatschek, 1942. Gletscherkunde. Wien.

Ferber, A.C.F. 1907. An exploration of the Mustagh Pass. Geogr. Journal 30, 630-643.

Finsterwalder, R. 1938. Die geodätischen, gletscherkundlichen und geographischen Ergebnisse der Deutschen Himalaja Expedition 1934 zum Nanga Parbat. Berlin.

Forbes, J.D. 1843. Travels through the Alps of Savoy...with observations on the phenomena of glaciers. Edinburgh.

Gardner, J.S. 1986. Recent fluctuations of Rakhiot Glacier. J. Glaciology 32 (112), 527-529.

Godwin-Austen, H.H. 1864. On the glaciers of the Mustagh Range. J. Roy. Geogr. Soc. 34, 19-56.

Goudie, A.S. 1984. Recent fluctuations in some glaciers of the Western Karakorum mountains. In: Miller, vol. 2, 411-455.

Kick, W. 1956. Der Chogo-Lungma-Gletscher im Karakorum, Part 1. Zeitschr. Gletscherkunde 3 (3), 335-347.

Kick, W. 1958. Auf den Spuren Schlagintweits. Mitteilungen Dtsch. Alpenverein. 10; Helft 3, 37-40. München (with painting Panmah glacier, 1856).

Kick, W. 1960. The first glaciologists in Central Asia. J. Glaciology 3 (28), 686-692 (with painting Pindari glacier, 1855).

Kick, W. 1962. Variations of some Central Asiatic glaciers. Sympos. Obergurgl. IASH-Publ. no. 58, 223-229.

Kick, W. 1964. Der Chogo-Lungma-Gletscher im Karakorum, Part 2. Zeitschr. Gletscherkunde 5 (1), 1-59.

Kick, W. 1967. Schlagintweits Vermessungsarbeiten am Nanga Parbat 1856. Dtsch. Geodät. Komm., Bayer. Akad. Wiss., C97.

Kick, W. 1969. Alexander von Humboldts Wirken für die Hochgebirgsforschung in Asien, besonders über die Brüder Schlagintweit. Petermanns Geogr. Mitt. 113, (H.2), 91-99.

Kick, W. 1972. Auswertung photographischer Bilder für die Untersuchung und Messung von Gletscheränderungen. Zeitschr. f. Gletscherkunde und Glazialgeologie 8, 147-167.

Kick, W. 1975. Application of geodesy, photogrammetry, history and geography to the study of long-term mass balances of Central Asiatic glaciers. Moscow Symposium 1971. IASH-Publ. no. 104, 150-160.

Kinzl, H. 1932. Die grössten nacheiszeitlichen Gletschervorstösse in den Schweizer Alpen und in der Montblanc-Gruppe. Zeitschrift f. Gletscherkunde 20, 269-397.

Kotlyakov, V.M. and Krenke, A.N. 1979. The regime of the present-day glaciation of the Kaukasus. Zeitschrift f. Gletscherkunde 15 (1), 7-21.

Kraus, H. 1967. Freie und bedeckte Ablation. In: W. Hellmich (ed.), "Khumbu Himal." Heidelberg, 203-235.

Longstaff, T.G. 1910. Glacier exploration in the estern Karakorum. Geogr. Journal 35, 622-658.

Madden, E. 1847. Notes of an Excursion to the Pindree Glacier in Sept. 1846. J. Asiat. Soc. Bengal 16, (176), 226-266.

Mason, K. 1930. The Glaciers of the Karakorum and Neighbourhood. Rec. Geol. Survey India, 63, Part 2, 214-278.

Mayewski, P.A. and Jeschke, P.A. 1979. Himalayan and Trans-Himalayan Glacier Fluctuations since A.D. 1812. Arctic and Alpine Research 11 (3), 267-287.

Mayewski, P.A. and others 1980. Himalayan and Trans-Himalayan Glacier Fluctuations and the S-Asian Monsoon Record. Arctic and Alpine Research 12 (2), 171-182.

Mercer, J.H. 1963. Glacier variations in the Karakorum. In: American Geographical Society. IGY World Data Center A; Glaciology, Glaciol. Notes no. 14. New York, 19-38.

Miller, K.J. (ed.) 1984. International Karakorum Project. 2 Volumes, Cambridge.

Neve, E.F. 1910. Mt. Kolahoi and its northern glacier. Alpine Journal 25 (187), p. 40.

Neve, E.F. and Oliver, D.G. 1910. Notes on some Kashmir Glaciers. Rec. Geol. Survey India 40.

Odell, N.E. 1963. The Kolahoi northern glacier, Kashmir. J. Glaciology 4, 633-635.

Patzelt, G. 1970. Die Längemessungen an den Gletschern der österreichischen Ostalpen. Zeitschrift f. Gletscherkunde und Glazialgeologie 6, 151-159.

Patzelt, G. 1978. Gletscherkundliche Untersuchungen im "Grossen Pamir". In: Senarclens de Grancy & Kostka: Grosser Pamir. Graz, 132-149.

Röthlisberger, F. 1987. 10,000 Jahre Gletschergeschichte der Erde. Aarau.

Shi-Yafeng and Xiangsong, Zh. 1984. Some studies of the Batura Glacier. In: Miller (ed.), Vol. 2, 51-63.

Schlagintweit, A. 1856. Beobachtungsmanuskripte aus Indien und Hochasien. Unpublished. Collection of Manuscripts Bavarian State Library. Schlagintweitiana II a Vol. 21.

Strachey, R. 1847. A description of the glaciers of the Pindur and Kuphinee rivers in the Kumaon Himalaya. J. Asiat. Soc. Bengal 16, part 2, no. 181, 794-812.

Strachey, R. 1849. On the snow line in the Himalaya. J. Asiat. Soc. Bengal 18, Part 1, 287-310.

Strachey, R. 1851. Physical Geography of Kumâon and Garhwal. J. Roy. Geogr. Soc. 21, 57-85.

Tewari, A.P. 1973. Recent changes in the position of the snout of the Pindari Glacier. Proc. Banff-Sympos. 1972. WMO-IASH, 1114-1149.

Visser, Ph.C. 1933/34. Benennung der Vergletscherungstypen. Zeitschrift f. Gletscherkunde 21, 137-139.

Wiche, K. 1958. Die österreichische Karakorum-Expedition 1958. Mitteilungen Geogr. Ges. Wien 100, 280-294.

Wissmann, H. von, 1959. Die heutige Vergletscherung und Schneegrenze in Hochasien. Abhdlgen math.-nat. Kl. Akad. Wiss. u. Lit. Mainz, no. 14, 1-307.

Zolotarev, E.A. and Seinova, I.B. 1982. On the spatial position and fluctuations of Bolshoi Azau Glacier in the last century. Data of Glaciol. Studies. Publ.no. 46, Russ., Moskva.

VARIATIONS OF RIO PLOMO GLACIERS, ANDES CENTRALES ARGENTINOS

J.C. Leiva, L.E. Lenzano, G.A. Cabrera and J.A. Suarez
Instituto Argentino de Nivologia y Glaciologia
Mendoza , C.C. 330
Argentina

ABSTRACT

The Rio Plomo glaciers have been studied since 1909 (Helbling, 1919). The fluctuations of these glaciers show a general retreat from 1909 to 1974 (Espizúa, 1986) but there are two surging glaciers in the region, one of them was the source of an ice-dammed lake in 1984, similar to the one in 1934. A great amount of information was originated after the 1934 flood, caused by the outburst of the ice-dam produced by the surge of the Grande del Nevado Glacier. The maps drawn by Helbling in 1914, together with a new map made in 1985 by photogrammetric restitution of the 1974 aerial photos, were used to estimate the mass loss throughout that period. This was carried out using two different methods.

Between 1974 and 1986 some of the Plomo glaciers readvanced and, in 1984, the Grande del Nevado Glacier advanced three kilometres and dammed the Plomo river during its surge.

1. INTRODUCTION

Rio Plomo is the main tributary of Rio Tupungato, which in turn, along with Las Cuevas and Las Vacas rivers, forms Rio Mendoza. Its basin has one of the most important glacier systems of the Argentine Central Andes. Located at 32°57' to 33°12'S latitude, and 69°57' to 70°06W longitude, this glacier system has been studied since 1909 (Helbling, 1919). There are about 50 km^2 of debris free ice surface and 20 km^2 of debris covered ice (Corte and Espizúa, 1981). After the 1934 flood, caused by the outburst of the ice-dammed lake, originated by the surge of Grande del Nevado Glacier, several expeditions were to study its source (King, 1935; Helbling, 1935; Razza, 1935). Unfortunately, they only recorded observations of the ice dam.

J. Oerlemans (ed.), Glacier Fluctuations and Climatic Change, 143–151.

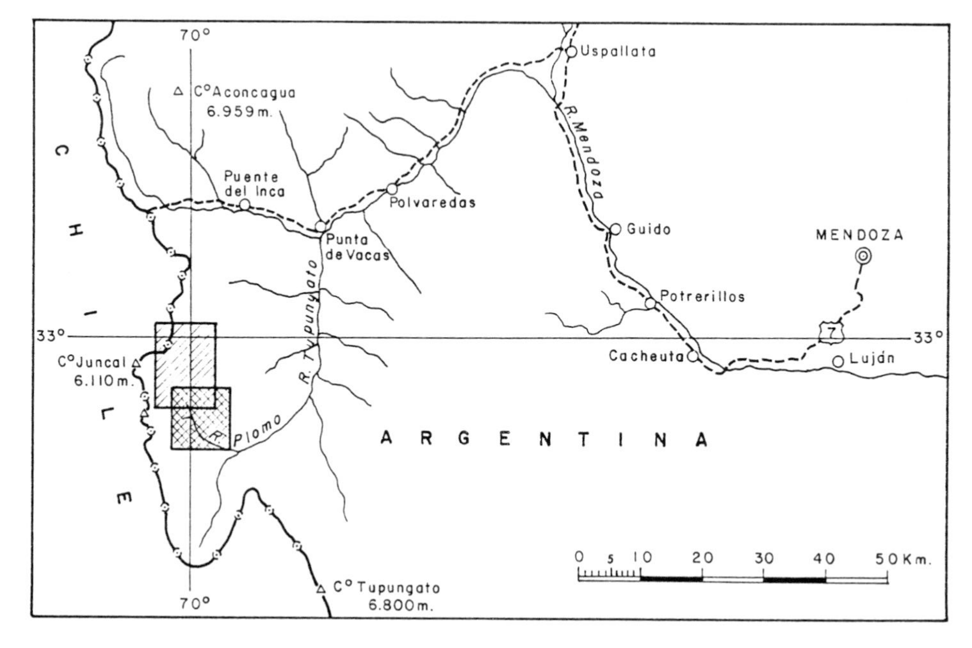

Figure 1. Location map.

2. GLACIER MASS VARIATION 1914-1974.

Front fluctuations of Rio Plomo glaciers show a general retreat from 1909 to 1974 (Espizúa, 1986), but there are two surging glaciers: one of them originated and ice-dammed lake in 1984 (Leiva, 1986) similar to the one in 1934 (King, op.cit.).

In 1985 there was obtained a contour map of the Plomo river headwaters by means of a photogrammetric restitution of 1974 aerial photos. It only covers part of the glacier system but it allows for comparison with the 1914 Helbling's maps in order to calculate the 1914-1974 glacier mass variations.

Firstly, volume changes over 60 years are evaluated following Finsterwalder (1953) and Hofmann (1958). Figure 2 shows the contour lines and ice boundary for both dates. If A_i are the areas created by the contour lines displacement, and dH the contour line interval, which is in this case chosen equal to 50 m, the volume changes calculated by

$$dV = \frac{1}{3} dH \ (A_1 + A_2 + A_1 \cdot A_2) \qquad (1)$$

are shown in Table I.

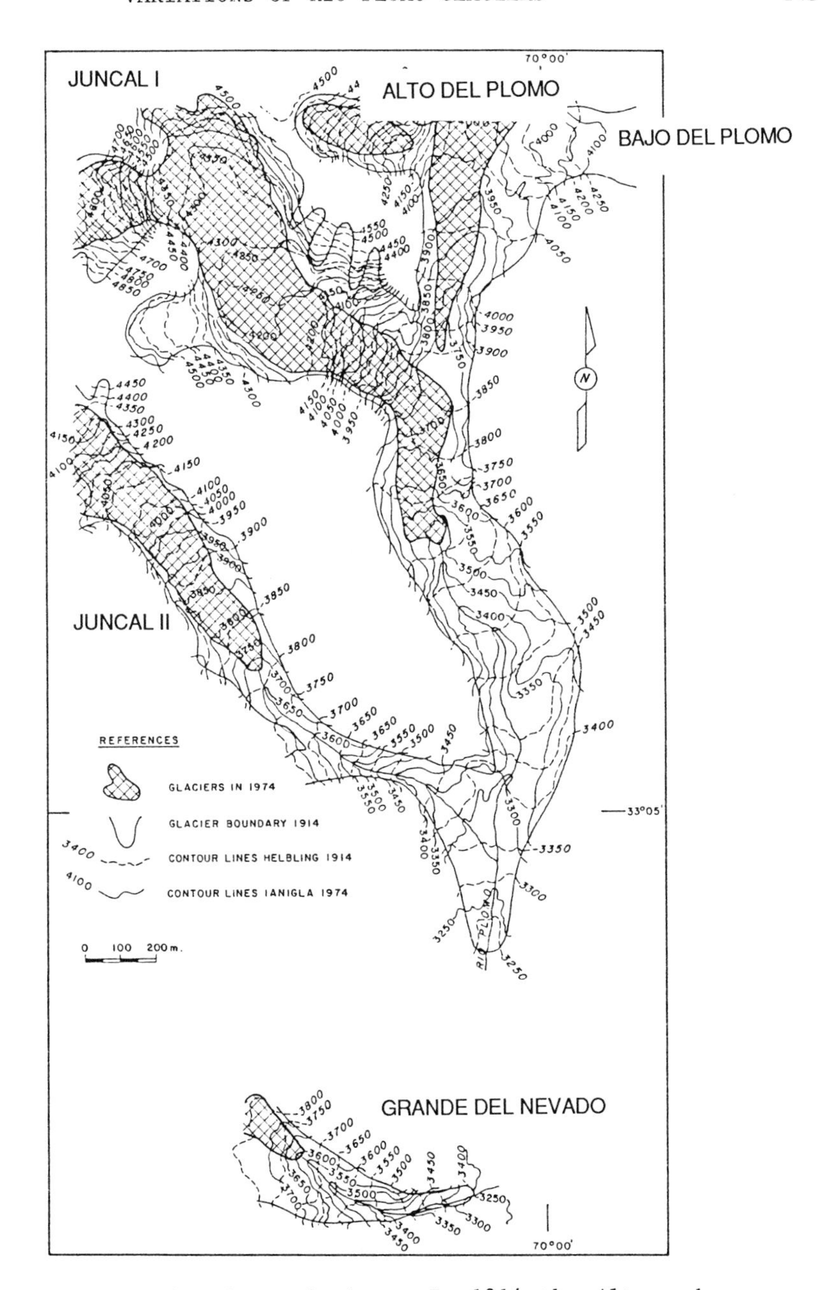

Figure 2. Río Plomo glaciers. In 1914 the Alto and Bajo del Plomo glaciers, together with Juncal I and Juncall II glaciers, formed only one Plomo glacier that reached 3,230 m a.s.l. (the figure shows only the region common to both, the 1914 and 1985 maps).

Table I. Volume of ice mass loss calculated following Finsterwalder (1953) and Hofman (1958).

Glacier	Volume of ice mass loss 1914-1974 ($m^3 \times 10^6$)
Alto and Bajo del Plomo	1,073.32
Juncal II	227.36
Juncal I	186.98
Grande del Nevado del Plomo	61.76
Totals	1,549.43

The lack of coincidence between the totals and additions is due to the fact that computations were carried out with a higher number of digits.

Afterwards, several profiles were drawn to calculate the ice level changes between 1914 and 1974. The vertical cross sections along two profiles define two faces of a prismatoid that contains the ice loss volume. Said volumes are calculated using the general prismatoid equation (1) but now A_1, A_2 are the cross section areas and dH is the planimmetric distance between two consecutive profiles. Results are presented in Table II.

Table II.

Glacier	Total basin surface (km^2)	Common surface in both maps used	Percentage of total area studied	Vol. of ice mass loss 1914 1974 $m^3 \times 10^6$
Alto del Plomo (including Contrafuerte del Juncal)	6.661	2.562	38.47	1,021.0
Bajo del Plomo	4.555	0.136	2.99	46.85
Juncal II	2.437	1.290	52.92	216.30
Juncal I	3.196	0.268	8.41	162.24
Grande del Nevado del Plomo	2.466	0.525	21.29	66.97
Totals	19.316	4.782	24.76	1,513.37

3. CLIMATIC INDICATORS AND GLACIERS

The most frequent synoptic situation is defined by the presence to two anticyclons, one over the Pacific Ocean, around 200°, and another on the Atlantic Ocean, around 0°. According to Goske et al. (1957) their position changes along the year; the high pressure maxima is around 30°S

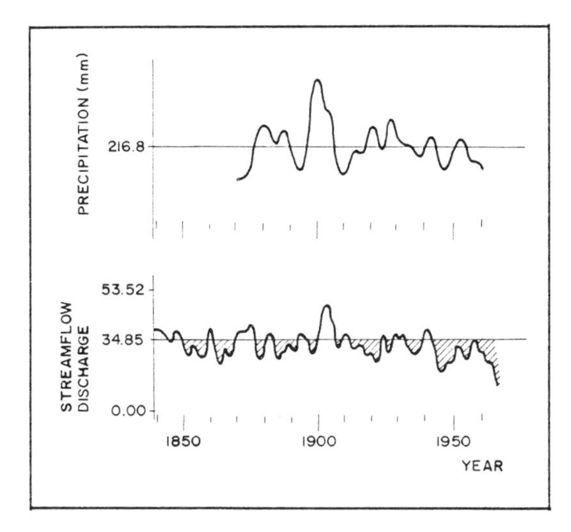

Figure 3. Santiago de Chile winter precipitation series (above) and Atuel river streamflow reconstruction (filtered with 5 years low-pass digital filter).

Figure 4. The ice-dammed lake created by the surge of Grande del Nevado del Plomo glacier. The maximum level attained by the lake can be seen on the right slope.

in January and 20°S in July. Their latitude fluctuates, remaining at lower latitudes from May to September, thus allowing the S-W perturbations to reach the Andes Centrales and climbing higher latitudes from October to April (Lliboutry, 1982).

Hence, the winter precipitations regime of the Central Andes is controlled by weather systems coming from the Pacific Ocean. The Andes height constitues a blockage to airflow, weather systems and humidity (Barry, 1981). The semi-desertic climate of northern San Juan and the semiarid conditions in the province of Mendoza are a consequence of this mechanism.

Mountain rivers of this region are mainly snowfed, and glacier contribution to runoff becomes relevant only when winter precipitations are scarce (Leiva et al., 1986).

The longest meteorological record on the eastern side of the Central Andes comes from Cristo Redentor (32°50'S, 70°05'W; 3832 m a.s.l.), temperature data from 1941 to 1983 are shown in Table III. No trend is found (slopes of 1×10^{-3} or less) by fitting a straight lines, both by rms and L1 techniques to the monthly mean summer temperature data. This is also true when using the Observatorio Mendoza records (32°50'S, 70°05'W; 3832 m a.s.l.) from 1906 to 1980, annual winter and summer monthly mean temperatures.

The winter precipitaiton series of Santiago de Chile and the long term streamflow reconstruction for Rio Atuel (Cobos and Boninsegna, 1983) may be used as climatic indicators. The snow- and course measurements of Polleras (33°09'S, 69°52'W, 1900 m a.s.l.) and Toscas (33°10'S, 69°56'W, 3000 m a.s.l.) were not considered because these records were begun in 1951.

The long term streamflow reconstruction series for Rio Atuel, based on four tree-ring width chronologies (LaMarche et al., 1979) obtained between 32°40' and 34°00'S latitude on the western side of Cordillera de Los Andes, show - when filtered with a five year low-pass digital filter - periods of high or low runoff. There is a marked period with runoff greater than the mean, especially at the beginning of the century. This period is the most important since 1900, except for very weak peaks around 1930 and 1940 (Fig. 3).

The winter precipitation series of Santiago de Chile (since 1865) show a similar behaviour when likewise filtered (Fig. 3). The same peaks appear when going through the Rio Mendoza streamflow record, started in 1909.

The glaciers of Rio Plomo basin have receded since 1909 until 1974. This general recession was interrupted by sudden advances of the Juncal I glacier (also called Grande del Juncal) in 1910 and sometime between 1934 and 1955, and the Grande del Nevado glacier surges in 1933 and 1984 (Fig. 4).

This behaviour gives significance to the general and weak tendency of precipitation data of Santiago and discharge measurements of all the snow-fed rivers of the Argentine Central Andes towards lower values, even if, during the 80s some of the Plomo glaciers have experienced small readvances.

The total ice-mass loss between 1914 and 1974 could be used to estimate the hydrological importance of this climatic tendency.

Table III. Cristo Redentor monthly mean temperature 1941-1983.
Note: The serie has many blanks. References: M = month, Y = year, T = temperature

M	Y	T	M	Y	T	M	Y	T	M	Y	T	M	Y	T	M	Y	T	M	Y	T	M	Y	T
01	41	6.3	10	45	-1.0	07	55	-9.0	11	60	1.1	11	65	0.1	08	70	-6.3	05	75	-3.4	11	80	-0.5
02	41	4.1	11	45	-1.4	08	55	-8.3	12	60	3.8	12	65	0.4	09	70	-3.9	06	75	-4.2	12	80	3.5
03	41	0.8	12	45	1.9	09	55	-4.7	01	61	5.1	01	66	4.4	10	70	-3.6	07	75	-7.2	03	81	3.7
04	41	-2.4	01	46	2.4	10	55	-5.9	02	61	3.7	02	66	2.3	11	70	-0.6	08	75	-5.9	04	81	0.5
05	41	-6.1	02	46	4.8	11	55	-0.5	03	61	1.9	03	66	1.7	12	70	2.0	09	75	-5.9	05	81	-2.0
06	41	-7.5	03	46	2.4	02	56	3.6	04	61	-0.7	04	66	-1.2	01	71	1.9	10	75	-3.4	06	81	-5.6
07	41	-7.9	02	47	3.3	03	56	0.2	05	61	-1.4	05	66	-1.5	02	71	3.1	11	75	-1.8	08	81	-4.3
08	41	-7.6	03	47	2.5	04	56	-2.9	06	61	-8.2	06	66	-8.9	03	71	1.3	12	75	4.7	09	81	-4.1
09	41	-6.8	04	47	-0.3	05	56	-6.0	07	61	-5.8	08	66	-7.9	04	71	-2.4	01	76	5.6	10	81	-2.4
10	41	-0.1	05	47	-1.5	06	56	-7.3	08	61	-4.4	09	66	-4.3	05	71	-3.5	02	76	3.3	11	81	1.3
11	41	-3.5	06	47	-5.1	07	56	-6.4	09	61	-8.0	10	66	-2.8	06	71	-9.1	03	76	2.8	12	81	4.7
12	41	2.0	07	47	-6.9	08	56	-5.7	10	61	-1.8	11	66	0.8	07	71	-5.2	04	76	0.4	01	82	5.6
01	42	4.7	08	47	-5.9	09	56	-4.9	11	61	0.6	12	66	0.1	08	71	-5.5	05	76	-2.7	02	82	4.6
02	42	4.8	09	47	-6.1	10	56	-3.4	12	61	3.0	01	67	2.8	09	71	-4.3	06	76	-6.4	03	82	3.4
03	42	2.4	10	47	-3.9	11	56	0.6	01	62	2.2	02	67	2.7	10	71	-1.2	07	76	-6.2	04	82	2.9
04	42	-0.5	11	47	1.2	12	56	2.0	02	62	3.0	03	67	1.6	11	71	1.6	08	76	-7.1	05	82	-2.0
05	42	-3.3	12	47	2.4	01	57	4.2	03	62	3.4	04	67	2.9	12	71	3.5	09	76	-5.7	12	82	5.4
06	42	-9.3	01	48	4.8	02	57	3.6	04	62	0.9	05	67	-2.0	01	72	5.1	10	76	-4.5	01	83	5.4
07	42	-7.4	02	48	3.9	03	57	4.0	05	62	-2.5	06	67	-9.7	02	72	4.8	11	76	-0.2	02	83	6.0
08	42	-6.8	03	48	2.0	04	57	-1.1	06	62	-6.7	07	67	-8.3	03	72	1.4	12	76	2.8	03	83	4.8
09	42	-5.1	04	48	-1.5	05	57	-3.2	07	62	-6.4	08	67	-6.9	04	72	-0.6	01	77	4.6	04	83	-0.1
10	42	-2.9	05	48	-4.5	06	57	-7.2	08	62	-4.7	09	67	-6.3	05	72	-2.6	02	77	5.2	05	83	-5.3
11	42	-0.1	06	48	-5.4	07	57	-9.0	09	62	-4.9	10	67	-2.6	06	72	-6.0	03	77	5.5	06	83	-8.9
12	42	2.8	07	48	-8.9	08	57	-7.0	10	62	-3.1	11	67	-1.0	08	72	-8.3	04	77	1.5	07	83	-7.9
01	43	5.2	08	48	-5.8	09	57	-6.4	11	62	1.4	12	67	4.1	09	72	-5.7	05	77	-0.8	08	83	-6.4
02	43	4.6	09	48	-5.6	10	57	-4.1	12	62	3.3	01	68	4.3	10	72	-5.7	06	77	-3.1	09	83	-8.8
03	43	0.5	01	49	3.6	11	57	0.0	01	63	3.6	02	68	4.5	11	72	-2.9	07	77	-9.0	10	83	-1.7
04	43	1.2	09	49	-8.1	12	57	3.2	02	63	3.3	03	68	0.7	12	72	2.1	08	77	-5.7	11	83	1.5
05	43	-3.2	10	49	-3.3	01	58	4.5	03	63	2.3	04	68	-1.3	01	73	4.1	09	77	-2.0	12	83	5.3
07	43	-6.1	11	49	-0.7	02	58	4.1	04	63	2.3	05	68	-3.0	02	73	2.9	10	77	-0.6			
08	43	-8.6	12	49	0.8	04	58	0.1	05	63	-2.0	06	68	-6.3	03	73	3.2	11	77	1.2			
09	43	-4.7	01	50	1.6	05	58	-4.8	06	63	-5.2	07	68	-4.4	04	73	0.4	12	77	4.7			
10	43	-2.4	07	50	-7.0	07	58	-4.2	07	63	-6.6	08	68	-4.9	05	73	-3.7	01	78	4.3			
11	43	-0.3	08	50	-7.3	09	58	-5.4	08	63	-7.2	09	68	-4.1	06	73	-6.0	02	78	5.2			
12	43	3.1	10	50	-5.6	01	59	3.7	09	63	-8.1	10	68	-4.4	07	73	-7.9	03	78	3.9			
01	44	4.2	11	50	-4.8	02	59	5.1	10	63	-5.4	11	68	2.0	08	73	-6.5	04	78	2.2			
02	44	3.8	12	50	2.1	03	59	3.2	11	63	-5.4	12	68	1.8	09	73	-5.0	05	78	-1.1			
03	44	1.3	01	51	1.6	04	59	-3.2	12	63	2.3	01	69	5.5	10	73	-3.9	02	79	3.7			
04	44	1.3	02	51	1.9	05	59	-3.3	01	64	3.9	02	69	4.5	11	73	0.3	03	79	3.5			
05	44	-3.8	03	51	2.3	06	59	-6.2	02	64	3.9	03	69	2.4	12	73	0.9	04	79	-0.4			
06	44	-5.2	06	51	-5.9	07	59	-5.7	03	64	2.0	04	69	0.5	01	74	5.8	05	79	-3.2			
07	44	-3.1	11	51	0.9	08	59	-7.3	04	64	-0.5	05	69	-2.1	02	74	3.2	06	79	-4.2			
08	44	-7.7	12	51	3.3	09	59	-2.7	05	64	0.6	06	69	-7.0	03	74	3.0	07	79	-3.6			
09	44	-1.8	05	54	-6.3	10	59	-3.6	06	64	-8.1	07	69	-5.7	04	74	2.4	10	79	-1.6			
10	44	-3.7	06	54	-8.7	11	59	-0.4	07	64	-7.2	08	69	-6.1	05	74	-3.5	11	79	-0.4			
11	44	-0.5	07	54	-8.5	12	59	4.0	08	64	-7.3	09	69	-3.6	06	74	-8.2	12	79	2.6			
12	44	3.9	08	54	-4.5	01	60	5.0	09	64	-3.5	10	69	-4.1	07	74	-6.0	01	80	5.4			
01	45	4.4	09	54	-5.7	02	60	4.3	10	64	-2.3	11	69	-0.8	08	74	-6.0	02	80	3.4			
02	45	0.8	10	54	-4.4	03	60	1.9	11	64	0.2	12	69	4.2	09	74	-6.0	03	80	5.5			
03	45	2.3	11	54	0.7	04	60	1.8	12	64	0.7	01	70	3.1	10	74	-2.8	04	80	-2.2			
04	45	2.3	12	54	2.3	05	60	-1.9	01	65	3.0	02	70	4.8	11	74	-0.6	05	80	-3.7			
05	45	-1.4	01	55	5.3	06	60	-5.4	02	65	3.5	03	70	3.0	12	74	1.3	06	80	-5.3			
06	45	-3.7	03	55	1.3	07	60	-8.9	03	65	3.9	04	70	2.8	01	75	5.8	07	80	-6.8			
07	45	-8.0	04	55	-2.3	08	60	-6.1	04	65	-2.3	05	70	-3.9	02	75	3.9	08	80	-5.5			
08	45	-4.1	05	55	-4.9	09	60	-5.5	05	65	-5.0	06	70	-7.9	03	75	2.3	09	80	-4.4			
09	45	-3.6	06	55	-6.4	10	60	-3.0	06	65	-3.5	07	70	-7.0	04	75	0.0	10	80	-2.6			

After 1974 some of the Rio Plomo glaciers readvanced (Espizúa, 1986; Leiva, 1986, op.cit.) and this behaviour is also reflected in the total discharge of Rio Tupungato (it climbs up to 800 × 10^6 m^3 for 1970-1986). Nevertheless, the evidence is insufficient to consider this situation as an inflection point on the tendency observed before 1974. Future research based on our glaciological observations between 28° and 42°S latitude may yield more information.

4. RESULTS AND COMMENTS

The hydrological importance of the climatic trend observed and pointed out by the ice mass loss since 1974 is reflected by the general glacier recession in the Plomo region between 1910 and 1974. Glacier fluctuations from a more regional point of view, including both sides of the Central Andes, will be discussed in a future paper.

The coincidence between the results of both methods of calculation gives an ice mass loss of about 1,500.00 × 10^6 m^3, to the mean total annual discharge of Rio Tupungato (1954 - 1974) which is 608 × 10^6 m^3 of water.

ACKNOWLEDGEMENTS

This research was supported by the Argentine National Research Council (CONICET).
Gratitude is expressed to the Mendoza Government that partially supported the field missions during 1985/1986. To Víctor Videla and Roberto Bruce for their valuable field work, to H. Miranda for the drawings and the C.E. Moyano and M. Silva for their help with the data analysis.

REFERENCES

Barry, R.G. 1981. Mountain weather and climate. Methuen, London and New York, 312 pp.

Cobos, D. and Boninsegna, J. 1983. Fluctuation of some glacier in the upper Atuel river basin. Mendoza, Argentina. Quaternary of South America and Antarctic Peninsula. AA. Balkema, Rotterdam 1, 61-82.

Corte, A. and Espizúa, L. 1981. Inventario de glaciares de la cuenca del Rio Mendoza. IANIGLA-CONICET, Mendoza, Argentina, 62 pp.

Espizúa, L.E. 1986. Fluctuations of Rio del Plomo glaciers. Geografiska Annaler, 68A (4), 317-327.

Finsterwalder, R. 1953. De zahlenmassige Erfassung des Gletscherrückgang und Ostalpengletschern. Zeit. f. Gletscherkunde und Glazialgeologie, II, hft. 2, 189-239.

Godske, C.L., Bergeron, T., Bjerknes, J. and Bundgaard, R.C. 1957. Dynamic meteorology and weather forecasting. Published by the American Meteorological Society and Carnegie Institute of Washington, 800 pp.

Helbling, R. 1919. Beitrage zur topografischen Erschliessung der Cordillera de los Andes zwischen Aconcagua und Tupungato. Sonderabdruck aus dem XXII Jahresbericht des Akademischen Alpenclubs, Zürich, 77 pp.

Helbling,R. 1935. The origin of the Rio Plomo ice-dam. The Geographical Journal 8 (1), 41-49.

Hofmann, W. 1958. Der Vortoss des Nisqually. Gletschers am Mt. Rainer, USA, von 1952 bis 1956. Zeit. f. Gletscherkunde und Glazialgeologie, IV, hft. 1-2, 47-60.

King, W.D.V.O. 1935. El aluvión del Rio Mendoza de enero de 1934. "La ingeniería", Buenos Aires, Argentina, 309-313, 389-399; and Observaciones adicionales sobre la obstruccíon en el valle Plomo recogidas en febrero de 1935. "La Ingeniería", Buenos Aires, Argentina, 514-518.

LaMarche, V.C. et al. 1979. Tree-ring chronologies of the Southern Hemisphere: Chile. Chronology series V, Laboratory of Tree-Ring Research, University of Arizona, Tucson, 69 pp.

Leiva, J.C. 1986. El surge del glaciar Grande del Nevado del Plomo. Technical report submitted to the Mendoza Government, 31 pp.

Leiva, J.C., Cabrera, G. and Lenzano, L.E. 1986. Glacier mass balances in the Cajón del Rubio, Andes Centrales, Argentina. Cold Regions Science and Technology 13, 83-90.

Lliboutry, L. 1982. Chapter 11B. South America, southern part. In: Satellite Atlas of Glaciers, R.S. Williams and J.G. Ferrigno (Eds.). US Geological Survey Professional Paper.

Razza, L. 1935. El glaciar del Nevado del Plomo. Revista Geográfica Americana IV (25), 22-238.

EFFECTS OF TOPOGRAPHIC AND CLIMATIC CONTROLS ON 19TH AND 20TH CENTURY GLACIER CHANGES IN THE LYNGEN AND BERGSFJORD AREAS, NORTH NORWAY.

W.B. Whalley, J.E. Gordon[1] and A.F. Gellatly[2]
Dept. of Geography, Queen's University
Belfast BT7 1NN, U.K.

ABSTRACT

Evidence of glacier extent and retreat in two areas of mountain plateaus in north Norway (70°N) are described. Glaciers on the Bergsfjord Peninsula are on rock plateaus about 900 m a.s.l. with a maximum summit altitude of 1200 m and ice thickness believed to be at least 100 m over substantial areas. The outlet glaciers flow down to about 400 m a.s.l., are receding at present and have been since the turn of the century.

The southern Lyngen Alps lie 100 km to the southwest of Bergsfjord. Three plateau areas with glaciers are ca. 1700 m, 1600 m and 1450 m a.s.l. respectively. Depths of ice vary across each plateau with greatest thickness reaching 70-150 m. Glaciers on the highest summits are roughly in equilibrium but below 1500 m are rapidly receding. Although less extensive in area than the Bergsfjord group, the two higher plateau ice caps in Lyngen nourish valley glaciers which descend to similar altitudes (i.e. ca. 400 m).

In both regions, some high plateau summits do not support permanent ice, reflecting the control of a threshold topographic area necessary for glacier growth. There is a distinct relationship between altitude and thermal regime of the glaciers; of the main glaciers it appears that the two highest ice caps are certainly cold-based.

Climatic data available for three stations in the study which suggest that there have been higher temperatures since about 1910 and increased precipitation from about 1930 than during the late 19th and early 20th centuries.

Photographic and sketched records show positions of both plateau ice caps and valley glaciers which can be related to present day ice limits and the positions of Neoglacial moraines. Where surface lowering of Øksfjordjøkelen is shown at the northern edge of the plateau, bedrock topography may be influential in shifting the location of the ice-divides. Hence the complexity of valley terminal moraine sequences is a response to both topographic and climatic controls at higher altitudes.

[1] Nature Conservancy Council, Peterborough PE1 1VA, U.K.
[2] Dept. of Geography, Plymouth Polytechnic, Plymouth PL4 8AA, U.K.

J. Oerlemans (ed.), Glacier Fluctuations and Climatic Change, 153–172.

1. INTRODUCTION

There is a close linkage between the snout fluctuations of a glacier, its mass balance and the general climatological environment. Thus, Fig. 1, a diagram from Meier (1965), is frequently used to illustrate the way in which ice marginal positions relate to climatic effects. However, more complex feedbacks are present (e.g. Andrews, 1975) and the behaviour of one glacier in an area may not be mirrored by its neighbours. Reconstructions can be made of conditions on the right from data on the left and vice-versa.

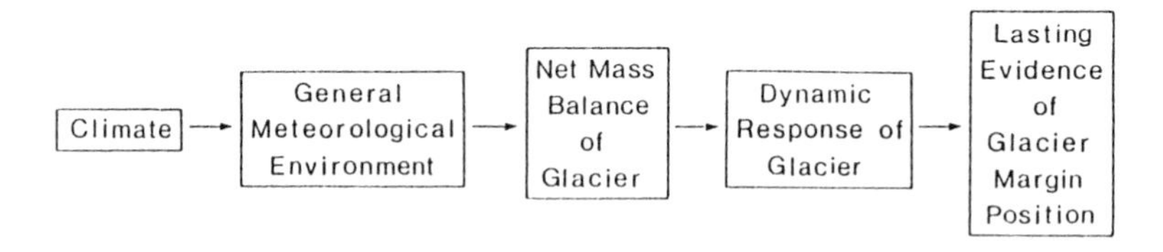

Figure 1. Linear relationship of climate-glacier interactions (after Meier, 1965).

Following Paterson (1981, p. 260; discussing the work of Nye and Lliboutry in particular) the reponse of a glacier can be related to perturbations in the mass balance. Conversely, given variations in snout position, it is possible to find variations in the mass balance that produced them. However, to explain the current behaviour of a glacier we need to know the mass balance over, maybe, several hundred years but to calculate the current mass balance only a few years record of terminus position data are needed.

It is well known that glaciers respond to climate in different ways because of disparities of size. Steepness, and thus velocity, as well as thermal regime will also have an effect. Apart from thermal effects, the application of Nye's models of glacier variation (Nye, 1965) can be used to explain how glaciers respond to climate or, more strictly, to mass balance changes.

Modelling glacier behaviour and the relationships with climate thus necessitates complex consideration, not suggested by the apparent simplicity of Fig. 1.

For instance, we may also need to consider the effects of topography as factors related to, but distinct from, altitude. The ideas of Manley (1955) suggest certain ways in which complications on the general scheme (Fig. 1) can be produced. He stressed the need to consider the topographic area available for glacier or ice cap growth in relation to the regional snowline. Whilst a mountain may rise well above the snowline, glacier development may be impeded by the amplitude of relief preventing ice accumulation. Even for recent fluctuations such complicating factors need to be accounted for, e.g. in plotting glacierization limits (altitudes) (Østrem, 1966) and snowline estimates (Finsterwalder, 1962; Osmaston, 1975). The same applies for palaeoclimatic studies such as reconstructing former ice sheet extents (e.g. Thorp, 1986), estimating timings of climatic events (e.g. Grove, 1985), elucidating patterns of deglaciation and till deposition (Drozdowski, 1979) or the problem of mountain refugia (Markgren, 1964).

We consider that north Troms - south Finnmark, north Norway (Fig. 2) provides a suitable test area for investigating some of these complexities of glacier behaviour in reponse to topographic controls as well as those of climate. Nevertheless, this does present considerable difficulties with regard to:

1. mapping present day glacier extents and glacierization
2. determination of the recent glacial history of an area
3. modelling the extent of ice limits from the relict terminal positions

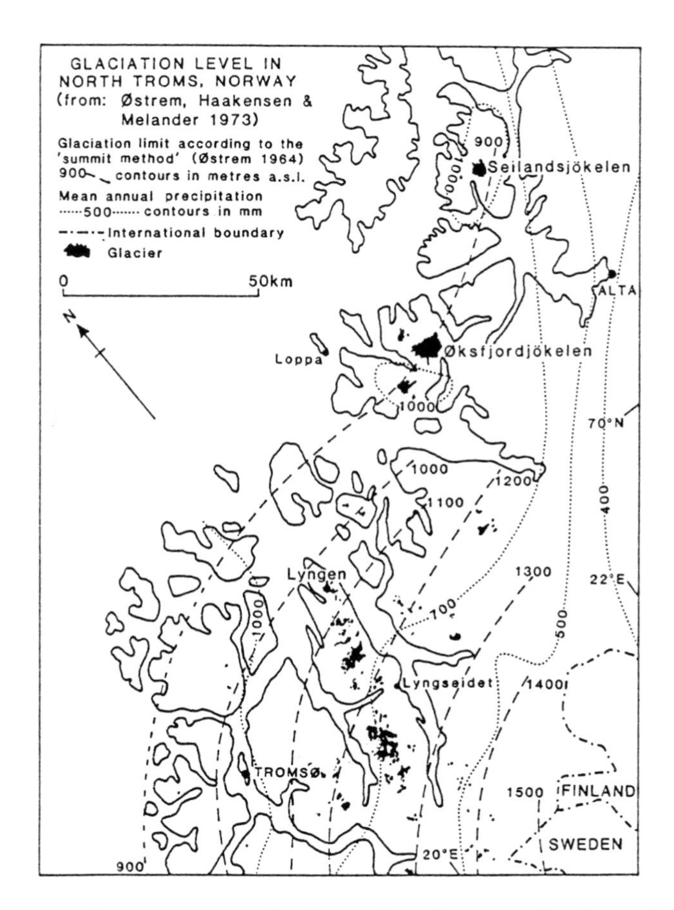

Figure 2. Location of the main plateau glaciers in north Norway with isoglacihypses and precipitation data (Østrem et al., 1973) and Meteorological Stations. The Bergsfjord peninsula, containing Øksfjordjøkelen, is to the west of Alta.

2. METHODOLOGIES OF RECONSTRUCTING FORMER ICE EXTENTS

In his review of the ways in which glacier variations can be used to indicate changes in the climatic environment, Porter (1981) included moraines and equilibrium line altitudes (ELAs). Such surrogate measures which have correlatives in Meier's early diagram (cf. Fig. 1) have been adopted by other workers in similar reconstruction-based studies: (Andrews, 1975; Andrews and Miller, 1972; Porter, 1977; Humlum 1985, 1986). However, their use does depend upon the recognition of a regular partitioning of the accumulation and ablation areas of a glacier.

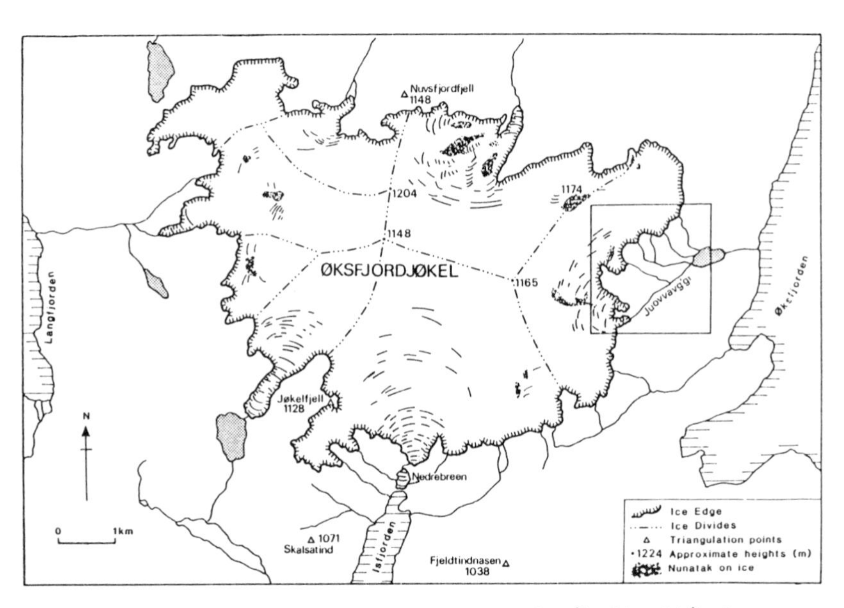

Figure 3. The main features of Øksfjordjøkelen showing nunataks and ice divides. See Fig. 4 for detail.

Similarly, equilibrium profiles of glaciers and ice caps can also provide a means of interpreting the past, gross history of an ice mass. However, when we consider the means of nourishment and ablation of many glacier systems with juxtaposed valley glaciers and plateau ice caps, we might suspect, and indeed find, that interpretation by simple models such as these is far from easy. Arising from the present study are a number of pertinent issues such as the interpretation of depositional sequences where the ablation and accumulation zones are physically separate and where the 'ELA' is postioned between two glacierized zones, but is, itself ice-free. An example here is Nedrebreen on Øksfjordjøkelen (Fig. 3) and the plateau-valley glacier systems of Lyngen (Gordon et al., 1987).

It is necessary to distinguish, albeit roughly, between 'local' and 'general' climate. Following Paterson (1981), local climate signifies the climate on the glacier and the immediate surroundings. General climate refers to the average conditions in a large area, indicated e.g. by the data from a network of standard weather stations. Procedures for reconstructing glacier events may fail to recognize this distinction and thereby blend methodological approaches as follows:

1. Glacier positions are mapped from year to year.
2. Glacier snout positions are reconstructed from ground evidence by dating moraines or associated glacier-marginal events.
3. General environments are reconstructed by reference to local responses to climate e.g. pollen and faunal evidence from an area.
4. General climatic environments are reconstructed by reference to documentary data, either directly e.g. meteorological data, or indirectly, e.g. intense geomorphological activity (Grove, 1972)

It should be noted that 1 and 2 are both local, site specific and distinctly located in time (even if averaged over more than one year). However, 3 and 4 are more general in both place and time. Furthermore, the response times (lags) and rates between the reaction of glaciers and vegetation to a climatic perturbation are different (e.g. Bryson and Wendland, 1967; Moran and Blasing, 1972). This provides further problems for estimates of long-term glacier mass balance.

The control on ice accumulation in mountain areas is a function not only of temperature and precipitation but also of topography. Hence Manley (1955), and later Humlum (1986) amongst others have, shown the relevance of precipitation shadows, altitudinal and spatial controls to glacier accumulation and survival. Clearly, glacier growth and glacier waning periods will be affected and, in turn, will affect the local climate differently.

Remarkably little attention has been paid to these considerations and where it has, attention has usually been drawn to valley glaciers. In this paper we examine two main areas of ice accumulation as discrete, high ice caps (900 to 1600 m a.s.l.) and their associated valley glaciers (descending to 400 m or below). Their dynamic and temperate characteristics over the last 150 years can be estimated from a variety of sources.

3. FIELD AREAS

Two glacierized areas, about 100 km apart, have been examined in North Norway (Fig. 2): the Bergsfjord Peninsula (70°10'N) which contains the large Øksfjordjøkelen, and smaller plateau glaciers in the Lyngen Peninsula (69°30'N). Within this area some portions have been examined in detail (Fig. 3).

3.1 Bergsfjord Peninsula

The Bergsfjord Peninsula supports a group of small plateau ice caps, with the three most extensive being Øksfjordjøkelen (40 km^2), Langfjordjökul (9 km^2) and Svartfelljökel (6 km^2) with maximum altitudes respectively 1204 m, 1062 m and 1162 m a.s.l. (AMS 1735, Øksfjordjøkelen 1:50,000, 1976). All three are associated with a glaciation level of c. 900 m a.s.l. (Østrem et al., 1973). Comparison of ice with revealed bedrock on the plateaus suggests that the thickness of the ice varies greatly in each case and reflects the general slope of the underlying plateau surfaces. We have measured the thickness on profiles on eastern Øksfjordjøkelen being at least 100 m thick over large areas. The highest surface elevation of Øksfjordjøkelen is 1200 m a.s.l. (Fig. 2). Mean annual precipitation is c. 1000 mm (Østrem et al., 1973). Øksfjordjøkelen, at least in part, is inferred to be warm-based from the presence of regelation ice and ice-moulded bedrock on nunataks and the plateau edge. Localized deposits of Late-glacial/Holocene and Neoglacial moraines occur around the plateau rim (Fig. 4). Outlet glaciers flow to altitudes of about 400 m a.s.l. and are retreating at present. One distributary (Nedrebreen on Øksfjordjøkelen) reaches sea level as a regenerated glacier (Hoel and Werenskiold, 1962).

In general, the break of slope around the respective bedrock plateau-rims is between 900-1000 m. Small sections of blockfield are exposed beyond the ice margin providing a 'promenade' in front of the retreating ice edge. Small areas of ice-free blockfield also exist on ice-free summits in the neighbourhood ranging between 800 m and 1000 m. For example, Fjeltindnasen (1041 m) has a surface area about 0.25 km^2 above 1000 m a.s.l. but with no present-day glacier.

3.2 South Lyngen, Troms

The mountains of the southern Lyngen Peninsula, 100 km to the southwest, support a range of small plateau ice caps, corries and valley glaciers (Gellatly et al. 1986a, 1986b; Gordon et al., 1987) associated with a glaciation level rising from 1200-1300 m a.s.l. south eastwards across the peninsula (Østrem et al., 1973 and Fig. 2). Mean annual precipitation is some 300 mm lower than at Øksfjordjøkelen. Two major plateau areas support summit ice caps (Fig. 5), Jiekk'varri (1833 m) and Balgesvarri (1625). Depths of ice vary across each with greatest thickness reaching c. 150 m. Unlike Øksfjordjøkelen, the glaciers on Jiekk'varri and Balgesvarri are cold-based and no moraines have been found on the fringing ice-free plateau surfaces (Whalley et al., 1981). A third plateau, Bredalsfjell (1538 m) supports a relatively thin ice field only a few metres thick while lower plateau surfaces are largely free of permanent ice (Gellatly et al., 1986a, 1986b). This glacier is not considered to be cold-based. No known deposits exist in relation to the higher and cold-based ice caps of the Lyngen Alps. Although less extensive than Øksfjordjøkelen, Jiekk'varri and Balgesvarri support outlet glaciers which also descend to around 400 m a.s.l.

4. GLACIER FLUCTUATIONS

4.1 Moraine deposition associated with the ice caps

Small fragmentary moraines have survived on the eastern plateaus in association with the ice caps of the Bergsfjord Peninsula, particularly Øksfjordjøkelen. The presence of intermittent deposits around the ice-cap perimeter affords speculation about glacier extent and climatic response during the last few millenia. As yet no means of establishing the ages of the older deposits exists, but some felicitous historical photographs have permitted the reconstruction of ice limits during the last 120 years (Gellatly et al., in preparation). Figure 4 details the location of plateau till deposits on Øksfjordjøkelen and show historical fluctuations of the main outlet valley glaciers.

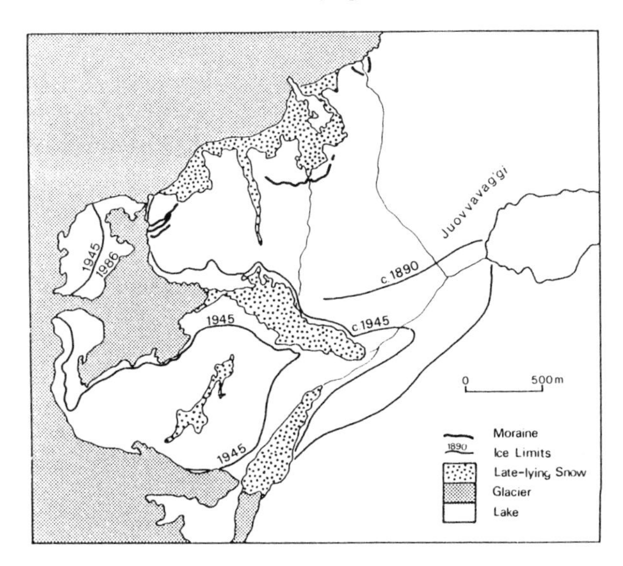

Figure 4. The eastern outlet glacier of Øksfjordjøkelen above Juovvavag'gi (Storelvdalen) showing moraines, present ice edge and the ice edge mapped from 1945 and ca. 1890 AD photographs (see also Fig. 7).

At Øksfjordjøkelen a complex configuration of plateau tills persist in two locations; above Nedrebreen - the main southern outlet of the glacier, and on the northern rim of Juovvavag'gi (formerly Storelvdalen), the chief catchment to the east of the plateau (Fig. 4). In both locations, three moraine groups of some antiquity could be distinguished and above Nedrebreen a fresh, unweathered moraine demarcated the recent maximum ice position which may also be confirmed by historical documents. Meltwater has probably been responsible for the destruction or resorting of many small deposits around the rest of the plateau rim although an almost intact moraine loop survives above Juovvavag'gi. Three moraine fragments have also survived on an adjacent rock buttress also overlooking Juovvavag'gi and their weathered appearance indicates that they are of considerable antiquity and possibly mid-early Holocene.

The ice cap deposits remain, as yet, uncorrelated with the sequences reported from the outlet valleys below the plateau, although here too, there are three well demarcated terminal moraines excellently preserved in most valleys and possibly dating from late-glacial advances (Skarpnes c. 12000-12500; Gaissa substage - often difficult to define; and the Tromsø-Lyngen substage 9000-8200) as documented by Sollid et al. (1973). Evidence of recent (19th and 20th century) moraine formation above the ice-fall of Nedrebreen indicates an hiatus in the valley deposition record, but it remains to see if the absence of moraines is due to erosion censoring or ice cap dynamics/behaviour.

4.2 Documentary evidence

Dating of moraines from both Lyngen and Bergsfjord presents difficulties back beyond a few tens of years. No precise dates have yet been determined and there are no calibrated lichen growth curves which can be used. However, we do know that glacier recession has been rapid at least since the turn of the century (e.g. Gellatly et al., 1986a). This recession applies, as would be expected, to the snouts in the valleys. In the present context it is important that some estimate of behaviour be obtained from the edges of the plateau ice caps. Elsewhere (Gellatly et al., 1986b) it has been suggested that there had been very little downwasting/recession of the plateau ice on the highest Lyngen summits (> 1500 m a.s.l.).

Evidence from Bergsfjord Peninsula includes photographs of Nedrebreen (south Øksfjordjøkelen) taken in 1898 that show it somewhat larger than today. As this is a regenerated glacier, fed directly from the plateau some 400 m above, it could well be maintained as a substantial ice body because of direct feeding. Important evidence from the north of the Øksfjordjøkelen comes from a composite sketch made by J.D. Forbes in 1851. The detail on the sketch is sufficiently accurate to show the ice cap skyline in comparison with that of the present day. For the most part, there is remarkable little difference in estimated elevation and profile. This similarity suggests that little change has taken place over the last 35 years. On the northeast edge of the plateau the ice cap apparently covered the area which is today exposed a nunatak (triangulation point 1174 m grid ref. 430863; map: ASTM 1735 II). Although this survey station was emplaced by the first survey of the area in 1897 insufficient accuracy prevents making precise comparisons of surface profile changes with the more recent aerial surveys of 1976. However, comparison of survey station photographs taken in 1961, by an expedition from Imperial College London, with the present day show no noticeable ice depletion in the last 25 years - a time during which valley snouts have been receding rapidly.

It is possible that, because of the somewhat complex surface topography of Øksfjordjøkelen, what slow thinning has taken place has been accompanied by a shifing of the ice divides (Fig. 3). Until repeat surveys are finished it will not be possible to assess this effect in detail. Nevertheless, it is yet one more complicating factor in the way the valley glaciers are supplied with ice.

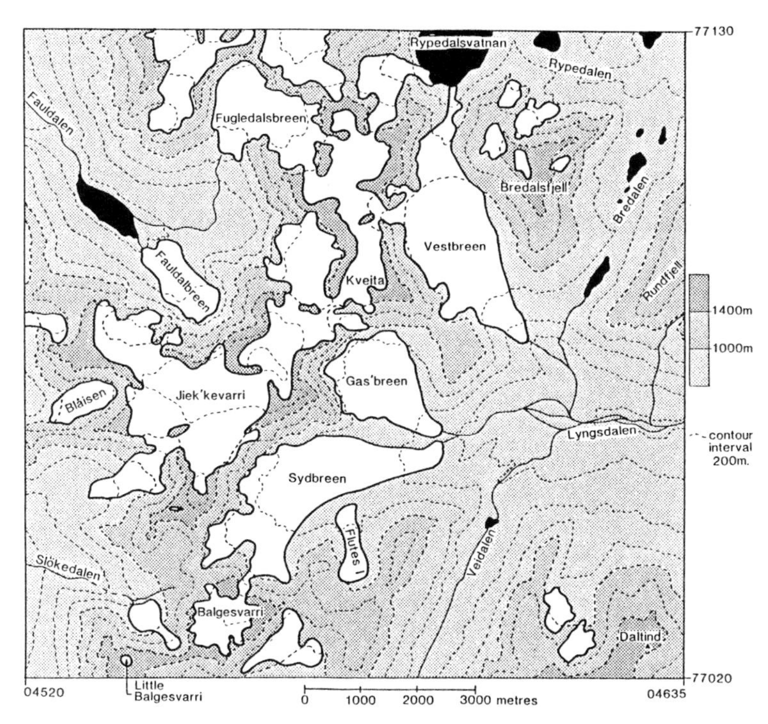

Figure 5. Map showing the main glaciers of the Jiekk'varri-Balgesvarri area, south Lyngen.

5. CLIMATIC TRENDS

In any investigation of glacier-climate interactions from a remote area, a substantial problem is the collection of adequate climate data which are representative of the area.

Relatively long temperature records are available from Tromsø (69°39'N; 18°58'E) beginning in 1868 and Alta (69°58'N; 23°15E) beginning in 1870. The longest precipitation records are from Alta, beginning in 1872 and Lyngseidet (69°35'N; 20°14'E) from 1896. The temperature and precipitation data series for these stations have been smoothed with a low-pass filter with wavelength determined from the ordinates of a Gaussian probability curve (Mitchell et al., 1966). In each case, the climatic variations with wavelengths greater than ten years are emphasized.

Selected data are presented in Fig. 6: mean summer and annual temperatures from Tromsø and Alta; and winter and annual precipitation for Lyngseidet and Alta. This selection is useful as it reflects the variables thought most important as estimators for summer and winter glacier balances (Liestøl, 1967). This approach is also taken by Anda et al. (1985) in their investigations of the behaviour of Holocene glacier variations on Jan Mayen.

Given that detailed snout recession data are only available for one or two glaciers in the whole of the area considered (Fig. 2), only general trends will be considered in both recession and climatic data. However, we may adduce certain broad trends in the climate.

5.1 Temperature

Seasonal and annual fluctuations follow similar patterns at both stations, although they are greater in magnitude at Alta. Three main trends are represented in mean summer temperatures. From the start of the records to around 1910, there was an overall cooling and remained relatively lower than during the rest of the observation period. From 1910 to the mid 1930's, there was a marked rise in temperature. From the late 1930s to the present, summer temperatures fell overall, although remaining above the average of the last few decades of the 19th century. Similar trends are apparent in mean autumn and winter temperatures, although the warming trend early this century began about five years later than for mean summer temperature. Mean spring temperatures show less strong overall trends than for other seasons.

Despite marked variablity, both annual and summer mean values show rises about 1910 AD. Some slight fall since late 1930s is shown but present day values show much warmer summers than in the previous century. Thus for Tromsø, 1970-80 had four times as many degree-days above 10°C for June to August as 1870-80. It is therefore no suprise to see the vast recession of valley glacier snouts since the turn of the century (Gellatly et al., 1986a; Whalley, 1973 and Fig. 7). Historic photographs do not show marked recession on the plateau ice caps. Taking comparative time periods, mean summer temperatures at Alta are slightly higher (+12.2°C) than at Tromsø (+10.7°C) but mean annual temperatures are slightly higher at Tromsø (+2.9°C) than at Alta (+1.7°C). For Lyngseidet and Loppa, which are slightly closer to Balgesvarri and Øksfjordjøkelen respectively, the figures are: Lyngseidet, +3.3°C (mean annual) and +11.5°C (mean summer); Loppa, +4.0°C (mean annual) and +10.9°C (mean summer).

Some estimates can be made of the temperatures on the summit plateaus, although confirming measurements in situ have yet to be obtained. On Balgesvarri, mean annual temperature is given as -4.8°C, using Lyngseidet (1931-1961) data and a lapse rate of 0.5°C/100 m (Laaksonen, 1976). Using similar Alta data and a similar lapse rate, the estimated mean annual temperature on Øksfjordjøkelen summit (1204 m) is -4.3°C. Alta may be rather more continental and use of data from the station at Loppa, on the west coast of the peninsula, gives an equivalent value of -2.0°C.

5.2 Precipitation

For the two stations, Lyngseidet and Alta, there is slightly less correspondence in the detail of the fluctuations than for temperature. Nevertheless, the overall trends for the two station are broadly similar (Fig. 6). The main feature of the data is a marked shift in mean spring, autumn, winter and annual totals around 1930, with relatively higher values in the decades after that date than in those preceding it.

Although the precipitation data from Lyngseidet do not go as far back as at Tromsø, inspection of Fig. 6 suggests that winter precipitation was some 50 mm greater per year in 1970-80 than in the first decade of the century. A higher precipitation is indicated at Lyngseidet than at Alta, although isohyets shown in Østrem et al. (1973) indicate higher precipitation on Øksfjordjøkelen (just less than 1000 mm) than on the Lyngen Peninsula (700 mm). The difference probably reflects the location of Alta in a slight rainshadow.

Snowpits have been dug on both Balgesvarri and near the highest point of Øksfjordjøkelen; the water equivalent depths are shown in Table I (more detail is given in Gordon et al. 1987). These suggest a reasonable steady input on to Balgesvarri but it is unclear what an 'average' input to Øksfjordjøkelen might be. These correspond to snow thicknesses (ca. 1 ma^{-1}) on Balgesvarri which may allow sufficient penetration of a winter cold wave to give a cold-based glacier; this is by analogy with Hardangejøkul (B. Wold, personal communication).

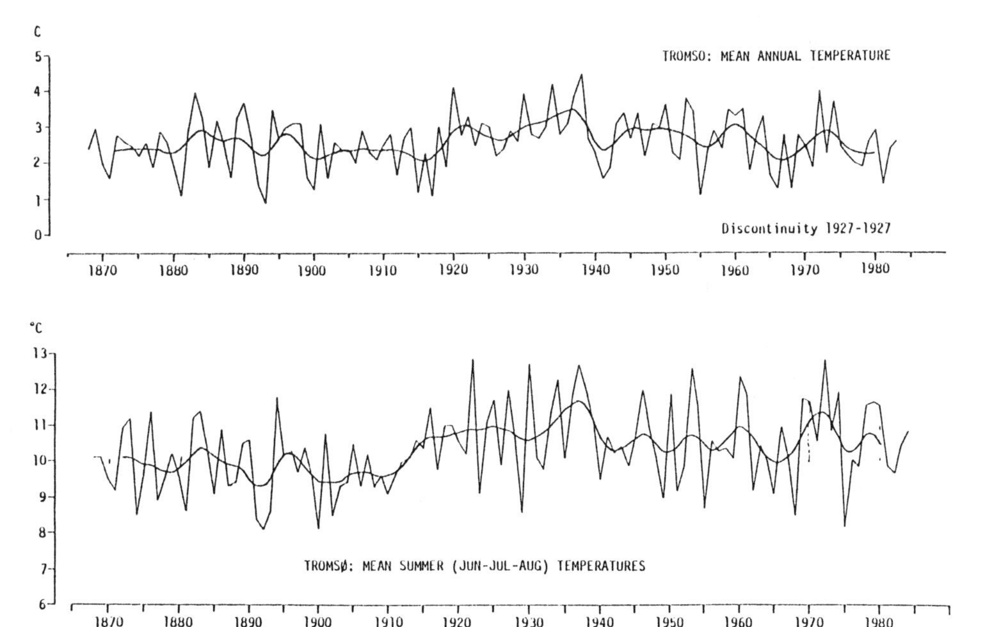

(Fig. 6 continued)

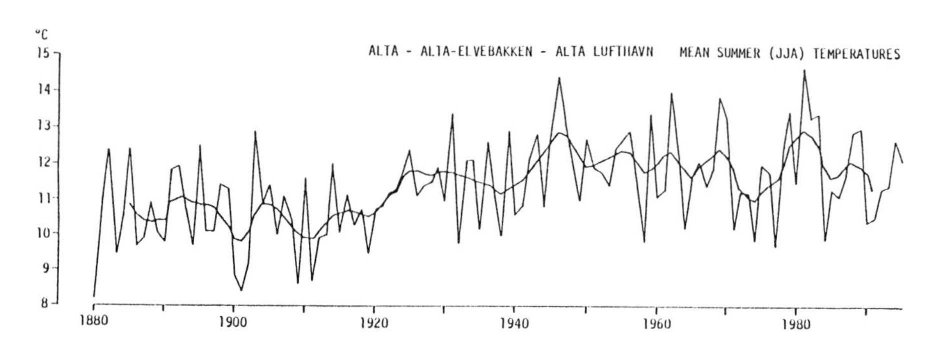

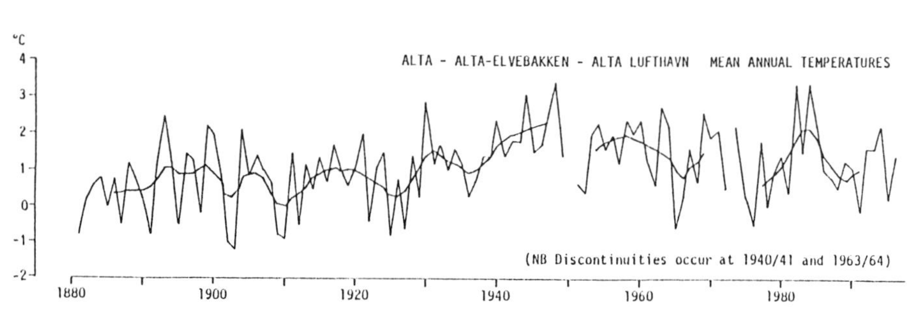

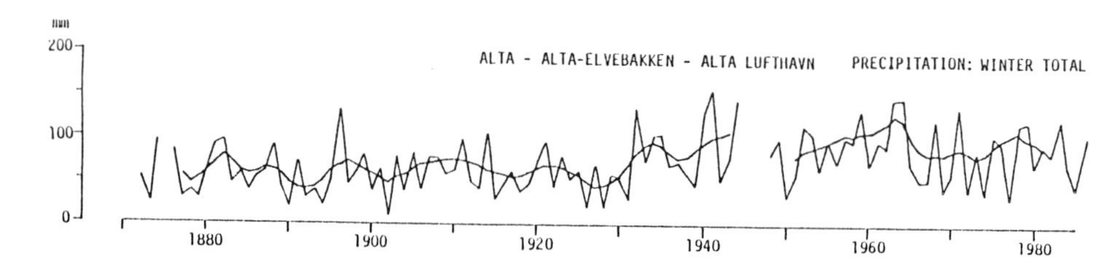

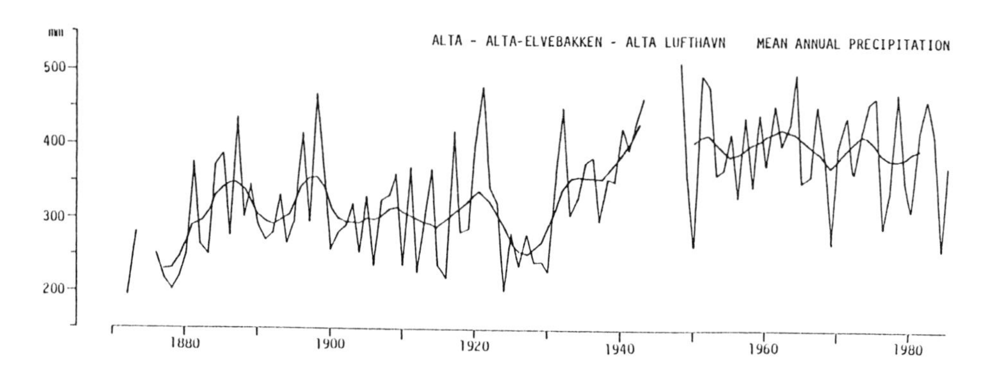

(Fig. 6 continued.)

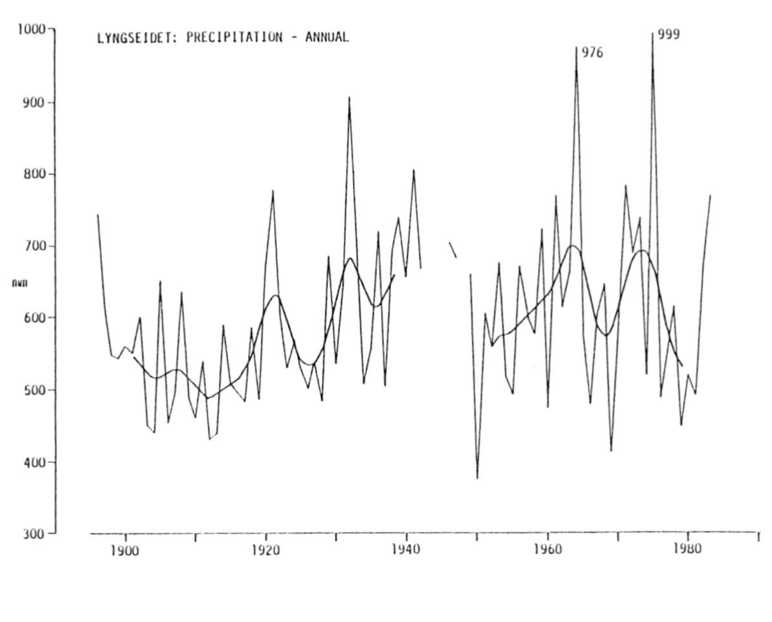

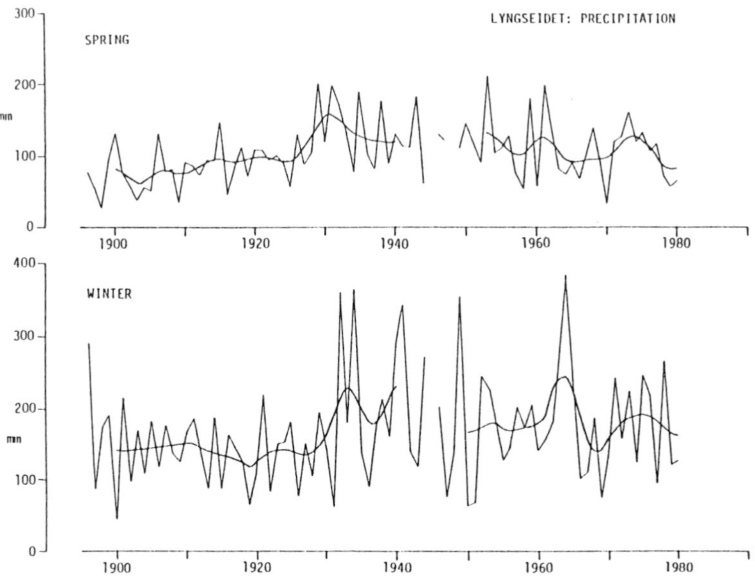

Figure 6. Meteorological data for Alta, Lyngseidet and Tromsø; for discussion, see the text.

6. GLACIER THERMAL REGIME

Amongst the variables to be considered in predicting mass balance changes from glacier variations is that of the glacier thermal regime. Indeed, it is of great importance to know this when concerned with glacier modelling (e.g. Gordon, 1979). Data for distinguishing between thermal regimes are not easy to obtain in the light of a lack of detailed, on-site temperature measurements. The recent use of remotely sensed evidence of naledi and their interpretation (Humlum and Svensson, 1982) is not applicable in this case; as far as we can judge there is no evidence of pronounced naledi in the study area with the exception of some thin aufeis near the Bredalsfjell glacier. Two lines of evidence, in addition to the geomorphological, suggest that the Balgesvarri is cold based. This is presented in more detail in Gordon et al. (in press) but, in summary, is shown by simple temperature modelling using the lapse rate data given above and from stable isotopes ($\delta D - \delta^{18}O$). The latter shows that ice on the edge of Balgesvarri has been on a flowline with no refreezing (Gordon et al., in press); conversely, ice at the edge of Øksfjordjøkelen has been sujected to refreezing conditions (Gordon and Darling, unpubl. data).

With this information, it is possible that our investigations in North Norway can shed some light onto the problem of identification of thermal regimes, albeit by way of a possible model which requires further field testing. This could be carried out both for our field area and for other plateau ice caps in Scandinavia (Østrem et al., 1973).

Geomorphic evidence (frozen and undeformed sorted stone circles), as well as that mentioned above, suggests that the present glaciers on Jiekk'varri and Balgesvarri are cold based (Whalley et al., 1981). However, the remnant body of ice on Bredalsfjell is no more than 2 m thick, and is receding rapidly. On Bredalsfjell, permafrost probably exists below the ice mass itself but we have not yet been able to examine this. We believe that the glaciers on high plateaus in Lyngen have formed on top of permafrost-blockfield. The very highest still maintain cold bases but with gradual warming the thinnest (e.g. Balgesvarri) have just about lost this condition, although some permafrost still remains in the blockfield. On current rates of shrinkage, we expect the ice on Bredalsfjell to have vanished with a few years.

On Øksfjordjøkelen, we found no evidence of either cold based glaciers or of permafrost, despite there being some stone sorting. Taken on its own, this evidence is not greatly informative but combined with observations on the presence or absence of moraines in the study area does shed more light on the problem. As demonstrated in a previous section, moraines are associated with the plateau glacier of Øksfjordjøkelen (and also with Seilandsjøkel, 900 m a.s.l. to the north) but not with any of the higher plateaus in south Lyngen. In the latter case, the cold based glaciers seem to be overriding frozen sorted circles rather than deforming them (Whalley et al., 1981). We suggest therefore, that sliding, warm-based glaciers could have removed any weathered blockfield material, formed before an ice advance, by pushing. If this line of reasoning holds then we can suggest the following distinction:

1. High plateau glaciers (Jiekk'varri, Balgesvarri) have been cold based during the Holocene, any Neoglacial advance having ridden over blockfields.
2. Certain, slightly lower plateau glaciers (Bredalsfjell) have been cold based but with Neoglacial/Holocene warming affecting them more (than 1.) such that they no longer have sub-pressure melting point ice bodies.
3. Low plateau glaciers (Øksfjordjøkelen, Seilandsjøkel) which during the Holocene have never been cold based.

The implication is that the threshold between basal freezing and melting may be a fine one and is modulated by altitude. The detection of permafrost, although its decay gives a lagged response to climatic warming, may locally affect the temperature characteristics of nearby glaciers. Both precipitation and temperature related to altitude seem to be slightly higher in the case of Øksfjordjøkelen and Seilandsjøkelen than the highest plateaus of south Lyngen. Distinction between these types in terms of geomorphic features may also allow differentiation on climatic grounds. Here, however, it is necessary to investigate further the relationships between altitude and 'controlling' meteorological variables such as winter precipitation and summer temperatures as well as critical summit area controls. These ideas could also be tested at other ice caps near the Norwegian-Swedish border (Østrem et al., 1973).

7. DISCUSSION

Reconstructions of former glaciers has been accomplished by a variety of methods (e.g. Porter, 1981). However, the generalized use of valley glacier systems should be tempered in view of the differences in response to climatic variables (specifically, summer temperatures and winter precipitation) for plateau ice caps with extensive valley outlets. A further complicating factor concerns the thermal regime of the system. Crucially, but most difficult to answer; if these effects do present problems then what can we use to describe past glacier dynamics and reponse? Is it possible, for example to use Nye's (1965) model for inferring the mass balance of a glacier from available advance and retreat figures? Thus, does the interpretation of moraine sequences or historically-determined ice extents reflect what actually happened to the mass balance?
From observations in north Norway, the answer to such questions appears to be 'no'. The current pattern of glacier response on the plateau ice caps of north Norway reflect regional climatic factors and local topographic influences. An appreciation of the long-term responses to general climate is essential for the interpretation of ice cap behaviour. Local, small scale amplitudes of climatic changes are manifest by valley glacier oscillations irrespective of patterns of nourishment and ice redistribution. The dual functioning of topography and climate needs to be more fully appreciated in order to interpret the dynamic responses of both the cold-based and warm-based ice caps. To this end, observations over several years have been started on both

Figure 7 (opposite page). Photograph taken ca. 1890 compared with a similar view in 1986 showing the marked glacier recession in the valley of Juovvavag'gi (Storelvdalen).

Balgesvarri and Øksfjordjøkelen with levelling across the ice caps which should allow long-term changes to be related to climate (Østrem and Tvede, 1986).

In the Lyngen Peninsula, it appears that the altitudinal transition between cold-based and warm-based glaciers today is between about 1500 and 1600 m a.s.l. This corresponds to a glacierization altitude of 1250 m for this area (Østrem et al, 1973). We do not know what corresponding elevations were over recent (cooler, and possible wetter) centuries and thus plateau ice caps are likely to have been modified at lower levels than at present.

Given the inferred asymmetry of accumulation and ablation vis-à-vis mass balance and climate, it would seem reasonable to infer that regional glacier history is unlikely to exhibit much apparent synchroneity. Similarly, the depositional record and traditional, trim-line, methods of glacier reconstruction (Osmaston, 1975; Porter, 1981) are unlikely to be applicable on a regional basis, but rather in association with individual ice-caps. Despite this apparent paradox, the plateau glaciers of north Norway present both a sensitive record of climatic change as it occurs at present and a unique response pattern. The inability to find positive correspondence with well-documented sequences further south in Scandinavia is well known and it has been recently suggested by Grove (1985) with reference to the Little Ice Age (16th-17th centuries) that 'unless we turn to a hypothesis that the ice caps in (north) Norway required a century longer than the alpine glaciers before frontal advances took place, a surprisingly long lag period, we are forced to turn to meteorological explanations of the difference in timing'. Given the complexity brought about by topographic control and the disparity arising through separation of the accumulation and ablation zones, the asynchronous behaviour is perhaps both explicable and of value in the interpretation of recent climatic trends in NW Europe. Furthermore, it is significant that in a recent study of Jan Mayen (Anda et al., 1985) similar trends in recent glacier behaviour to those of the Lyngen and Bergsfjord Peninsulas have been observed at approximately the same latitude (69°-71°N).

8. CONCLUSIONS

Knowledge of the past and present climatology of the northeastern Atlantic and Norwegian Sea is important in assessing trends in glacier activity and for the better modelling of climate and glacial events. Currently, there is a gap in our knowledge about the behaviour of the small plateau glaciers of north Norway. The examples discussed here show that useful data can be obtained with a range of glacier activities and thermal regimes. There are also useful analogies with some glaciated mountains in the British Isles.

ACKNOWLEDGEMENTS

We thank the Royal Society for funding fieldwork involving members of the British Schools Exploring Society Expeditions to Lyngen (1979, 1984) and the University of Sheffield Øksfjordjøkelen Expedition (1986). We also thank the members, sponsors and donors to those expeditions, in particular: The New York Explorer's Club, The Mount Everest Foundation and the Manchester Geographical Society. We should also like to thank the University of Sheffield, Plymouth Polytechnic and the Queen's University of Belfast for financial and other assitance. We gratefully acknowledge help from the Alpine Club in tracking down early photographs and members of the Imperial College, London, 1961 Expedition to Øksfjordjøkelen for unpublished data as well as Dr. George Darling for stable isotope analyses.

REFERENCES

Anda, E., Orheim, O. and Mangerud, J. 1985. Late Holocene glacier variations and climate at Jan Mayen. Polar Research 3 n.s., 129-140.

Andrews, J.T. 1975. "Glacier systems: an approach to glaciers and their environments". Duxbury Press, North Scituate, Mass., 191 pp.

Andrews, J.T. and Miller, G.H. 1972. Quaternary history of northern Cumberland peninsula, Baffin Island, N.W.T. Canada: Part IV: Maps of the present glaciation limits and lowest equilibrium line altitude for north and south Baffin Island. Arctic and Alpine Research 4, 45-59.

Bryson, R.A. and Wendland, W.M. 1967. Tentative climatic patterns for some late-glacial and post-glacial episodes in central North America. In: W.J. Meyer-Oakes (ed.), "Life, Land and Water". University of Manitoba Press, Winnipeg, 271-298.

Drozdowski, E. 1979. The patterns of deglaciation and associated depositional environments of till. In: C. Schluchter (ed.), "Moraines and Varves". Balkema, Rotterdam, 237-248.

Finsterwalder, R. 1962. Measurement of glacier variations in the eastern Alps, particularly in the Gurgl area. International Association of Scientific Hydrology, Symposium of Obergurgl, 58, S 7-15.

Gellatly, A.F., Whalley, W.B. and Gordon, J.E. 1986a. Topographic control over recent glacial changes in southern Lyngen Peninsula, North Norway. Norsk Geografisk Tidsskrift 41, 211-218.

Gellatly, A.F., Whalley, W.B., Gordon, J.E. and Ferguson, R.I. 1986b. Movement at the ice front. Geographical Magazine, June, 284-299.

Gellatly, A.F., Gordon, J.E., Whalley, W.B. and Hansom, J. Reconstruction of ice limits for some glaciers in North Norway since 1850. In preparation.

Gordon, J.E. 1979. Reconstructed Pleistocene ice sheet temperatures and glacial erosion in northern Scotland. Journal of Glaciology 22, 331-344.

Gordon, J.E., Whalley, W.B., Gellatly, A.F. and Ferguson, R.I. 1987. Glaciers of the southern Lyngen Peninsula, Norway. In: V. Gardiner (ed.), "International Geomorphology 1986, Part II". John Wiley, 743-758.

Gordon, J.E., Darling, G., Whalley, W.B. and Gellatly, A.F. $\delta D - \delta^{18}O$ relationships and the thermal history of basal ice near the margins of two glaciers in Lyngen, North Norway. Journal of Glaciology, in press.

Grove, J.M. 1972. The incidence of landslides, avalanches and floods in western Norway during the Little Ice Age. Arctic and Alpine Research 4, 131-138.

Grove, J.M. 1985. The timing of the Little Ice Age in Scandinavia. In M.J. Tooley and G.M. Sheail (eds.), "The climatic scene". George Allen & Unwin, 132-153.

Hoel, A. and Werenskiold, W. 1962. Glaciers and snowfields in Norway. Norsk Polarinstitutt Skrifter nr. 114.

Humlum, O. 1985. The glaciation level in West Greenland. Arctic and Alpine Research 17, 311-319.

Humlum, O. 1986. Mapping of glaciation levels: comments on the effect of sampling area size. Arctic and Alpine Research 18 (4), 407-414.

Humlum, O. and Svensson, H. 1982. Naledi i Gronland. Flyfotografisk inventering af perennerde flod- og kildeis i gronlandske permafrostomrader; forelobige resultater. Geografisk Tidsskrift 82, 51-59.

Laaksonen, K. 1976. The dependence of mean air temperatures upon latitude and altitude in Fennoscandia (1921-1950). Annales Academiae Scientiarum Fennicae, Series A III, Tom. 119.

Liestøl, O. 1967. Storbreen glacier in Jotunheimen, Norway. Norsk Polarinstitutt Skrifter nr. 141, 63 pp.

Manley, G. 1955. On the occurrence of ice domes and permanently snow-covered summits. Journal of Glaciology 2, 453-456.

Markgren, M. 1964. Geomorphological studies in Fennoscandia. Lund Studies in Geography, Series A, no. 28.

Meier, M.F. 1965. Glaciers and climate. In: H.E. Wright Jr. and D.G. Frey (eds.), "The Quaternary of the United States". Princeton University Press, Princeton, 795-805.

Mitchell, J.M., Dzerdzeevski, B., Flohn, H., Hofmeyr, W.L., Lamb, H.H., Rao, K.N. and Wallén, C.C. 1966. "Climate change". World Meteorological Organisation, Geneva, Technical note 79.

Moran, J.M. and Blasing, T.J. 1972. Statistical approaches to 'alpine' glacier response. Palaeogeography, Palaeoclimatology and Palaeoecology 11, 235-236.

Nye, J.F. 1965. A numerical method of inferring the budget history of a glacier from its advance and retreat. Journal of Glaciology 5, 589-607.

Osmaston, H. 1975. Models for the estimation of firnlines of present and Pleistocene glacier. In: R. Peel, M. Chisholm and P. Haggett (eds.), "Processes in Physical and Human Geography". Heinemann, London, 218-245.

Østrem, G. 1966. The height of the glaciation level in southern British Columbia and Alberta. Geografiska Annaler 48A, 126-138.

Østrem, G., Haakensen, N. and Melander, O. 1973. Atlas over breer i Nord Skandinavia. Norges Vassdrags- of Elektrisitetsvesen, Meddelese frå Hydrologisk Avdeling 22, 315 pp.

Østrem, G. and Tvede, A. 1986. Comparison of glacier maps - a source of climatological information. Geografiska Annaler 68A, 225-231.

Paterson, W.S.B. 1981. "The physics of glaciers" (2nd edition). Pergamon Press, Oxford, 380 pp.

Porter, S. 1977. Present and past glaciation threshold in the Cascade Range, Washington, USA: Topographic and climatic controls, and palaeoclimatic implications. Journal of Glaciology 18, 101-116.

Porter, S.C. 1981. Glaciological evidence of Holocene climatic change. In: T.M.L. Wigley, M.J. Ingram and G. Farmer (eds.), "Climate and History". Cambridge University Press, 82-110.

Sollid, J.L., Andersen, S., Hamre, N., Kjeldsen, O., Salvigsen, O., Sturød, S., Tveitå, T. and Wilhelmsen, A. 1973. Deglaciation of Finnmark, North Norway. Norsk Geografisk Tidsskrift 27, 233-325.

Thorp, P. 1986. A mountain icefield of Loch Lomond Stadial age, western Grampians, Scotland. Boreas 15, 83-97.

Whalley, W.B. 1973. A note on the fluctuations of the level and size of Strupvatnet, Lyngen, Troms and the interpretation of ice loss on Strupbreen. Norsk Geografisk Tidsskrift 27, 39-45.

Whalley, W.B., Gordon, J.E. and Thompson, D.L. 1981. Periglacial features on the margins of a receding plateau ice cap. Journal of Glaciology 27, 492-496.

A HISTORICAL PERSPECTIVE OF THE NINETEENTH CENTURY ICE TRADE

R.A. Smith
Cambridge University
Engineering Department
Trumpington Street
Cambridge, CB2 1PZ
United Kingdom

ABSTRACT

The use of naturally occurring ice as a refrigerating medium became widespread in north-west Europe from the beginning of the nineteenth century. There are records of Norwegian glacier ice being exported to Britain from c. 1820 onwards. In 1844, the first cargo of "Wenham Lake Ice" from the Boston area of America arrived in Liverpool. High pressure advertising ensued, with the result that a fashionable market in ice was created. The increase in demand soon outstripped, and overpriced, the limited American supplies with the result that from the 1850's the Norwegian lake ice trade increased extremely rapidly.

The supply to Britain rose steadily, presumably because of the limited possibilities of home based sources. Germany was, in normal years, self-supporting, but she needed to import after mild winters. High average winter temperatures in Hamburg corresponded to huge peaks in the Norwegian export tonnage.

The decline of the lake ice trade was abrupt and dramatic just before the First World War, when mechanical/chemical refrigerating plant became widely available. In northern Norway, however, the glacier ice trade lasted until as late as 1949, in order to supply fishing vessels operating into the Arctic Ocean.

The paper ends with a short description of a rare example of a present-day glacier ice trade in the Hunza region of the Karakoram mountains. The methods of collecting, protecting and distributing glacier ice must be echos of the widespread, but little known, ice trade of the last century.

The Industrial Revolution brought a rapid increase in wealth and luxury to Western Europe from the early years of the nineteenth century. Foods were required to be fresher, drinks and room temperatures were desired to be comfortably cool. Ice was the agent used to produce these effects. It must not be thought that its use was new. Snow and ice had been stored in classical times; Soloman in the book of Proverbs, Chapter 15 v. 13, mentions:

J. Oerlemans (ed.), Glacier Fluctuations and Climatic Change, 173–182.

> As the cold of snow in the time of harvest, so is a faithful messenger to them that send him; for he refresheth the soul of his masters!

In a recently published short history (Ellis, 1982) many early examples of the use of ice are quoted. In southern countries such as Egypt, snow came from the mountains of Syria on relays of fast horses. Oran, a Spanish possession in North Africa, imported snow from Spain by sea. Many Mediterranean towns were supplied by mule transported snow and ice gathered in the mountain areas and stored in caves and pits until required in the hot seasons. Grand country houses had their own ice storage houses of which many examples still exist. In northern Europe ice could be obtained, in cold winters, from local lakes and ponds, but with increasing demand and concentration of the population in towns this form of supply became inadequate and ice was moved great distances to meet the increasing need. Although this short paper concentrates mainly on the British experience, it is clear that similar developments took place in the other North European countries.

One of the earliest written references (Blackwoods Magazine, 1823) marks the beginning of this export trade:

> Two or three mild winters, of late, have brought a new article of foreign trade to England. Ice,...now comes to us all the way from Norway, where a gentleman, we understand, is making arrangements to send over even snow, at a far cheaper rate than it can afford to fall in this country... This imported ice is the foremost in our streets now of a morning, moving along in huge cartloads...and looking, as it lies in bulk, like so much coagulated Epsom salts.

The beginning of this trade was not without its bureaucratic difficulties. It is said that when the first cargo arrived in June 1822, it was so novel that the Customs officers could not classify it for dury. By the time it was decided that ice was 'dry goods', the whole 300 tons had melted away! The perishable nature of ice dictated the methods available for its transportation and storage.

The source of this first consignment of ice was most probably Bondhusbreen, an outflow glacier on the west side of Folgefonni in southern Norway. Hoel & Wevenskiold (1962) have given and interesting account, but some of his dates (understandably in view of the confusing evidence) conflict each other. He tells of local men hacking ice loose, rowing it across a lake 1400 m long, carrying it on their back, up to 1201 lbs (or 55 kg) at a time, across 7.5 km of scree and loading it onto the schooner, Albion, on which the ice was taken to Scotland for preserving salmon. The trade later became regular enough for a road, Isveien, to be build across the scree. Ice was more easily obtained from the eastern rim of Folgefonni, were regenerated blocks calved from the vertical front of Støkken were located just over one mile from the waiting ships. A regular trade developed from this source also over the next decade.

In the north eastern states of America, the long cold winters made lake or pond ice a commonly available commodity. As early as 1805, Frederick Tudor of Boston, Massachusetts, recognised its export potential. His first venture was to send 130 tons of ice on the 'Favorite' to Martinique, where what quantity remained on arrival, was used to alleviate the effects of yellow fever which then raged through the towns of the West Indies. From these small beginnings, Tudor rapidly became known as the 'Ice-King', and went on to establish a monopoly throughout the Caribbean by 1816. Markets in the southern United States followed, and astonishingly, by 1833, India was beginning to get cargoes from Boston! Of course, losses were considerable, but with insulation provided by boards, tarpaulins, sawdust and wood shavings, remarkable results were achieved. Up to two thirds of the ice perished between Boston and Calcutta, but even if only as little as 10% remained on arrival in India, a handsome profit could be made. New England apples, packed on board a ship in ice, could be sold in China for their weight in silver (Augustine Heard papers). The great advantage of lake ice over irregularly shaped pieces of glacier ice was that rectangular shaped blocks could be used to fill the storage space available in a ship hold and the surface area/volume ratio could be reduced thus decreasing the melt rate (Figs. 1 and 2).

Figure 1. Fresh Pond, Cambridge, Mass., USA. Horse drawn grooving prior to hand sawing of ice blocks (Scribner's Monthly, 1875)

Inevitably, the American ice trade spread to Europe. The first shipment of 1842 was not a financial success, but American ice soon established itself as a fashionable product. In 1844 Charles Lander began to ship his product, Wenham Lake ice, to Britain. Wenham Lake had a surface area of a mere 320 acres (1.3 km^2) but could produce two or more harvests in a cold winter. The product was advertised with enthusiasm in The Times of 8 July 1845:

> Wenham Ice, from America, can be obtained from the principal offices (of the Company) at 164A, Strand or at any of the principal agents throughout the Kingdom. Frozen from pure mountain lake water, this ice is suited for table use, for mixing with liquids, or placing in direct contact with provisions. Owing to its solidity, it is easy of preservation and is dispatched without difficulty into the country, in large or small quantities, packed in a hamper or blanket. The Company's vans leave the principal office at 8 a.m. and p.m. daily, delivering the ice to all parts of town. Orders should be sent in a least an hour before the time of departure. Many fishmongers and dealers in common rough ice profess to supply American ice to their customers, but it is only to be obtained from the Company's advertised agents, no other American ice than theirs being now in the country, and the present scarcity of ice in America, rendering the purchase of it there, for export, impossible during this season.

Figure 2. Technological development! This remarkable ice-sawing machine was constructed in the early 1880's by an American, Sauncry A. Sager.

This was advertising hype of the first order, for only two weeks later, from the Time of 18 July, we read that a cargo of American ice from a rival source was being discharged on the Clyde and that:

> The novelty of import has attracted a considerable degree of curiosity and interest. The cargo was processed at the Rockland Lake, a fine sheet of water situated about 40 miles up the Hudson river, (from New York) and the ice is packed in beautifully sawn blocks of about 2 cwt (100 kg) each, with all the regularity of a cargo of square-dressed stones. The only protective covering is a layer of rice-chaff and sawdust, and we learn that the loss during the passage has been exceedingly little.

The trade flourished, aided by an article in the Illustrated London News (1845), which pictured the various tools used at Wenham Lake for splitting and cutting the ice, the large warehouses erected for its storage and a branch railway built for its transportation. This was clearly an established business. The article also introduced to the public the term 'refrigerator', being a small insulated box (portable ice-house!) into which ice could be packed round the articles to be preserved.
The Times of the following year (1846) charted the progress of the trade. On 28 March it was noted that 460 tons (4.6×10^5 kg) had arrived from Boston at St. Katherine's Dock, London. Since the frosts of the previous winter had been slight, it was thought that the imported arcticle would be in great demand. On 17 June, the arrival of a further 600 tons (6×10^5 kg) from Boston was recorded, together with the statement:

> Since the arrivals of ice which recently took place from Norway and other parts of the north of Europe have ceased from the time they were last noticed, this is the first importation of the article which has taken place from any foreign country.

On 24 August, another 460 tons arrived from the United States, but:

> The wonder which at first existed as to the manner in which so extraordinary an article of importation could be collected and brought in a perfect state of preservation to this country from so distant a quarter of the globe, has now ceased.

Accordingly the paper stated that it would not report future arrivals in full detail. However, on 29 August we read that a further 900 tons of Wenham Lake ice had arrived on the ship Tiberius and that:

> Its market price, in consequence of the frequent importations is now reduced to 8 l per ton.

This price of £8 per ton would be approximately equal to £270 per ton or 27p per kg based on equivalent purchasing power today (1987). The current price of artificially manufactured block ice is around £20 per ton.)

In the two following years other sources of supply became available. On 6 July 1847, The Time records the arrival of 291 tons of ice at Liverpool from Boston, whilst noting one or two arrivals in the three months previously form the 'northern regions of Europe'. On 31 March 1848, the same newspaper announced that further imports from Hanover and other parts of Prussia had arrived at St. Katherine's Dock where:

> a particular place has been set aside for the landing and housing of the article on the southern side of the docks, of a cool temperature and protected from the heat of the sun. (These arrivals)...announce the possibility of further supplies of this novel article of commerce from a European state which has hitherto failed or declined to traffic in this commodity.

What effect did these importations have on the Norwegian glacier ice trade? It was Hoel's (1962) opinion that the invention, in the period 1850-1860, of methods of storing ice in sawdust and peat, made it easier and cheaper to cut ice on tarns and lakes near the sea. Glacier ice could not compete. Whilst this is clearly true, it is also obvious that other lessons were being learned from the American trade because the lake ice business in Norway started and extremely rapid development in the 1850's. It might also be supposed that the American trade had created a receptive market, which the Norwegians were keen to exploit. Developments such as the building of large ice warehouses for storage and shoots to load the ice onto waiting ships took place. By the latter half of the century over 90 ice works were located on the coast of Telemark (Ouren, 1981). However the name 'Wenham' became a generic term for all ice. Its use was so universal that, for example, in a work of popular fiction in 1865 (Chambers Journal, 1985), the hero:

> was to go (to) where I could get my Wenham Lake ice, as in London, with my seltzer and sherry

Thus ice was, by now, an important home comfort. However, his Wenham ice might well have come from a host of other sources in America or even from a Norwegian lake. Sometime later, in 1899, Lake Oppegard, in Norway, was indeed commercially christened Lake Wenham, to add to the confusion!

Short reviews of both the American and Norwegian ice trades were published in 1981 (Proctor, 1981), the paper on Norway by Ouren (1981) containing some particularly interesting quantitative data. It is clear that from 1860 on, Britain took the largest part of Norway's exported ice whilst the American supply dwindled, presumably because of transportation costs. The trade build up steadily from 1860 to reach a peak just before 1900. Germany, on the other hand, was nearly self sufficient and only imported large quantities of ice after a particularly mild winter (see Fig. 3). It is probable that the exports from Germany to Britain in 1848 were a curiosity: no further records of imports from this quarter can be dound, so it can be assumed that the ice produced in Germany was mostly consumed domestically. The peaks in Norwegian exported tonnage of 1884, 1898, 1899 and 1910 correspond to

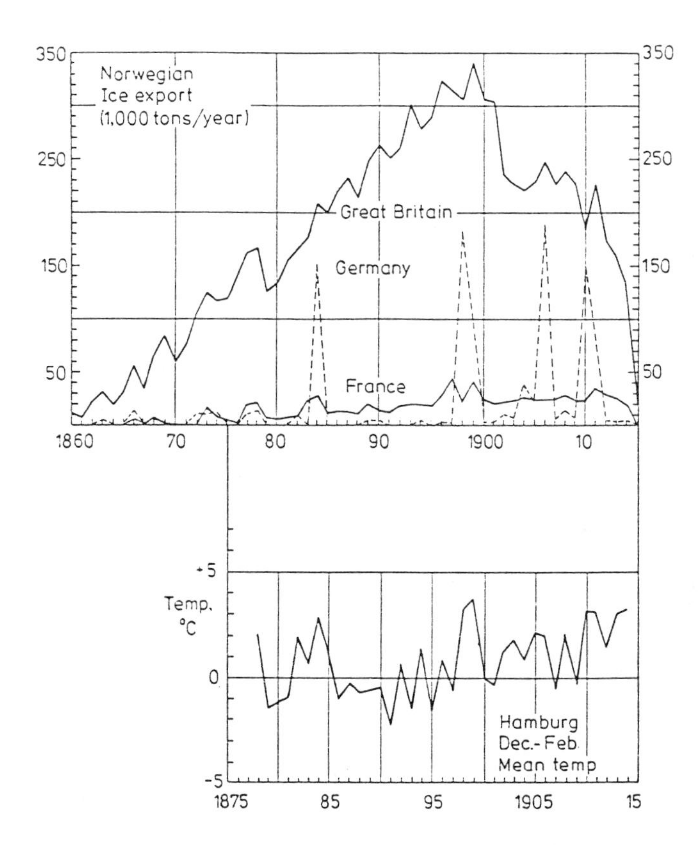

Figure 3. Statistics of the Norwegian lake-ice trade (Ouren, 1981).

high average winter temperatures in Hamburg. These peaks very nearly double the normal rate of Norwegian supply, the elasticity of which was quite remarkable. It is also interesting to note the huge quantities involved. The peak of 340,000 tons in 1899, shows a remarkable increase on the relatively small cargoes of the 1850's. The accelerating decline from the turn of the century to the death of the trade by the start of the First World War can be attributed to the introduction of mechanical heat extraction machines which made the small scale production of artificial ice much more convenient.

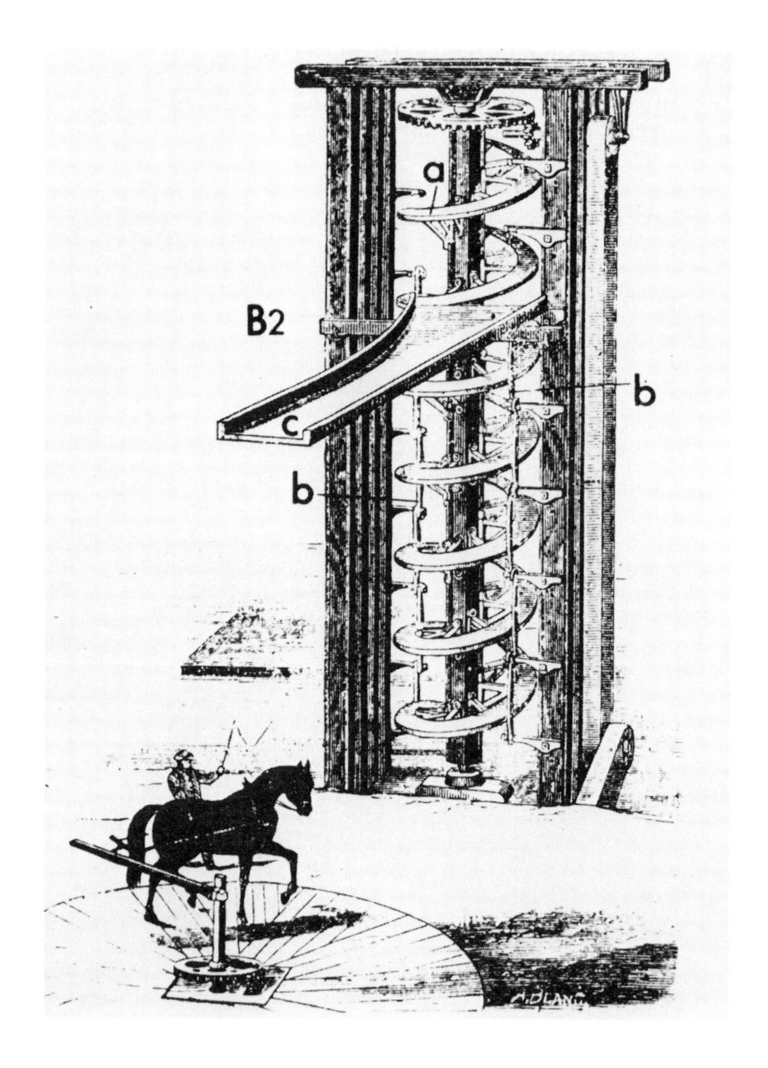

Figure 4. Ice was stored in large buildings where the blocks were placed some 50 mm. apart, to give room for drainage and ventilation and were covered with hay. The device illustrated brought the ice up to the correct level in the store-house by a horse powered lift. The ice blocks were placed at the bottom of the screw (a), behind vertical rods (b). When the screw rotated to the left, the blocks slid upwards until they reached a vertically adjustable slide (c), on which they slid into the store-house.

An interesting survival of the glacier trade described by Hoel & Wevenskiold (1962) was the collection up to 1949 of ice from Øksfjordjøkulen in Finnmark and Troms. Apparently the use was for the preservation of fish in vessels operating in the Arctic Ocean. When natural calving of the glacier failed, explosives were used on a cone of ice which descended into the sea:

> the ice was collected in iron-covered ice-chutes, which were 21 inches wide and 9 inches deep. These were pushed further and further into the ice as blasting proceeded. Eventually these chutes might be as much as 150 m long. They were laid on a dogleg. In order to reduce the speed of the ice in the chute, bolts were, where necessary, placed in the bottom of the chute. It was possible to load as much as 30 tons in one hour from a chute. This work was extremely dangerous (!). In 1949 the authorities build a refrigeration plant and the sale of ice ceased.

In the summer of 1986, the author was able to witness a present day remnant of the glacier ice trade. In that summer the Karakoram Highway was opened to 'tourists' over the Kunjerab Pass, thus linking north Pakistan with China. The upper reaches of this road pass through the near-desert regions of Chilas and the Hunza Valley. All the water in this region, both for irrigation and drinking, is obtained by tapping off glacial melt streams, which also power tiny medieval-style corn mills. Ice is sawn from accessible glacier snouts, of which the Minnapin, flowing down from Rakaposhi, is only a few kilometers from the Highway. Every day, the ice is transported first by mule then by truck, protected from the fierce sun by sacking and sawdust. The town of Gilgit, some 3 or 4 hours away, is a distribution centre. The ice is stored in a shaded wooden shed, the price is halved at regular intervals througout the day. The main uses appear to be the cooling of fruit in roadside shops, cooling bottles of the ubiquitous 'Coke' or as the 'rocks' in fruit cordial sold for half a rupee a glass. The recent development of local hydro-electric power will probably render this trade obsolete within a few years.

REFERENCES AND NOTES

Augustine Heard Papers, Baker Library, Harvard Business School, quoted in Ref. 8, Foster Smith.

Blackwood's Magazine, (1823), Now., p 509.

Chambers Journal (1865), No 89, Sept 9, p 561.

Ellis, M. (1982), 'Ice and Icehouses Through the Ages', Southampton University Industrial Archaeology Group, Southampton University, U.K. Traces the history if the Icehouse and gives a short account of the transport and storage of ice through the centuries and of its later manufacture. The reference to Proverbs comes from this source.

Hoel, A & Wevenskiold, (1962), 'Glaciers and Snowfields in Norway', Norsk Polarinstitutt Skrifter, Oslo, 114, p 99-117. Hoel's account of Bondhusbreen is from a reference in the Christiana 'Morgenbladet' of Oct 4, 1925. This must be 1825.

Illustrated London News, (1845), May 17, p 315-316.

Ouren, T., (1981), 'The Norwegian Ice Trade', p 31-41 (in last ref.)

Proctor, D.V. (ed), (1981), 'Ice Carrying Trade At Sea', Proceedings of a Symposium held at the National Maritime Museum, Sept. 1979, Maritime Monographs & Reports No 49. Of particular interest in this context are the papers on the Norwegian Ice Trade, (Ouren, 1981) and 'Concentrated Wenham, New England Ice Trade in Albion, Foster Smith, P.C., p 43-52.

Scribner's Monthly, X, 4, Aug., 1875, p. 467.

Contemporary references from The Times, (London), Include:

1845	8 July	(Supplement)	
1845	18 July	page 6	column 5
1846	28 March	6	4
1846	17 June	6	4
1846	24 Aug	3	6
1846	29 Aug	7	5
1847	5 May	7	4
1847	17 May	7	4
1847	6 July	7	4
1848	31 March	5	4
1848	8 June	6	5

MASS BALANCE HISTORY OF BLUE GLACIER, WASHINGTON, USA

Richard L. Armstrong
CIRES & World Data Center-A for Glaciology
University of Colorado, Boulder
Colorado USA

ABSTRACT

Mass balance and weather data have been collected for Blue Glacier since 1957. Annual net balance has been determined by combining the total mass loss from ice and firn melt with the residual snow remaining at the end of the summer season. In order to define confidence limits for the historical mass balance data set, it is necessary to quantify several sources of potential error. Preliminary analysis of the spatial variance of ice melt within the ablation zone and the variance of snow density within the residual snow of the accumulation zone indicate that neither of these parameters are likely to be a source of significant error. The largest potential source of error in computing the mass balance of the glacier as a whole is the migration of an ice divide.

We investigate the possibility of developing a simplified, cost efficient, measurement technique which can be used to extend the mass balance record into the future to a time period adequate for the testing of glacier dynamics models. Initial analysis indicates a good correlation between mass balance and equilibrium line altitude (ELA).

1. INTRODUCTION

Blue Glacier is located on the north-east slope of Mount Olympus, the highest peak (2424 m) in the Olympic Mountains of western Washington State (Figs. 1 and 2). It is the largest of the more than 250 glaciers in the Olympic Mountain Range. It is 4.3 km in length, has a volume of approximately 0.5 km^3, a total area of 5.3 km^2, and an elevation range of 1275 m to 2350 m. Two separate accumulation zones, the Cirque and the Snowdome, discharge through ice falls into the valley tongue or Lower Glacier. The local climate is strongly maritime with the Pacific Ocean located only 50 km to the west. Mount Olympus receives the greatest precipitation of any area in the United States, excluding Alaska and Hawaii. In general, glaciers in the Olympic Mountain Range also experience very high rates of summer ablation due to their low elevation. Annual mass balance measurements were begun on Blue Glacier

J. Oerlemans (ed.), Glacier Fluctuations and Climatic Change, 183–192.

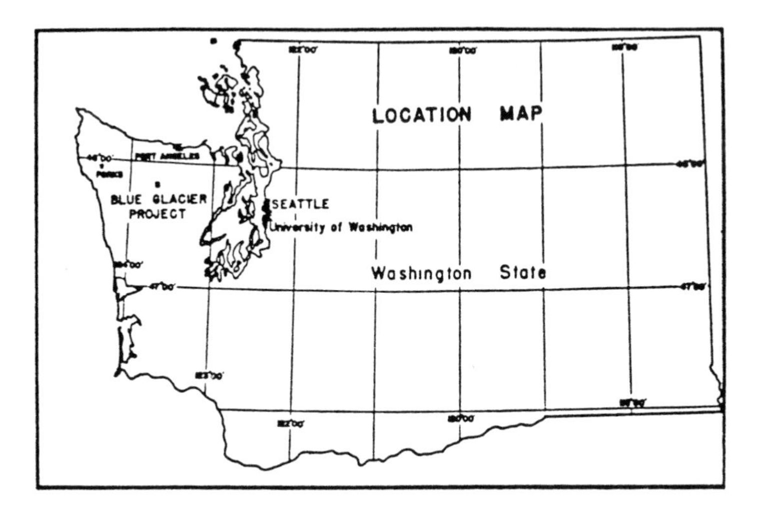

Figure 1. Location map.

Figure 2. Oblique aerial photograph of Blue Glacier from the north north-east (Photograph by A.S. Post).

in 1957 during the International Geophysical Year (IGY) and have been maintained almost continuously up to the present. These data indicate that Blue Glacier is in approximate equilibrium with the present climate. Terminus location data have been obtained by direct measurement over the past 48 years.

This paper presents a preliminary analysis of the recently compiled weather, mass balance and terminus location data from Blue Glacier since 1963. Data for the period 1957 to 1963 have previously been presented by LaChapelle (1959, 1960, 1965).

Meteorological data are collected at the Blue Glacier research building site (Fig. 3) at an elevation of 2025 m during the period approximately 15 June to 15 September. For this period the average air temperature is 7.6°C, average precipitation per month is 97.6 mm and the mean wind speed is 3.1 m/sec. Data for all months of the year were collected only during the IGY of 1957-1958 (LaChapelle, 1959) when a total annual precipitatin of 3782 mm was measured. Based on data from mass balance measurements the average annual precipitation for Blue Glacier is estimated to be between 3500 and 5000 mm. Eighty percent of

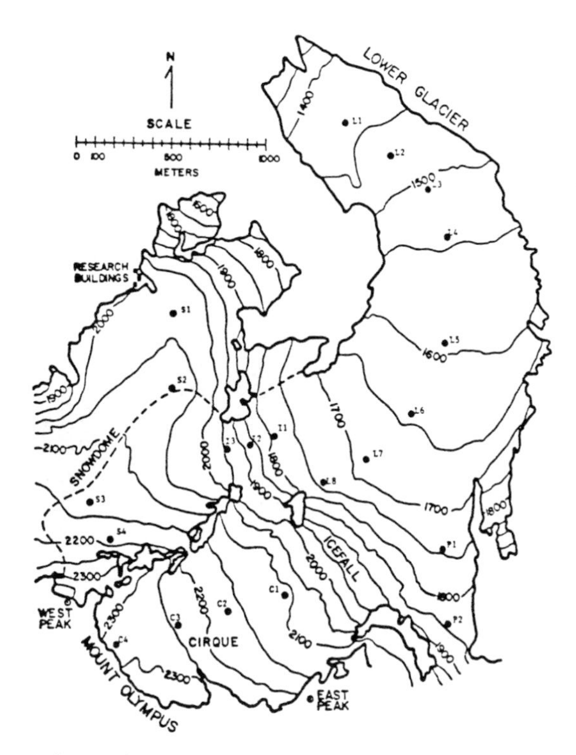

Figure 3. Blue Glacier elevation contour map including location of ablation stake network, 1981-1986.

the total annual precipitation falls in 6 months from October through March and less than 5 percent falls during July and August. The nearest location from which annual weather data are available is the National Park Service at the Hoh River Ranger Station located 20 km north-west of Blue Glacier at an elevation of 200 m. The average annual precipitation at this site is 3500 mm with a standard deviation of 518 mm.

Annual net mass balance for Blue Glacier is computed by measuring total ice and firn ablation combined with measurements of residual snow accumulation at the end of the melt season (LaChapelle, 1965; Hubley, 1956). In order to remain in a near-equilibrium condition a large winter accumulation is required to offset the significant summer ablation which has prevailed during the period of record. During the melt season ablation rates average 30 to 40 mm per day over the surface of the glacier. Mass exchange during the average balance year is large, equivalent to the gain and loss of a layer of water 3.0 to 4.0 m in thickness over the total glacier surface. Figure 4 contains the area-averaged net balance for the period of measurement. The summation of net mass balance values for the period of record is 9.2 meters water equivalent which corresponds to a gain in total mass of approximately 7 percent. (Data for the years 1976-1978 are currently estimates only, with an expected accuracy of 0.2-0.3 m water equivalent. These data will be finalized in the near future.)

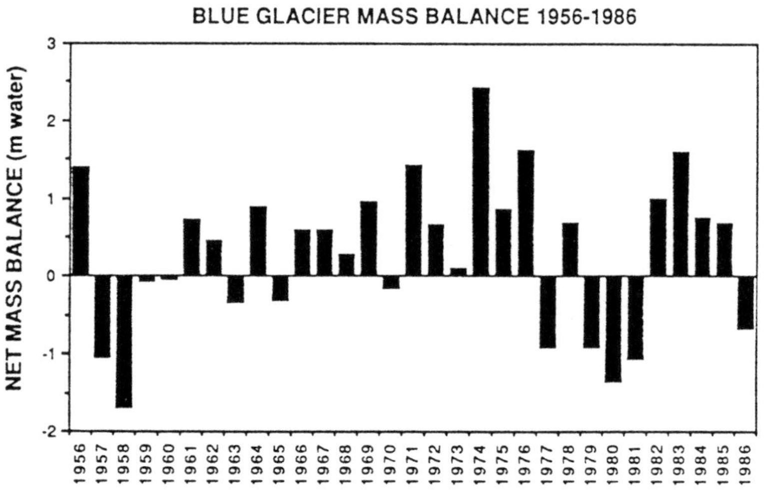

BLUE GLACIER MASS BALANCE 1956-1986

Figure 4. Blue Glacier mass budget history, 1956-1986.

2. POTENTIAL ERRORS ASSOCIATED WITH MASS BALANCE CALCULATIONS

In order to define a confidence limit for the annual net mass balance values it was necessary to evaluate several potential sources of error.

The centre-line ablation stake network currently used on Blue Glacier provides a measurement density of about 0.25 km^2/stake (Fig. 3). In order to evaluate the extent to which the individual stakes truly represent the average ice ablation over a 0.25 km^2 area, data were analyzed from stake networks with ten times this density. These networks included cross-glacier transects of seven stakes (100 m spacing) at two elevations as well as networks of 40-50 stakes arranged in a grid over the entire exposed ice surface of the Lower Glacier. The coefficient of variation within an area currently represented by one ablation stake is 10 to 15 percent. Data from the larger network covering the entire Lower Glacier provided the same range of variation. Therefore, a single stake placed on an ice surface would, hypothetically, monitor the melt averaged over the total surface of exposed ice with a probable error range of 10 to 15 percent regardless of the total area involved. There was no evidence of a systematic elevation or cross-glacier effect.

Snow ablation data, while not currently used as input to the net mass balance calculations, can be used to reconstruct the winter balance (LaChapelle, 1965). It had been assumed that snow ablation would indicate a lower spatial variance compared with ice ablation due to a more homogeneous surface texture, albedo and topography. A dense stake network simular to that used on the ice surface was installed near stake S1 on the Snowdome (Fig. 3) and the coefficient of variation of ablation on the snow surface was about 10 percent.

Depending on prevailing weather conditions the seasonal termination of ice melt may occur at any time between mid-September and early November. Direct observations have normally not been obtained beyond late September. The most reliable method to monitor the final increment of ice melt during the ablation season is to leave ablation stakes in place throughout the winter season. As soon as these stakes appear above the melting snow the following summer the distance to the ice surface below is measured and any additional melt subsequent to the final observation of the previous fall is noted. Extrapolation of late-season ablation based on meteorological data from nearby surface weather and radiosonde stations is a possible approach but it is not likely to be as accurate as the direct measurement of winter stakes.

The thickness of the residual snow remaining at the end of the ablation season can be sampled by depth probing. For the period of record it has been assumed that the residual snow layer has an average density of 580 kg m^{-3} throughout the accumulation area. Data from snow profiles through the layer of residual snow at four elevations in the accumulation zone support this assumption. Average snow density was 577 kg m^{-3} with a standard deviation of 17.5. There were no apparent elevation or other spatial effects. Snow layer thickness values obtained from probing were generally within 0.1 to 0.2 m of the values obtained from direct observations of the snow profile (dirt layers and sharp density boundaries).

The boundary of Blue Glacier for mass balance purposes is clearly defined by surrounding non-glaciated topography except along the western edge of the Snowdome (dashed line in Fig. 3). For the mass balance calculations the boundary for this area has been assumed to be the topographic divide represented by the snow surface. Surveys of a network of velocity markers showed this to be true; however, a series of topographic maps produced over the past 30 years show that the flow divide migrates within a zone up to 350 m wide (Waddington and Marriott, 1986). The area of the migration zone corresponds to about 10 percent of the accumulation area for a year of zero net mass balance.

3. INDIRECT METHODS TO COMPUTE MASS BALANCE

The equilibrium line altitude (ELA) has long been considered a useful parameter for the indirect computation of glacier mass balance (LaChapelle, 1962; Gross et al., 1976; Kuhn, 1980; Braithwaite and Mueller, 1980; Braithwaite, 1984). Such data compiled for Blue Glacier are shown in Fig. 5 along with the 95 percent confidence interval for the prediction of mass budget given ELA. The r^2 value for the regression is 0.90 with a standard error of 0.31.

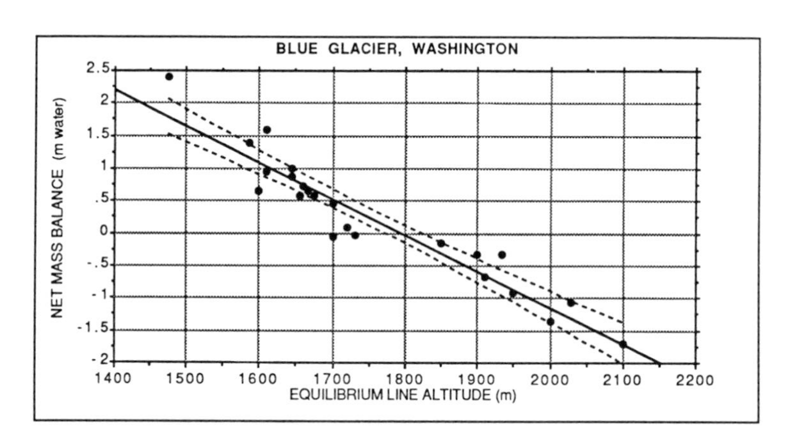

Figure 5. Specific net mass balance versus equilibrium line altitude for selected years of highest quality data (95% confidence interval shown).

The full 30-year data set could not be used for this test due to uncertainties in the exact location of the ELA at the end of some ablation seasons, especially those when ablation continued long after direct observations had ceased. On Blue Glacier the contour of the retreating snow line often does not correspond with a single elevation contour. A single ELA value will actually be an average altitude with a typical range of 25 to 50 m. This range would correspond to a variation in the computed specific mass balance of as much as 0.3 m. Therefore, assuming the location of the snowline at the end of the ablation season could be accurately located, either by air photography or by direct observation, it would be of greater practical and statistical value to make use of the accumulation area ratio (AAR) rather than the ELA.

An AAR of about 0.7 often corresponds to a mean specific balance of zero (Hoinkes, 1970; Gross et al. 1976; Braithwaite and Mueller, 1980). For a zero balance on Blue Glacier this ratio has averaged 0.48. This result is consistent with the nearly linear relation of net balance versus elevation found for the glacier surface (Fig. 6) and indicates the significance of the extraordinarily large winter accumulation on Blue Glacier.

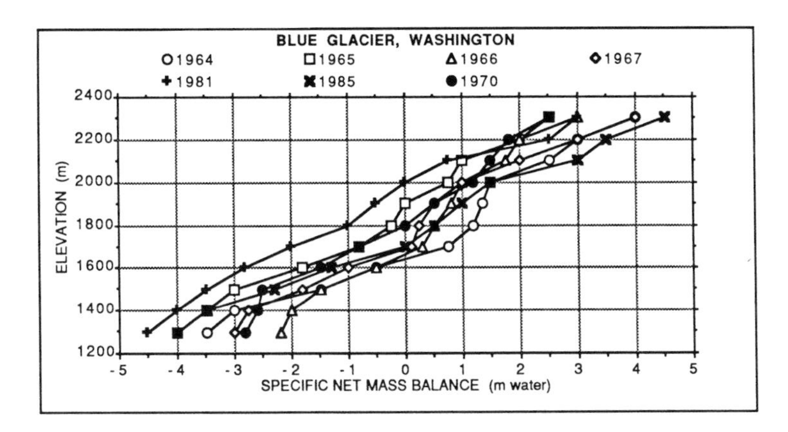

Figure 6. Specific net mass budget versus elevation for seven selected years on Blue Glacier

Figure 6 shows the relationship between net balance and elevation for seven years of Blue Glacier data. Applying this same relationship to South Cascade Glacier, Meier and Tangborn (1965) noted that while such curves may be displaced horizontally from year to year they tend to be nearly parallel for different years. Therefore, if sufficient data have been collected on one glacier, the net mass balance for a given elevation can be estimated by assuming the parallel nature of the curves. Data analyzed thus far for Blue Glacier do not show such a consistent parallel relationship with elevation compared with other studies. This may be due to the fact that the atmospheric freezing level, especially during fall and early winter, often moves up and down through an elevation band which includes the Lower Glacier. As a result there may be extended periods of time when the lower glacier is experiencing net ablation during rainfall while snowfall is occurring at the higher elevations. This would result in a mass balance versus elevation curve with a much different shape than that produced when conditions of ablation and accumulation occurred simultaneously at all elevations on the glacier.

4. GLACIER RESPONSE TO RECENT CLIMATE

Within the time frame of the annual balance measurements, Blue Glacier appears to be in near equilibrium with the local climate, with a slightly positive average mass budget (0.3 m) for the 30 years of record. Terminus location data do, however, indicate some major fluctuations on the longer time scale. The earliest advance of Blue Glacier which has been dated (Heusser, 1957) took place about 1650. Old-growth forest bordering the 1650 moraines is more than 700 years old. Heusser's study indicates a second advance with a maximum at about 1815 followed by a major recession. It is estimated that the terminus retreated 1400 m from 1815 to 1938, when direct measurement of the terminus location began. Preliminary investigation of relationships between terminus location and climate trends have been undertaken by LaChapelle (1965), Armstrong (1979), and Fountain (1987).

Figure 7 shows the 48-year history of terminus location presented as the cumulative distance from the location of 1939 when measurements began (Spicer, 1986). When the indirect measurements of Heusser, based on photography, are added to these data, a general pattern of retreat is indicated between 1900 and 1960. This is followed by a short period of moderate advance between 1969 and 1979 and then little change up to the present.

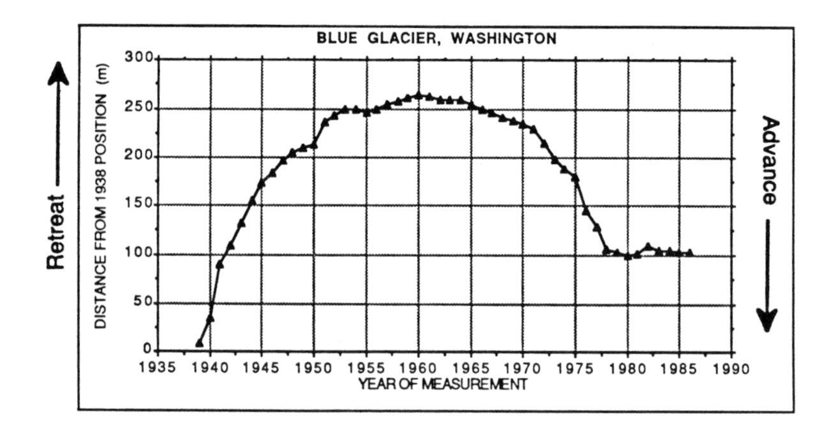

Figure 7. Blue Glacier terminus history shown as cumulative distance from the 1938 position (data from Spicer, 1986).

Measured mass balance and terminus location data indicate that Blue Glacier is in approximate equilibrium with the present local climate. Based on interpretation of aerial photographs (Spicer, 1986) it is likely that this condition is typical for other glaciers in the coastal range of the Olympic Mountains. This is in sharp contrast with conditions on South Cascade Glacier, the only other glacier in the northwestern United States with a long-term mass balance record. South Cascade Glacier has experienced significant mass loss during the period 1958-1985. The summation of net balance values is -10.06 meters water equivalent and the terminus has retreated a total distance of 388 m during this period (Krimmel, 1987). South Cascade Glacier is located in a drier region inland from Blue Glacier, 200 km to the north-east in the Cascade Mountain Range. The maximum elevation for the Blue Glacier exceeds South Cascade by about 125 m but the terminus extends to an elevation nearly 400 m below that of South Cascade.

Explanation for this contrasting response to recent climate fluctuations in the northwestern United States may be tied to variations in the large-scale weather patterns between maritime and more continental locations. It is the opinion of Yarnal (1984), who analyzed data from a maritime and continental glacier in southwestern Canada, that the explanation for these differences lies in the synoptic-scale circulation patterns. It is his opinion that the mid-tropospheric oscillations embedded within long-wave disturbances may be dampened by the topography of the coastal ranges, thus reducing the precipitation reaching the next mountain range to the east.

ACKNOWLEDGEMENTS

The evaluation and analysis of the historic Blue Glacier data set is a joint project between the University of Washington and the University of Colorado. Drs. C.F. Raymond and S.G. Warren are the principal investigators at the University of Washington and I thank them for their review of this manuscript. This work is supported by National Science Foundation Grant No. DPP-8412461.

REFERENCES

Armstrong, B.R. 1979. An attempt to reconstruct Blue Glacier mass balance from climatic records for the period 1904-1956. Unpublished research report, Dept. of Geography, University of Colorado, Boulder, Colorado, 26 pp.

Braithwaite, R.J. 1984. Can the mass balance of a glacier be estimated from its equilibrium line altitude? Journal of Glaciology, vol. 30, 106, 364-368.

Braithwaite, R.J. and Mueller, F. 1980. On the parameterization of glacier equilibrium line altitude. IAHS-AISH Publ. 126, 263-271.

Fountain, A. 1987. U.S. Geological Survey, Tacoma, Washington, personal communication.

Gross, G., Kerschner, H. and Patzelt, G. 1976. Methodische Untersuchungen über die Schneegrenze in alpinen Gletschergebieten. Z. Gletscherk. Glazialgeologie 12, 2, 223-251.

Heusser, C.J. 1957. Variations of Blue, Hoh, and While Glaciers during recent centuries. Arctic, vol. 10, 3, 139-150.

Hoinkes, H. 1970. Methoden und Möglichkeiten von Massenhaushaltsstudien auf Gletschern. Z. Gletscherk. Glazialgeologie 6, 1-2, 37-89.

Hubley, R.C. 1956. Glaciers of the Washington Cascade and Olympic Mountains; their present activity and its relation to local climate trends. Journal of Glaciology, vol. 2, 19, 669-675.

Krimmel, R.M. 1987. Mass balance and volume of South Cascade Glacier, Washington, 1958-1985 (this volume).

Kuhn, M. 1980. Die Reaktion der Schneegrenze auf Klimaschwankungen. Z. Gletscherk. Glazialgeologie, vol. 16, 2, 241-254.

LaChapelle, E.R. 1959. Annual mass and energy exchange on the Blue Glacier. Journal of Geophysical Research, vol. 64, 4 , 443-449.

LaChapelle, E.R. 1962. Assessing glacier mass budgets by reconnaissance aerial photography. Journal of Glaciology, vol. 4, 33, 290-296.

LaChapelle, E.R. 1960. Energy exchange measurements on the Blue Glacier, Washington. IAHS-AISH Publ. 54, 302-311.

LaChapelle, E.R. 1965. The mass budget of Blue Glacier, Washington. Journal of Glaciology, vol. 5, 41, 609-623.

Meier, M.F. & Tangborn, W.V. 1965. Net budget and flow of South Cascade Glacier, Washington. Journal of Glaciology, vol. 5, 41, 547-566.

Spicer, R.C. 1986. Glaciers of the Olympic Mountains, Washington: present distribution and recent variations. Master of Science Thesis, Department of Geological Sciences, University of Washington, Seattle, 158 pp.

Waddington, E.D. & Marriot, R. 1986. Ice divide at Blue Glacier, U.S.A. International Glaciological Society, Annals of Glaciology, vol. 8, 175-177.

Yarnal, B. 1984. Relationships between synoptic-scale atmospheric circulation and glacier mass balance in southwestern Canada during the International Hydrological Decade, 1965-1974. Journal of Glaciology, vol. 30, 105, 188-198.

MASS BALANCE AND VOLUME OF SOUTH CASCADE GLACIER, WASHINGTON 1958-1985

Robert M. Krimmel
U.S. Geological Survey
Tacoma, Washington, 98402, U.S.A.

ABSTRACT

The mass balance of South Cascade Glacier has been calculated by means of snow accumulation and ice ablation measurements since 1958. The summation of the net balance values for the 1958-85 period is -10.0 meters water equivalent. Topographic maps of the glacier were made from late summer vertical aerial photography in 1958, 1961, 1964, 1977, 1980 and 1985. In addition, the topography in the late summer of 1970 is well known from the theodolite survey of about 10 points on the glacier surface. The mass balance and photogrammetry are independent measurements of the cumulative volume change. The volume change between 1958 and 1985, measured from maps and adjusted for the density of ice, is -12.7 m, multiplied by the area of the glacier. This discrepancy can be reconciled by using the density of firn in the accumulation zone, rather than ice, to convert the volume change to water equivalent. Differences between the methods within the 27 year period are probably due to map errors, and indicate the need for caution in this type of comparison.

Surveying, rather than photogrammetric, methods for determining the volume change of glaciers is preferable, and density profiles are necessary if the glacier is not in equilibrium.

1. INTRODUCTION

Cumulative net balance for several years, measured using traditional (e.g. the measurement of material gain or loss at discrete spots on the glacier) mass balance methods, integrated over an entire glacier, should be in agreement with the density adjusted change in volume (measured from maps produced photogrammetrically) of the glacier over the same time period (Paterson, 1981). This comparison, using virtually independent methods, should be useful to check long-term mass balance results. Meier and Tangborn (1965) claimed excellent agreement for a 1958-61 balance-volume comparison at South Cascade Glacier. Haakensen (1986) obtained excellent agreement for the 1968-80 period for Hellstugubreen, but for Grasubreen during the 1968-84 period traditional

J. Oerlemans (ed.), Glacier Fluctuations and Climatic Change, 193–206.

methods gave an 8.1 m mass loss while the volumetric method gave a loss of 5.6 m. The discrepancy is attributed to difficulties in measuring internal accumulation. The purpose of this paper is to compare the results of traditional and volumetric methods at South Cascade Glacier for the period 1958-85.

South Cascade is a small (2.5 km^2) valley glacier in the North Cascade Mountains of Washington State. It is in a maritime climate, with an annual mass exchange of about 6 m water equivalent. It is isothermal throughout, and superimposed ice and internal accumulation are both minor factors in the balance calculations.

2. GLACIER VOLUME

2.1 Surface maps

Topographic maps were produced from vertical photographs taken in 1958, 1961, 1964, 1977, 1980 and 1985 (Table I). The 1964 to 1977 gap was filled with data from a 1970 survey of 112 points well distributed on the glacier surface (Tangborn et al., 1975). This survey was converted to a topographic map by using the approximate shape of contours on the 1977 map, but shifted in position to fit the 1970 surveyed altitudes.

For this paper, it is assumed that map error in the horizontal dimensions is negligible compared with errors in the vertical. A 1:5000 scale was used, because it produced a convenient working size. Scale differences between maps were not more than 1 mm over a 500 mm length (5 m over 2500 m at 1:5000).

The accuracy of the map in the verical is critical. In order to meet the U.S. National Map Accuracy Standard, at least 90% of the altitudes tested on a map must agree with the field checked altitudes within one-half the contour interval (Slama, 1980). The contour intervals on the maps used in this study were either 20 feet (6.1 m) or 5m. Thus, a 2.5 m error may occur at any point, but if the error were random, it would not be a serious detriment to the balance-volume comparison. A systematic error of 2.5 m magnitude that affected the entire glacier would be serious. Two methods were used to test the accuracy of the maps.

Near the time that each set of vertical photography was obtained the glacier surface was surveyed at a few points using 1 second or 0.0001 grad theodolites. These altitudes have an accuaracy of 0.1 m, but must be corrected to the date of the photography by considering surface lowering at ablation stakes. This correction adds an error of 0.2 m. Meier and Tangborn (1965) determined that for the 1958 and 1961 maps the systematic error was less than 0.15 m. However, for later years the discrepancy between the maps and field checked altitudes are somewhat greater (Table I).

The second method examined the representation of the terrain adjacent to the glacier. The altitudes of 15 calibration points (A-O on Fig. 1) around the glacier perimeter were taken from each map. The points were chosen to be ice free on all maps, on less steep terrain, and distributed around the glacier as much as possible. Calibration

Table I. The 1958-85 maps of South Cascade Glacier. The photogrammetry for each map was field c ecked, and the error suggested by the field check shown, but a datum adjustment was warranted only or the 1964 map (-4.1 m). The 1970 map was synthesized from 112 surveyed altitudes. The glacier volume was calculated by diffencing the surface and bed topographic maps. Change in volume, map to map, divided by the average glacier area between the two maps, is the average surface level change. These are not density corrected in this table. The 1958 and 1961 maps were field checked by Meier and Tangborn (1965).

Date	contour interval	number of field checked points	error (m)	datum corr. used (m)	glacier area $\times 10^4$ m^3	glacier volume $\times 10^4$ m^3	Δ volume/area (m)
21 Aug 1958	20 ft	-	-	0.0	281	23758	-3.50
12 Sep 1961	20 ft	-	-	0.0	273	22789	-0.09
11 Sep 1964	20 ft	5	4.1	-4.1	267	22766	-5.51
9 Sep 1970	5 m	(112)	-	0.0	263	21305	-0.25
13 Sep 1977	5 m	7	0.8	0.0	257	21242	-4.58
3 Oct 1980	5 m	3	-1.9	0.0	255	20070	-0.27
24 Sep 1985	5 m	21	1.0	0.0	252	20002	

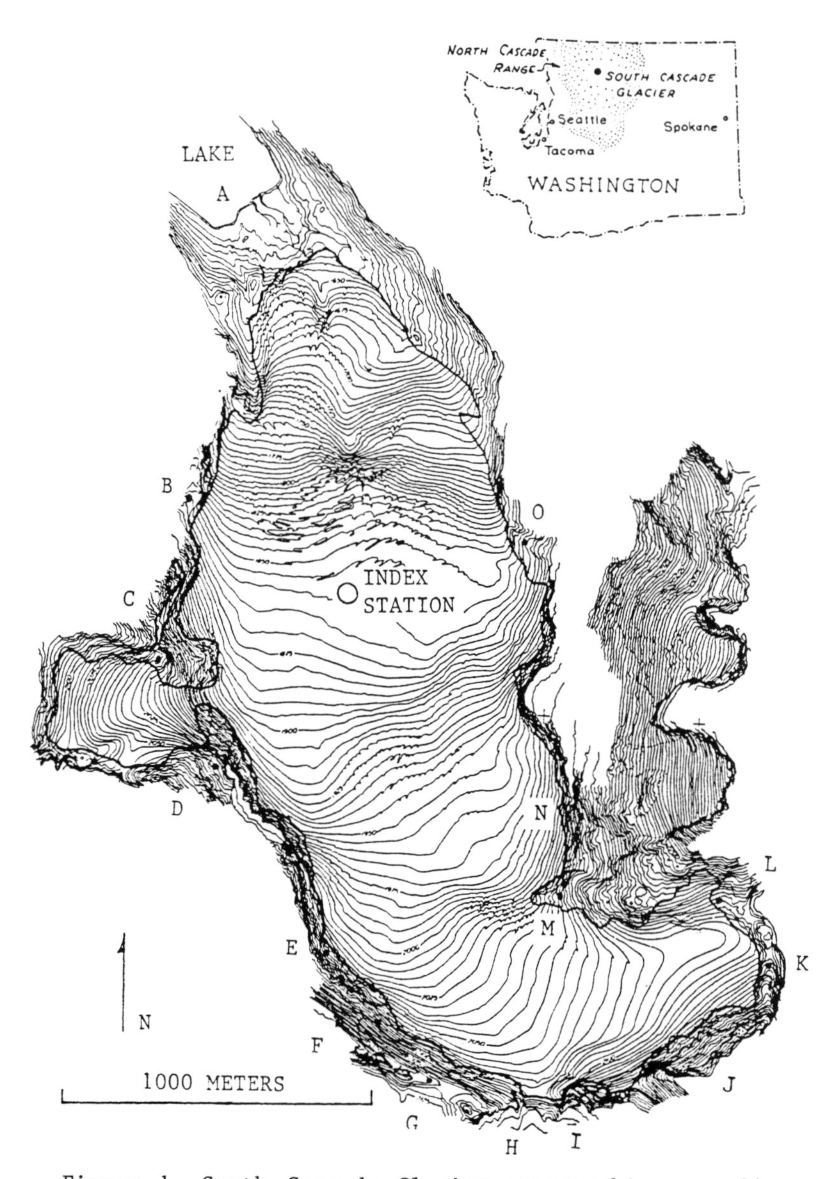

Figure 1. South Cascade Glacier topographic map, 24 September 1985, contour interval 5 m. The glacier margin is indicated by a heavy line and the 1958 terminus position is shown as a dotted line. Heavy dots A-O around the perimeter of the glacier were used as calibration points to compare the altitude of terrain adjacent to the glacier from map to map. Dot A is the lake surface.

point A was the lake surface, which for fair weather in late summer has a stage range of less than 0.4 m. The deviation from the average altitude of the 6 values (one from each map) for each calibration point is shown in Fig. 2. The most striking feature of Fig. 2 is the large deviation of the 1964 map, it is often several meters higher than the other maps, but for some calibration points is lower. The discrepancy of the 1964 map, which averages 4.1 m high, is suggested by the field altitude check as well. The 1964 map tends to contaminate the means and standard deviations of calibration points for other maps, so these were recalculated without 1964; results are given in Table II.

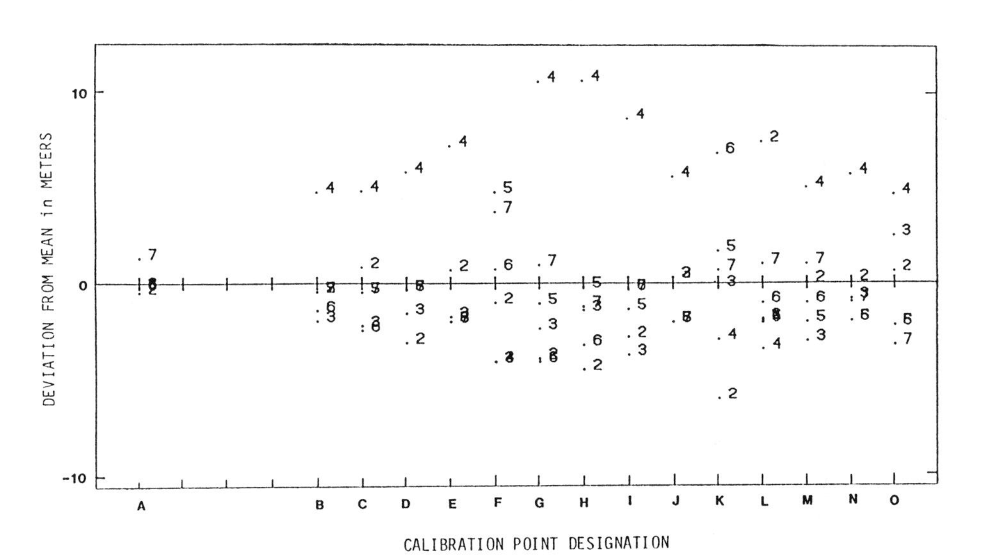

Figure 2. The deviation from the mean of the altitudes for each map for all 15 calibration points A-O (Fig. 1). The horizontal scale is arbitrary and the numerals forming columns above the point designators indicate the date of the map: 2 = 21 August 1958, 3 = 11 Sept 1961, 4 = 10 Sept 1964, 5 = 13 Sept 1977, 6 = 3 Oct 1980 and 7 = 24 Sept 1985. Each calibration point on each map is represented, but in some cases the digits are drawn over one another.

No trends (except for 1964) are readily apparent between the field altitude check and deviations from the mean of the calibration. For this reason the only datum correction applied was −4.1 m for 1964. It is estimated that each map datum is accurate to 1 m. The random error on each map is also estimated to be 1 m, but is less significant to the results. The datum accuracy of the 1970 synthesized map is estimated at 0.3 m.

Table II. Statistics of altitudes of calibration points on terrain around the perimeter of the glacier. The altitudes of 15 fixed points were compared on all maps to test for altitude datum consistency. The means and standard deviations from the mean altitudes at the calibration points are given here. The 1964 map was the only one to stand out as especially different (see Fig. 2), thus the calculations were made twice, including and excluding the 1964 map. After excluding the 1964 map, no other map was deemed to need a datum correction.

year	with 1964		without 1964	
	average dev. of mean	standard deviation	average dev. of mean	standard deviation
1958	-0.82	2.98	0.01	2.81
1961	-1.49	1.59	-0.67	1.81
1964	4.12	4.53		
1977	-0.66	1.77	0.16	1.50
1980	-0.99	2.37	-0.17	1.88
1985	-0.16	1.63	0.66	1.34

2.2 End of year correction

Ideally, the volume of the glacier would be measured at the end of the balance year. In practice, the photography or surveying required to obtain volume information must occur in fair weather before an autumn storm ends the balance year. The volume loss between the time of the volume measurement and the end of the balance year must be considered. This correction diminishes in importance as the number of years between measurements increases, and can also be minimized by making measurements as near as possible to the end of the balance year. All but one of the volume measurements at South Cascade were made 9 September or later (Table I), after which time it is unlikely that more than 0.5 m of ablation occurs on the lower, or more than 0.3 m on the upper glacier (these values are material thickness). Thus no corrections for subsequent ablation were made on the maps other than for 1958. The 1958 volume was determined from 21 August photography. Not only was this date much earlier than the others, but the ablation rate during the late summer and fall of 1958 was exceptionally high. An end of year correction was thus made, and is discussed under the mass balance section.

2.3 The volume determination from the maps

A topographic map is a wealth of information, but it is difficult to obtain volume information directly. For this analysis a grid system was used to convert topographic contours to more easily handled numerical values. A 100 m grid, parallel to a local Universal Transverse Mercator coordinate system, was superimposed on each map. The altitude at each grid point over the glacier (281 in 1958, 252 in 1985) was considered to be representative of the surrounding 10^4 m^2; no mental or other averaging within the hectare represented by each grid point was attempted. The bedrock map, determined by gravimetric (Krimmel, 1970), borehole, and radar (Hodge, 1979) measurements, was similarly reduced to gridded values. Errors in the bed estimation have no effect on the outcome of the balance-volume comparison. This gridding method is similar to that used by Haakensen (1986) and was found to introduce insignificant error. An example of the gridded map data is shown in Fig. 3.

The volume, V, of the glacier on each date, was determined by

$$V = \sum (S-B)_{i,j} \times 10^4 \text{ m}^2$$

over the set of grid points i,j on the glacier, S is the altitude of the glacier surface, and B is the altitude of the glacier bed. The glacier volumes are given in Table I.

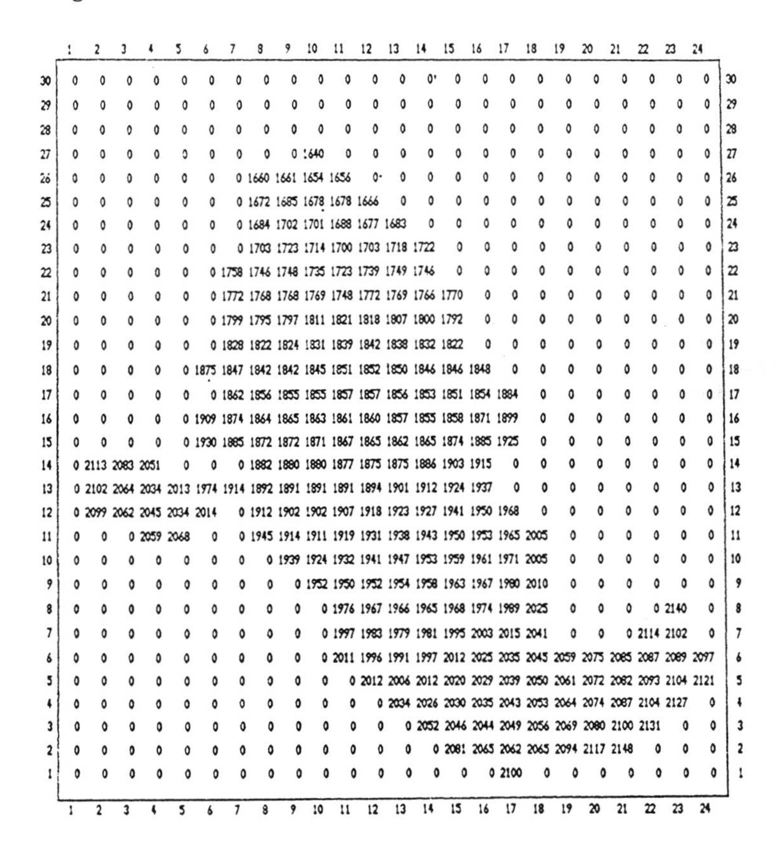

	1	2	3	4	5	6	7	8	9	10	11	12	13	14	15	16	17	18	19	20	21	22	23	24
30	0	0	0	0	0	0	0	0	0	0	0	0	0	0	0	0	0	0	0	0	0	0	0	0
29	0	0	0	0	0	0	0	0	0	0	0	0	0	0	0	0	0	0	0	0	0	0	0	0
28	0	0	0	0	0	0	0	0	0	0	0	0	0	0	0	0	0	0	0	0	0	0	0	0
27	0	0	0	0	0	0	0	0	0	1640	0	0	0	0	0	0	0	0	0	0	0	0	0	0
26	0	0	0	0	0	0	0	1660	1661	1654	1656	0	0	0	0	0	0	0	0	0	0	0	0	0
25	0	0	0	0	0	0	0	1672	1685	1678	1678	1666	0	0	0	0	0	0	0	0	0	0	0	0
24	0	0	0	0	0	0	0	1684	1702	1701	1688	1677	1683	0	0	0	0	0	0	0	0	0	0	0
23	0	0	0	0	0	0	0	1703	1723	1714	1700	1703	1718	1722	0	0	0	0	0	0	0	0	0	0
22	0	0	0	0	0	0	1758	1746	1748	1735	1723	1739	1749	1746	0	0	0	0	0	0	0	0	0	0
21	0	0	0	0	0	0	1772	1768	1768	1769	1748	1772	1769	1766	1770	0	0	0	0	0	0	0	0	0
20	0	0	0	0	0	0	1799	1795	1797	1811	1821	1818	1807	1800	1792	0	0	0	0	0	0	0	0	0
19	0	0	0	0	0	0	1828	1822	1824	1831	1839	1842	1838	1832	1822	0	0	0	0	0	0	0	0	0
18	0	0	0	0	0	1875	1847	1842	1842	1845	1851	1852	1850	1846	1846	1848	0	0	0	0	0	0	0	0
17	0	0	0	0	0	0	1862	1856	1855	1855	1857	1857	1856	1853	1851	1854	1884	0	0	0	0	0	0	0
16	0	0	0	0	0	1909	1874	1864	1865	1863	1861	1860	1857	1855	1858	1871	1899	0	0	0	0	0	0	0
15	0	0	0	0	0	1930	1885	1872	1872	1871	1867	1865	1862	1865	1874	1885	1925	0	0	0	0	0	0	0
14	0	2113	2083	2051	0	0	0	1882	1880	1880	1877	1875	1875	1886	1903	1915	0	0	0	0	0	0	0	0
13	0	2102	2064	2034	2013	1974	1914	1892	1891	1891	1891	1894	1901	1912	1924	1937	0	0	0	0	0	0	0	0
12	0	2099	2062	2045	2034	2014	0	1912	1902	1902	1907	1918	1923	1927	1941	1950	1968	0	0	0	0	0	0	0
11	0	0	0	2059	2068	0	0	1945	1914	1911	1919	1931	1938	1943	1950	1953	1965	2005	0	0	0	0	0	0
10	0	0	0	0	0	0	0	0	1939	1924	1932	1941	1947	1953	1959	1961	1971	2005	0	0	0	0	0	0
9	0	0	0	0	0	0	0	0	0	1952	1950	1952	1954	1958	1963	1967	1980	2010	0	0	0	0	0	0
8	0	0	0	0	0	0	0	0	0	0	1976	1967	1966	1965	1968	1974	1989	2025	0	0	0	0	2140	0
7	0	0	0	0	0	0	0	0	0	0	1997	1983	1979	1981	1995	2003	2015	2041	0	0	0	2114	2102	0
6	0	0	0	0	0	0	0	0	0	0	2011	1996	1991	1997	2012	2025	2035	2045	2059	2075	2085	2087	2089	2097
5	0	0	0	0	0	0	0	0	0	0	0	2012	2006	2012	2020	2029	2039	2050	2061	2072	2082	2093	2104	2121
4	0	0	0	0	0	0	0	0	0	0	0	0	2034	2026	2030	2035	2043	2053	2064	2074	2087	2104	2127	0
3	0	0	0	0	0	0	0	0	0	0	0	0	0	2052	2046	2044	2049	2056	2069	2080	2100	2131	0	0
2	0	0	0	0	0	0	0	0	0	0	0	0	0	0	2081	2065	2062	2065	2094	2117	2148	0	0	0
1	0	0	0	0	0	0	0	0	0	0	0	0	0	0	0	0	2100	0	0	0	0	0	0	0

Figure 3. Gridded map data for the 24 Sept 1985 South Cascade surface, to the nearest meter above sea level. The grid is oriented with north to the top. Grid cells are 100 × 100 m, and are compressed slightly in the north-south in this figure.

3. MASS BALANCE

The mass balance of South Cascade Glacier has been measured each year by traditional methods, that is, stake and snow density measurements, beginning in 1958. Results of the work are reported in Meier and Tangborn (1965), Meier et al. (1971), Tangborn et al. (1976), and Krimmel et al. (1980). The balances reported here are the glacier-integrated net balance, given as a one-dimensional quantity in meters of water equivalent (Mayo et al., 1972).

The methodology used has evolved over the period of record. The stake placement density from 1958 through 1973 was more than 10 km^2, and was reduced to 2 or 3 km^2 from 1973 through 1982. From 1958 through 1982 the net balance values at the stakes were contoured by hand using glaciological judgement and knowledge of accumulation and ablation patterns. The glacier average net balance was determined with a planimeter or a 100 m grid similar to that earlier described in this paper. For 1983-85 contoured balance maps have not been produced from the low density stake network, and the glacier net balances were determined from a single stake near the equilibrium line using a method of simple linear regression (Fig. 4).

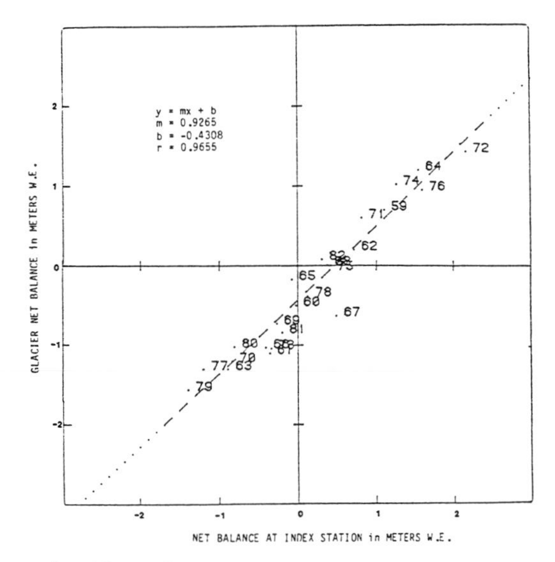

Figure 4. The glacier net balance as a function of the net balance at an index station (Fig. 1) near the equilibrium line of South Cascade Glacier. For the 24 years shown here, 1959-82, the glacier net balance was determined by measuring the area within balance contour intervals from maps produced using data from numerour measurements of balance. This regression equation was used to determine the glacier net balance from the index station balance only, for the years 1983-85.

The errors associated with balance values for the 27 years are estimated considering the method of balance calculation, the stake placement density, and the amount of snow or firn from the previous winter remaining at the end of the balance year. When there is a large amount of residual material (a positive net balance), the error is larger because of difficulties in determining the summer horizon. Balance values and errors are given in Table III.

Table III. The net balance of South Cascade Glacier, 1959-85. The year given refers to the year in which the balance year ends; thus, 1959 refers to the mass change from the autumn of 1958 to the autumn of 1959. The glacier area is determined from a grid cell count. The net balance is the area-integrated mass change from the end of one balance year to the end of the next balance year, the net balance. The error in net balance is estimated and includes consideration of the stake net density, the method to integrate point measurements, and the amount of residual snow. The cumulative balance is referenced to 21 August 1958, after which there was -1.5 m mass change before the end of the balance year. Assuming that the errors are uncorrelated, then + or -0.70 is the error to associate with the sum of the balances.

year	area $\times 10^4$ m^2	balance (m)	error (m)	cumulative balance
1959	278	0.70	0.10	-0.80
1960	275	-0.50	0.10	-1.30
1961	273	-1.10	0.10	-2.40
1962	271	0.20	0.10	-2.20
1963	269	-1.30	0.10	-3.50
1964	267	1.20	0.10	-2.30
1965	266	-0.17	0.12	-2.47
1966	265	-1.03	0.10	-3.50
1967	265	-0.63	0.10	-4.13
1968	264	0.01	0.10	-4.12
1969	263	-0.73	0.10	-4.85
1970	262	-1.20	0.10	-6.05
1971	261	0.60	0.10	-5.45
1972	261	1.43	0.20	-4.02
1973	260	-1.04	0.10	-5.06
1974	259	1.02	0.16	-4.04
1975	258	-0.05	0.10	-4.09
1976	258	0.95	0.16	-3.14
1977	257	-1.30	0.12	-4.44
1978	256	-0.38	0.12	-4.82
1979	256	-1.56	0.12	-6.38
1980	255	-1.02	0.12	-7.40
1981	254	-0.84	0.12	-8.24
1982	254	0.08	0.12	-8.16
1983	253	-0.76	0.23	-8.92
1984	253	0.13	0.23	-8.79
1985	252	-1.17	0.23	-9.96

This study is concerned with the balance and volume change from 21 August 1958 to 24 September 1985. There was insignificant balance change between 24 September 1985 and the end of the 1985 balance year. There was, however, major ablation between 21 August 1958 and the end of the 1958 balance year. Meier and Tangborn (1965) accounted for this ablation in their 1985-61 comparison, but were not explicit as to how it was calculated. Meier and Tangborn did give enough data to calculate the change in balance after 21 August 1958; this calculation requires a value for: the volume of ice loss from 21 August 1958 to 12 September 1961, 7.2×10^6 m^3 (which is equal to 6.5×10^6 m^3 when converted to mass change, ΔM, by multiplication by 0.9); the net balances of 1959-61, which were 0.7, -0.5, and -1.1 m respectively; and the area of the glacier, 2.7×10^6 m^2. The balance change, B, from 21 August 1958 to the end of the 1958 balance year can be estimated from

$$B = \Delta M/A + \sum_{1959}^{1961} b_t$$

where A is the area, and b_t are the mass balances for each year. B thus calculated is -1.5 m, a value which seems reasonable considering the weather of the summer and fall of 1958.

4. BALANCE-VOLUME COMPARISON

We shall relate the cumulative net balance and volume changes to each other by adjusting the volume changes to water equivalent averaged over the glacier area. This requires a division of the volume change by the average glacier area within the time interval and multiplication by a factor to convert ice to water equivalent. If the density-depth profile at the initial and final times is the same over the entire glacier, e.g. Sorge's Law is followed, it can be assumed that any volume change between the two times was due to change in volume of ice; thus, 0.9 is used to convert to water equivalent. The cumulative mass balance and density corrected volume changes since 21 August 1958 are shown in Fig. 5.

First, consider only the entire 1958-85 interval. Total loss estimated from traditional balance calculations is 10.0 m, while loss estimated by volumetric methods is 12.7 m. This difference could be explained if the net balance was systematically mis-measured by 0.1 m/year, or could be partially explained by a systematic map error of -1 m in 1958 and +1 m in 1985. There is, however, another possibility. Over the 27 year period, there has been a net loss of material, the density of which was assumed to be 0.9. This is probably a correct assumption for the ablation zone, where the surface material was ice initially and finally, and for the upper accumulation zone where over the 27 years there was a small increase in altitude of the glacier surface (a volume gain); in the upper accumulation zone Sorge's Law is followed. There is, however, a large area of the glacier between the ablation and the upper accumulation zones in which there has been a nearly continual mass loss;

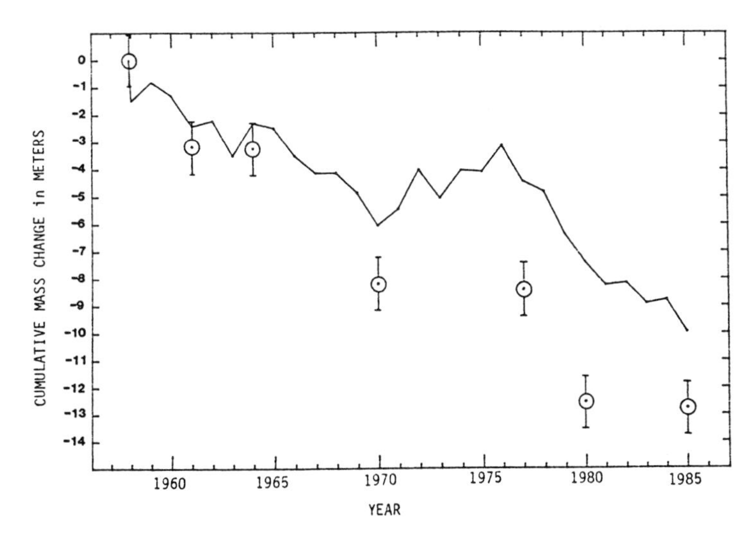

Figure 5. The cumulative net balance for South Cascade, in meters water equivalent, since 21 August 1958 (solid line). The balance change from 21 August 1958 to the end of the 1958 balance year, -1.5 m, is only a portion of the 1958 mass loss. For the years 1959-85 the mass change indicated is for the entire balance year.
The water equivalent volume loss, averaged over the glacier area, as determined from topographic maps since 21 August 1958 is shown as circles. Ice volume was converted to water volume by multiplication by 0.9. The 1964 map datum is shifted -4.1 m.

it has been firn that has melted in most years, and this firn has not been replaced. In the mid-zone the material lost had a density of something less than ice. A core through several years of material in the upper accumulation zone of South Cascade Glacier showed firn density from 0.6 to 0.8 (Fig. 6).

It is possible to estimate the density of material that was melted off, ρ_f, by allowing it to be residual, by using

$$B_T = V_T S_i \rho_i + V_T S_a \rho_i + V_T S_f \rho_f$$

$$\rho_f = [B_T/V_T - \rho_i(S_i + S_a)]/S_f$$

where B_T is the total balance change over the interval, S_i, S_a and S_f are fractional areas of the ablation, accumulation, and firn zones, V_T is the volume change over the interval, and ρ_i is the density of ice. Using values from South Cascade of B_t = -10.0 m, S_i = 1/4, S_a = 1/4,

$S_f = 1/2$, $V_T = -14.1$, and $\rho_i = 0.9$, then $\rho_f = 0.5$, a value only slightly below that expected for firn density.

The 27 year interval can be subdivided into six shorter intervals, and all the same comparisons made. Judging from figure 5, the first three intervals give reasonable results. But from 1970-77, a period during which 1.6 m of water equivalent of material was added to the glacier, the volume measurements show a 0.2 m loss. And given that it was low density material (young firn, ~0.6 density) added in the 1970-77 interval, the volume should have shown an even greater increase than if ice was added. The 1977-80 and 1980-85 intervals show fairly good agreement between the mass balance and volumetric changes.

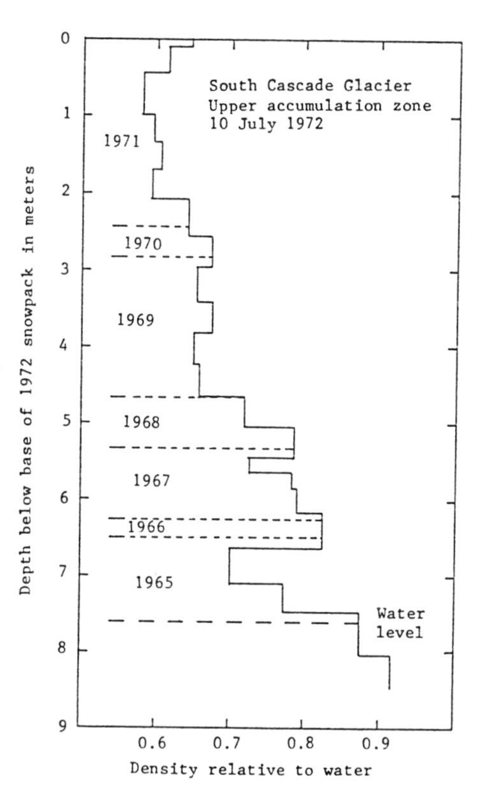

Figure 6. Firn-density profile, upper accumulation zone, South Cascade Glacier, 10 July 1972. The vertical scale is in meters below the base of the 1972 snowpack. The year the material was deposited is indicated. At the location of this core, there was some firn from each year after 1965; on much of the area above the ice (ablation) zone, but below the location of this core, material from several years was completely melted away.

5. CONCLUSIONS

In a gross sense, the traditional mass balance measurements and the photogrammetric volume measurements of South Cascade Glacier are consistent, as both indicate major mass loss for the 27 year period. Unfortunately, the errors are large enough that comparison of the two methods does not significantly improve the mass balance results. For the 27 year interval at South Cascade, there is good agreement when the volume change is adjusted for the density of ice and firn.

Comparisons for shorter intervals within the 27 year period are not as good, and tend to make the total interal comparison suspect. Several things are evident: 1) The cartography must be very carefully done, with special attention given to the absolute datum, 2) the ratio of map accuracy to balance change should be considered when investing in maps for the purpose of balance-volume comparison, and 3) density profiles should be obtained through the firn layers near the time of the aerial photography of surveying that is used to compile the maps.

A much higher accuracy can be obtained using theodolite surveys rather than photogrammetically produced maps, and is a preferable method, but obviously not retrospectively possible. The need for density profiles during periods when the glacier is not in equilibrium is evident, and density profiles will be especially important in conjunction with the anticipated satellite altimeter data (which may be repeatable to within a few centimeters) for the Greenland and Antarctic ice sheets.

REFERENCES

Haakensen, N. 1986. Glacier mapping to confirm results from mass-balance measurements. Annals of Glaciology 8, 73-77.

Hodge, S.M. 1979. Direct Measurement of basal water pressures: progress and problems. Journal of Glaciology 23, no. 89, 309-319.

Krimmel, R.M. 1970. Gravimetric ice thickness determination, South Cascade Glacier, Washington. Northwest Science 44, no. 3, 147-153.

Krimmel, R.M., Tangborn, W.V., Sikonia, W.G. and Meier, M.F. 1980. Ice and water balances, South Cascade Glacier, 1957-1977: Data of Glaciological Studies (Moscow, Academy of Sciences of the U.S.S.R., Section of Glaciology of the Soviet Geophysical Committee and Institute of Geography), Publication no. 38, Reports of the International Symposium on the Computation and Prediction of Runoff from Glaciers and Glacial Areas (Tbilisi, 1978), 143-147, 217-219.

Mayo, L.R., Meier, M.F. and Tangborn, W.V. 1972. A system to combine Stratigraphic and annual mass balance systems. Journal of Glaciology 61, no.11, 3-14.

Meier, M.F., Tangborn, W.V., Mayo, L.R. and Post, A. 1971. Combined ice and water balances of Gulkana and Wolverine Glaciers, Alaska, and South Cascade Glacier, Washington, 1965 and 1966 hydrologic years. U.S. Geological Survey Professional Paper 715-A, 23 pp.

Paterson, W.S.B., 1981. The Physics of Glaciers, 2nd edition, Pergamon Press Ltd. New York.

Slama, C.C. 1980. Manual of Photogrammetry, 4th Edition, Falls Church, VA, American Society of Photogammetry.

Tangborn, W.V., Krimmel, R.M. and Meier, M.F. 1975, A comparison of glacier mass balance measurements by glaciologic, hydrologic and mapping methods, South Cascade Glacier, Washington. International Association of Hydrological Sciences General Assembly 1971, IAHA-AISH Publication no. 104, 185-196.

Tangborn, W.V., Mayo, L.R., Scully, D.R. and Krimmel R.M. 1976. Combined ice and water balances of Maclure Glacier, California, South Cascade Glacier, Washington and Wolverine and Gulkana Glaciers, Alaska, 1967 hydrologic year. U.S. Geological Survey Professional Paper 715-B, 20 pp.

FIELD STATIONS FOR GLACIER-CLIMATE RESEARCH, WEST GREENLAND

Ole B. Olesen and Roger J. Braithwaite
The Geological Survey of Greenland
Copenhagen, Denmark

ABSTRACT

During the past ten years, the Geological Survey of Greenland (GGU) has measured glacier mass-balance parallel with basic climate elements. Field stations were operated near the margin of the Greenland ice sheet in Johan Dahl Land 1978-1983, at Qamanârssûp sermia 1980-1986 and at Tasersiaq since 1982. Measurements of snow accumulation and ice ablation are made at many stakes drilled into the ice. Ablation measurements are made daily at the stakes closest to the stations while more distant stakes are only measured a few times during the season. The climate measurements include air temperature, precipitation, wind, humidity, evaporation, sunshine duration and shortwave radiation. The field stations are expensive to operate but it will be possible to change over to cheaper field programmes in the future by relying on automatic instruments and by stressing "index" measurements.

1. INTRODUCTION

This paper outlines research which was started by the Geological Survey of Greenland as part of hydropower investigations but has also resulted in new information on relations between glaciers and climate in Greenland. The investigations include parallel mass balance and climate measurements at field stations established close to glaciers. The emphasis is therefore on the left-hand side of the diagram in Fig. 1 which illustrates the extent of the glacier-climate problem. As

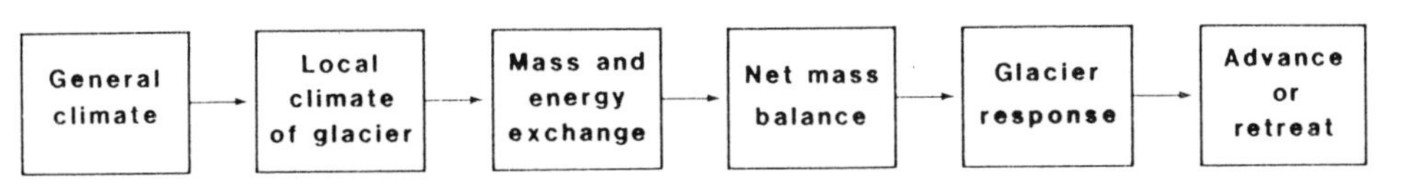

Figure 1. Block diagram illustrating glacier-climate relations, after Meier (1965) and Paterson (1981, p.299).

J. Oerlemans (ed.), Glacier Fluctuations and Climatic Change, 207–218.

conditions vary greatly from year to year, the measurements are made for periods of a few years at each site. This allows assessment of average conditions as well as the relations between glaciers and climate under varying conditions. The basic philosophy stresses simple measurements which can be repeated over a few years rather than more detailed studies which are difficult to maintain. The guide by Østrem & Stanley (1969) was useful for the initial planning of the work.

2. THE STATIONS

2.1 Locations

The locations and periods of operation of the three manned glacier-climate stations are shown in fig. 2 and Table I together with locations of newer unmanned stations.

Table I. Locations of glacier-climate stations operated in West Greenland by the Geological Survey of Greenland.

STATION	LAT. North	LONG. West	ELEVATION m a.s.l.	PERIOD
Johan Dahl Land	67°27'	45°22'	850	1978-1983
Isortuarsuup Tasia*	63°52'	49°32	1095	1984-
Qamanârssûp sermia	64°29'	49°29'	750	1980-1986
Qamanârssûp sermia*	64°29'	49°29'	760	1979-
Tasersiaq	66° 7'	50° 7'	950	1982-
Paakitsup Akuliarusersua*	69°29'	50° 8'	440	1984-
Paakitsup Akuliarusersua*	69°30'	49°47'	910	1986-

* Automatic climate station in cooperation with the Greenland Technical Organization.

The first manned station was established in Johan Dahl Land, South Greenland, in late summer 1977 as part of investigations for hydropower (Olesen, 1978). The station was located close to the margin of Nordbogletscher which is a tongue of the Greenland ice sheet and was manned for six summers of fieldwork until closed in late summer 1983 (Clement, 1981, 1982, 1983 and 1984).

The second manned station was established in late summer 1979 (Olesen, 1981) at the margin of Qamanârssûp sermia which is a tongue of the ice sheet, reaching nearly to sea level. The station was manned for seven summers 1980-1986 (Olesen and Braithwaite, 1982; Braithwaite, 1983, 1984, 1985 and 1986a) and will continue for several years more using an automatic climate station (Braithwaite, 1987).

The third manned station was established at Tasersiaq, West Greenland, in late 1981 (Olesen, 1982). The station is located close to

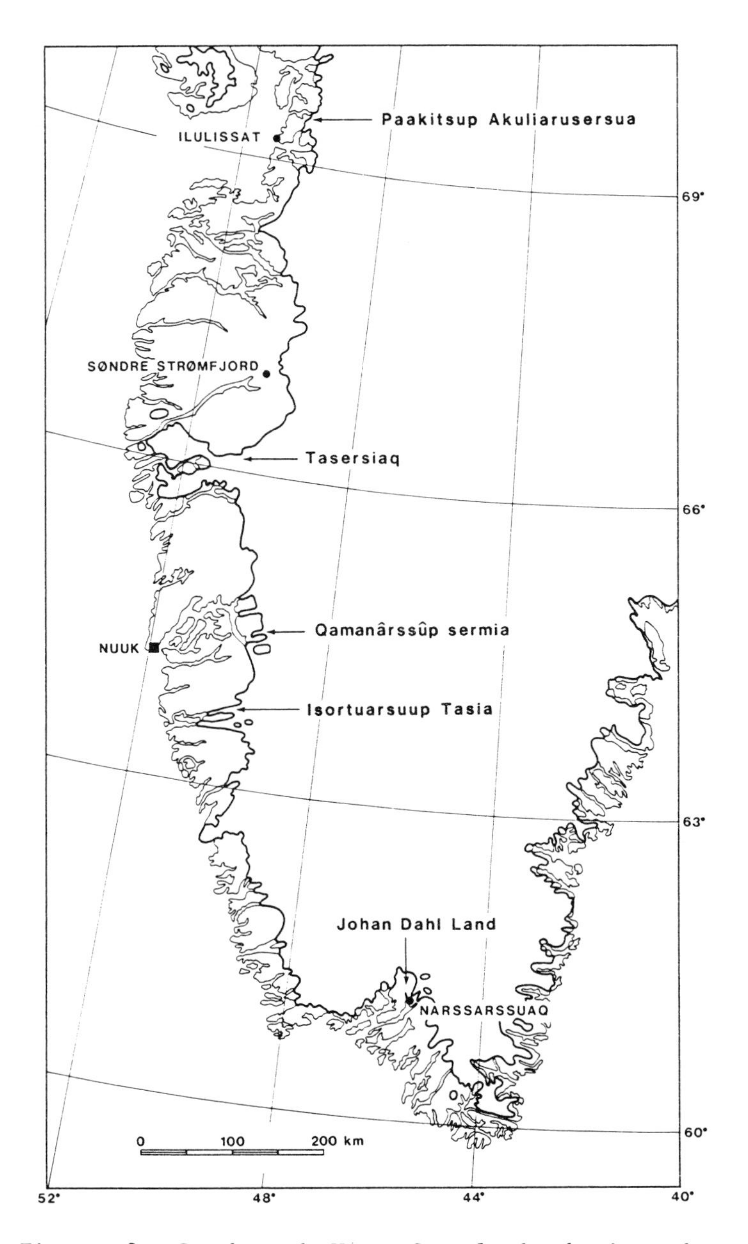

Figure 2. South and West Greenland showing the location of glacier-climate stations operated by the Geological Survey of Greenland. Johan Dahl Land, Qamanârssûp sermia and Tasersiaq are manned stations while Isortuarsuup Tasia and Paakitsup Akuliarusersua use automatic stations

to the margin of the Amitsulôq ice cap which drains into Tasersiaq, a lake which is one of the largest hydropower potentials in Greenland. The station has been manned every summer since 1982 (Olesen and Andreasen, 1983; Olesen, 1985 and 1986) and will continue in operations for several more years.

2.2 Facilities

The facilities at the manned stations provide comfortable accomodation for the field parties and thereby contribute to the quality of data collected. The older stations in Johan Dahl Land and at Qamanârssûp sermia are built around caravan trailers while the last station at Tasersiaq uses a pre-fabricated hut. In each station, there is accomodation for 3 to 5 persons, a small workshop for storing and repairing equipment and a latrine. The stations are equipped with radio-telephones for contacts with each other, and with marine radio stations.

2.3 Logistics

The stations are far from towns and existing infrastructure of roads, harbours and airports. The stations were established, and are operated, with the help of helicopters chartered from the nearest airport. This is expensive and logistics represent the largest item on the budget, see Table II.

The field programme at the manned stations requires activity over the whole ablation season repeated for a few years, e.g. six years in Johan Dahl Land and seven years at Qamanârssuûp sermia. With personnel, food, equipment and spare parts, the total weight of mobilizing a glacier climate station for one summer approaches two tons. This is transported to the station at the beginning of the season by a medium sized helicopter (Sikorski S-61N) from the nearest airport.

Smaller helicopters, e.g. Bell-206 Jetrangers, are used for local transport around the staions in areas which are otherwise inaccessible due to crevasses or slush fields. For example, crevasses on the Inland Ice were serious problems in Johan Dahl Land and at Qamanârssûp sermia whilst surface transport by snow machines is relatively safe at Tasersiaq.

2.4 Manpower

All three stations were established and lead initially by Olesen. The Johan Dahl station was lead by P. Clement in the four summers 1980-1983, Braithwaite lead the work at Qamanârssûp sermia 1981-1986, and Olesen continues to lead at Tasersiaq. The stations are manned every summer by 4-5 students each hired for about two months.

The field programme is designed to be carried out by 2-3 persons and the stations are dimensioned accordingly. Three is the minimum number for safety when travelling on glaciers but stations are operated in quieter periods by only two men. The station leaders always oversee the beginning and end of the field season in May and September although routine operations during the middle of the season are often delegated to students.

Personnel costs are the second biggest item on the budget, see Table II. The field work also involves a commitment of about two man-months by each station leader as well as time for data analysis after the field season.

Table II. Breakdown of running costs for a manned glacier-climate station in Greenland based on an average for two stations for the years 1982-1986.

ITEM	% OF TOTAL	
Logistics	44	
(Freight & helicopter charter)		
Personnel costs		40
(Wages & food)		
Travel		11
(Air tickets & hotels)		
Consumables		2
(Fuel & materials)		
Equipment	4	
(Replacement & repairs)		

2.5 Costs

A breakdown of the running costs of a typical glacier climate station is given in Table II. The figures are averages for the Qamanârssûp sermia and Tasersiaq stations for the years 1982 till 1986 inclusive and indicate the "normal" running costs as capital costs of the stations were paid in previous years, i.e. in 1979 for Qamanârssûp sermia and 1981 fro Tasersiaq. The largest items on the budget are respectively for logistics and for personnel with only a minor sum for consumables like fuel, stakes, flags as well as replacements and repair of equipment. The average running cost per station for one summer is about $ 54,000.-- to $ 81,000.-- respectively at 1985 and 1986 exchange rates and the capital cost of a new station would be about $ 17-25,000.-- depending on exchange rate.

3. GLACIOLOGICAL PROGRAMME

The glaciological programme includes measurements of snow accumulation, ablation and ice movement at stakes drilled into the ice at many points in the general neighbourhood of the stations. Changes in surface level due to accumulation and ablation are registered by measuring from the top of the stake to the ice or snow surface. Horizontal and vertical movements of the ice are measured by theodolite survey from fixed points on the ice-free land while longer term glacier variations are evaluated by repeated mapping using terrestrial photogrammetry.

3.1 Stakes and drills

The stakes are standard aluminium tubes 5 or 6 metres long with a diameter of 3.2 cm which are placed in holes drilled in the ice. The stakes are marked with orange or black flags to make them easier to find. The stakes can be fitted with 2 metre extensions for extra length, e.g. in the autumn so that the top of the stake will be visible in following spring after accumulation of 1-3 metres of snow.

The standard drilling equipment was designed by Olesen and comprises toothed and spoon bits, extensions, a two-stroke motor, tools and spare parts. The total equipment weighs about 30 kg and is carried on packframes by two persons or by helicopter in less accessible areas. A hole 5-6 metres deep can be drilled in 15-60 minutes depending on ice conditions, e.g. very wet ice is difficult to drill in. A newly constructed steam drill, tested at Tasersiaq, will do the same job in 10 to 15 minutes almost regardless of ice conditions. Stakes must be redrilled as they melt out and for most sites a redrilling once a year is enough although the high ablation rate on the lower tongue of Qamanârssûp sermia makes more frequent redrillings necessary, e.g. in the very warm summer of 1985 (Braithwaite, 1986a).

3.2 Snow accumulation and ablation

Accumulation of winter snow in water equivalent is determined from the depth of snow and its density as measured in the late spring, e.g. in May and ideally before the onset of melting. The snow depth around the stake is measured using a Swiss snow probe, taking care to measure possible ice accumulation by refreezing of meltwater from the snow cover. The snow density is determined by weighing snow samples from the wall of the snow-pit. In Greenland, as in other places, the snow density at any time does not vary greatly from place to place so it is only necessary to determine density at a few stakes and interpolate values for other stakes.

The ablation of snow is seldom observed directly in either the ablation or accumulation area. The ablation area is an uncomfortably wet and dangerous place during the period of snowmelt because of meltwater swamps, streams, water-filled cracks and weakened crevasse bridges. It is therefore natural to measure snow accumulation before melting starts and to set snow ablation in the lower ablation area equal to accumulation.

The above is valid for the melting of snow on ice. The situation for the melting of snow on firn, as in the lower accumulation area, is more difficult. This is because meltwater from the snow surface percolates down into the underlying firn and refreezes when it encounters temperatures below the freezing point. This process results in internal accumulation (Trabant and Mayo, 1985), sometimes to depths of several metres. The amount of refreezing is hard to measure, especially if one has a short time on-site as in a helicopter-borne operation. It is therefore difficult to determine annual balances in the lower accumulation area and some published results are incorrect. Our present opinion is that refreezing on the Greenland ice sheet is so strong that ablation, i.e. melting followed by runoff, is reduced to zero a little above the annual snow line.

3.3 Ice ablation

Ablation of ice is determined by measuring the lowering of the ice surface and assuming a constant density for ice. However, the latter assumption is not entirely correct as the density of the surface layer of the glacier fluctuates according to weather conditions. For example, a whitish ablation crust several centimetres deep often develops in sunny weather due to melting along grain boundaries and disappears again under rainy or cloudy conditions. Although important on a short-term basis, these density variations have little effect on seasonal ablation totals.

The areas covered by ablation stakes around the manned stations are relatively large, e.g. the ablation area at Qamanârssûp sermia covers about 150 km^2. This means that the more remote stakes, e.g. those closest to the equilibrium line on the ice sheet, can only be visited a few times during the summer or, in the worst case, only in spring and late summer. Such remote stakes are often visited by helicopter and measurements are sometimes missed because of bad weather and other problems. However, there are stakes close to the glacier edge and near enough to the base camps to be reached in a few minutes walk. These are the so-called "daily stakes" which are measured every day throughout the ablation season and have essentially the same climate as the base camps. Glacier climate relations can be studied therefore by comparing ablation at the daily stakes with climate at the base camps.

4. CLIMATOLOGICAL PROGRAMME

The programme at the manned stations is designed so that it can be carried out by a small field party. This means that the climatological observations are simplified to allow personnel to do other jobs requiring absences from the base camp. e.g. measuring balances at remote stakes or surveying ice movement. The programme at the base camps is based upon simple recording instruments, supplemented by hand observations twice daily (morning and evening).

The base camps are supplemented by nearby secondary stations consisting of temperature recorders and precipitation guages which are visited at intervals of a few days to several weeks to reset the recorders and to read the guages. For example, such stations are established on glaciers near to each of the three manned stations at roughly the same elevations to measure "glacier effects" or a few hundred metres higher to measure vertical temperature gradients.

4.1 Manned stations

Air temperature is recorded continuously by a Lambrecht thermohyrograph exposed in a standard instrument shelter at 2 metres above ground-level. The temperature recordings are checked and corrected by twice daily manual readings of present temperature as well as maximum and minimum temperatures. The temperature recorders are generally reliable but corrections of up to 1°C are often indicated by the hand readings.

The precipitation is measured by two Hellman precipitation guages mounted about 1 metre above ground and emptied twice daily. One guage is unshielded while the other one a few metres away is protected with a Woelfle shield.

A Lambrecht cup anemometer with a simple counter is mounted 5 metres above the ground. The run-of-wind is read from the counter twice daily so that average wind speeds are calculated for 12 and 24 hour intervals. An instantaneous wind direction, indicated by a wind vane, is observed twice daily.

Global radiation, i.e. shortwave radiation from sun and sky, is recorded with a Belfort actinograph. This is supplemented with daily measurements of sunshine duration using a Campbell-Stokes recorder mounted 1.5 metres above ground-level in a well-exposed site.

The relative humidity of the air is recorded in parallel to air temperature by thermohygrograph. Some checks have been made on the recorded relative humidity using a psychrometer but errors are generally small.

Potential evaporation is measured by two different methods. One uses a Lambrecht clockwork recorder to register the loss of weight of a small dish filled with water and exposed in the instrument shelter beside the thermohygrograph. The other method uses twice-daily readings of water level in a large water-filled pan similar to a Class A evaporation pan.

Table III. Data collection by manned and unmanned glacier-climate stations at Qamanârssûp sermia.

DATA	Time-scale	Manned Station	Unmanned Station
Ablation, near base camp	Daily	Yes	No
	Seasonal	Yes	Yes[1]
Ablation, other stakes	Seasonal	Yes	Yes[1]
Ice movement	Seasonal	Yes	No
Air temperature	Daily	Yes	Yes[2]
Wind speed	Daily	Yes	Yes[2]
Precipitation	Daily	Yes	Yes[2]
Humidity	Daily	Yes	Yes[2]
Global radiation	Daily	Yes	No
Sunshine duration	Daily	Yes	No
Potential evaporation	Daily	Yes	No

[1] Measured on brief visits in the spring and autumn
[2] Data collection by automatic climate station

4.2 Automatic stations

When the first station was established in 1977 in Johan Dahl Land, installation of an automatic climate station was contemplated but had to be rejected because of high capital cost. Furthermore, there is glaciological data that can not readily be measured by automatic instruments. It was therefore decided to rely upon the glaciological field personnel to collect climate data during the summer. However, a Danish-built automatic climate station was tested close to the manned station at Qamanârssûp sermia and has performed better than expected. It is not wholly fail-safe as complete data have only been recorded for four out of the seven years since September 1979 but this has been sufficient to establish year-round temperature variations which are important for understanding ablation variations.

The manned data collection at Qamanârssûp sermia was dicontinued in favour of the automatic station at the end of th 1986 field season (Braithwaite, 1987), see Table 3 for a comparison of the two systems. The latest stations to be established in Greenland are also unmanned. For example, three new stations have been established since 1984 on the ice sheet using the same type of automatic station as at Qamanârssûp sermia but equipped only with temperature sensors, see Table I and fig. 2 for the locations of these stations.

5. LESSONS LEARNT

5.1 Mass balance concepts

The establishment of well-defined and accepted concepts of mass balance study (Anonymous, 1969) was one of the big achievements of modern glaciology. The present studies in Greenland were therefore started with the good intention of following these concepts but they have been found too difficult to apply. For example, modern mass balance concepts include both the mass balance at a point, i.e. specific balance and the mass balance of the whole glacier. The second concept is difficult to determine under Greenland conditions because it requires knowledge of the variation of specific balance over the whole glacier and of the areal distribution of the glacier both of which are unknown for sectors of the ice sheet. For example, in Johan Dahl Land and at Qamanârssûp sermia, the specific balance measurements are essentially confined to the ablation area (and lower accumulation area where measurements are inaccurate) because the accumulation area is so far from the stations while the altitudinal distributions of ice sheet sectors are poorly known because of shortcomings of present maps. Mass balances cannot therefore be calculated for "glaciers" represented by sectors of the Greenland ice sheet.

5.2 Index stakes

Although the density of stakes in the ablation area is very low, i.e. a few square kilometres per stake instead of a few stakes per square kilometre, some useful regularities have been found. For example, specific balances approximately satisfy a simple version of the model by Lliboutry (1974) whereby space and time variations of balance can be separated (Braithwaite, 1986b). This means that time variations in the balance of a certain area can be determined even if the absolute values of the balance are unknown. It is in fact the former which are important for correlations with climate and with runoff.

The above leads to the concept of index stakes which are sparsely distributed over a glacier to measure the time variations in ablation and accumulation on the glacier although they are too few to measure the mass balance of the glacier as a whole (Braithwaite, 1986b, Reynaud et al., 1986). Further studies of the index stake concept are required to find out the optimum distribution of such stakes and to study the homogenuity of the data as the stakes move through varying local topography. Future index stakes should ideally be combined with automatic stations (see below).

5.3 Automatic climate stations

Although correct in its time, the decision to rely on manned stations has been re-examined in the light of the statisfactory performance of the automatic station at Qamanârssûp sermia. It is now clearer than before that the most important climate element for ablation is temperature which is easier to measure than precipitation, wind and radiation. The running cost of the unmanned station at Qamanârssûp sermia will be less than one-third than that of the manned station. The changeover from manned to unmanned operations is therefore favourable for costs versus benefits. However, in the long term, development of an automatic sensor for measuring ice ablation rate is also needed so that fully automatic stations can be deployed to measure ablation and climate in parallel. Recently developed solid-state recorders might be used for such stations as they are cheaper than the present automatic stations using tape recorders. They are also lighter so the old dream of a complete glacier climate station mounted on a single stake may soon be realized.

5.4 Glacier-climate relations

The most complete information gained by the present programme refers to relations between ablation and climate in the lower ablation area where there is little or no snow ablation. In particular there are strong correlations between ice ablation and positive temperature sums while increases in snow accumulation reduce ice ablation, see Braithwaite & Olesen (this volume). This finding is useful for planning hydropower projects as most runoff for proposed sites comes from the lower to middle ablation zone. However, the upper ablation zone is probably of greater importance for the mass balance of the Greenland ice sheet as a whole and it would be desirable to get more data from there. For example, the glacier-climate relation is frequently parameterized by variations of the equilibrium line altitude. However, ablation conditions in the upper ablation zone are complicated as snow ablation is of increasing importance compared with ice ablation while travel is more difficult during the snowmelt period. One future possibility would be to rely on remote sensing to monitor conditions during the snow ablation period although this will not be as easy as it sounds. For example, a first requirement would probably be to improve existing maps of the Greenland ice sheet which are greatly in error at elevations around the normal equilibrium line altitude.

ACKNOWLEDGEMENTS

This paper is published by permission of the Director, the Geological Survey of Greenland. The field station in Johan Dahl Land was partly funded by the European Economic Community (EEC) and by the Danish Energy Ministry. The Qamanârssûp sermia and Tasersiaq stations are wholly funded by the Geological Survey of Greenland. Poul Clement lead the work in Johan Dahl Lland 1980-1983. The automatic climate stations are operated in co-operation with the Greenland Technical Organization.

REFERENCES

Anonymous, 1969. Mass-balance terms. Journal of Glaciology 8, (52), 3-7.

Braithwaite, R.J. 1983. Glaciological and climatological investigations at Qamanârssûp sermia, West Greenland. Rapport Grønlands geologiske Undersøgelse 115, 111-114.

Braithwaite, R.J. 1984. Glaciological and climatological investigations as Qamanârssûp sermia, West Greenland. Rapport Grønlands geologiske Undersøgelse 120, 109-112.

Braithwaite, R.J. 1985. Glacier-climate investigations in 1984 at Qamanârssûp sermia, West Greenland. Rapport Grønlands geologiske Undersøgelse 125, 108-112.

Braithwaite, R.J. 1986a. Exceptionally high ablation in 1985 at Qamanârssûp sermia, West Greenland. Rapport Grønlands geologiske Undersøgelse 130, 126-129.

Braithwaite, R.J. 1986b. Assessment of mass-balance variations within a sparse stake network, Qamanârssûp sermia, West Greenland. Journal of Glaciology 32 (110), 50-53.

Braithwaite, R.J. 1987. Was 1986 the last year of glacier-climate investigations at Qamanârssûp sermia, West Greenland? Rapport Grønlands geologiske Undersøgelse.

Clement, P. 1981. Glaciological investigations in Johan Dahl Land 1980, South Greenland. Rapport Grønlands geologiske Undersøgelse 105,62-64.

Clement, P. 1982. Glaciological investigations in connection with hydropower, South Greenland. Rapport Grønlands geologiske Undersøgelse 110, 91-95.

Clement, P. 1983. Mass balance measurements on glaciers in South Greenland. Rapport Grønlands geologiske Undersøgelse 115, 118-123.

Clement, P. 1984. Glaciological activities in the Johan Dahl Land area, South Greenland, as a basis for mapping hydropower potential. Rapport Grønlands geologiske Undersøgelse 120, 113-121.

Lliboutry, L. 1974. Multivariate statistical analysis of glacier annual balances. Journal of Glaciology 13 (69), 371-392.

Meier, M.F. 1965. Glaciers and climate. In: Wright, H.E. & Frey, D.G. (editors) The Quaternary of the United States, 795-805, Princeton, New Jersey: Princeton U.P.

Paterson, W.S.B. 1981. The Physics of glaciers (2nd ed.), 380 pp. Oxford: Pergamon.

Olesen, O.B. 1978. Glaciological investigations in Johan Dahl Land, South Greenland, as a basis for hydroelectric power planning. Rapport Grønlands geologiske Undersøgelse 90, 84-86.

Olesen, O.B. 1981. Glaciological investigations at Qamanârssûp sermia, West Greenland. Rapport Grønlands geologiske Undersøgelse 105, 60-61.

Olesen, O.B. 1982. Establishment of a new survey station at Tasersiaq. Rapport Grønlands geologiske Undersøgelse 110, 86-88.

Olesen, O.B. 1985. Glaciological investigations in 1984 at Tasersiaq and Qapiarfiup sermia, West Greenland. Rapport Grønlands geologiske Undersøgelse 125, 104-107.

Olesen, O.B. 1986. Fourth year of glaciological field work at Tasersiaq and Qapiarfiup sermia, West Greenland. Rapport Grønlands geologiske Undersøgelse 130, 121-126.

Olesen, O.B. & Braithwaite, R.J. 1982. Glaciological investigations in 1981 at Qamanârssûp sermia, West Greenland. Rapport Grønlands geologiske Undersøgelse 110, 88-90.

Olesen, O.B. & Andreasen, J.O. 1983. Glaciological, glacier-hydrological and climatological investigations around 66°N, West Greenland. Rapport Grønlands geologiske Undersøgelse 115, 107-111.

Østrem, G. & Stanley, A. 1969. Glacier mass balance measurements, a manual for field and office work. Ottawa, Canadian Department of Energy, Mines and Resources, and Oslo, Norwegian Water Resources and Electricity Board, 127 pp.

Reynaud, L., Vallon, M. & Letréguilly, A. 1986. Mass-balance measurements: problems and two new methods for detemining variations. Journal of Glaciology 32 (112), 446-454.

Trabant, D.C. & Mayo, L.R. 1985. Estimation and effects of internal accumulation on five glaciers in Alaska. Annals of Glaciology 6, 113-117.

CALCULATION OF GLACIER ABLATION FROM AIR TEMPERATURE, WEST GREENLAND

Roger J. Braithwaite & Ole B. Olesen
The Geological Survey of Greenland
Copenhagen
Denmark

ABSTRACT

Recent measurements by the Geological Survey of Greenland (GGU) confirm a relation between glacier ablation and air temperature expressed in the form of positive temperature sums. The relation has already been used for the simulation of runoff from glacier-covered basins but might also be suitable for calculating ablation under alternative climates.

The degree-day factor, linking ice ablation to positive temperature sums, shows no distinct seasonal variation although it does fluctuate to some extent from year to year and possibly according to location. There is no evidence that the model can be improved by including incoming shortwave radiation but variations of winter snow accumulation have an important effect on summer ablation. This can be expressed by using a lower degree-day factor for snow than for ice.

Ablation can also be calculated from summer mean temperature although less accurately than using positive temperature sums. For example, ablation will increase substantially if summer temperatures rise due to greenhouse-effect warming of the atmosphere.

INTRODUCTION

Recent measurements in Greenland (Olesen & Braithwaite, this volume) have confirmed a relation between glacier ablation and positive air temperatures as suggested first by Finsterwalder & Schunk (1887). This relation has already been included in runoff models for highly glacierized basins (Braithwaite & Thomsen, 1984) but it is also of general interest for glacier-climate research. For example, the ablation-temperature relation is useful for both paleo-glaciological reconstructions and for prediction of future effects of climatic warming on the Greenland ice sheet. Application of such a simplified model to alternative climates is not without problems but it is a necessary first step to show that the model is valid under present climate. This is attempted in the following.

J. Oerlemans (ed.), Glacier Fluctuations and Climatic Change, 219–233.

The interpretation of the mechanisms underlying the relation between ablation and temperature will not be discussed at length here as the authors have already expressed opinions (Braithwaite, 1980 & 1981; Braithwaite & Olesen, 1985). Rather, the emphasis of the present paper is on results of recent measurements from Greenland which support or contradict various hypotheses on the ablation-temperature relation. It is, however, worth stressing that the present results refer to ablation facies with little or no snow accumulation, i.e. they refer essentially to ice ablation, while there is still too little information about snow ablation.

1. THE DATA

The locations and field programmes of the glaciological stations are described by Olesen & Braithwaite (this volume). The data used in the present paper are taken from two of the stations, i.e. from Nordbogletscher, Johan Dahl Land, South Greenland (two data sets) and from Qamanârssûp sermia, West Greenland (one data set). Brief descriptions of the three data sets are given below.

1.1 Nordbogletscher (1)

Ablation was measured almost daily at Stake 53 at 880 m a.s.l. near the edge of Nordbogletscher for nearly five summers, i.e. July-August 1979 and June-August for the four years 1980-1983. Parallel climatological measurements were made for the six summers 1978-1983 at the base camp which was located on bare ground only a few hundred metres from Stake 53. As the elevation difference is only 30 m, the climate conditions at the stake are essentially the same as measured at base camp. Ablation and temperature can therefore be compared for four complete summer 1980-1983.

The ablation readings from Stake 53 are averages of three sub-stakes and essentially refer to ablation of ice as there is little or no snow at the site in the period June-August.

1.2 Nordbogletscher (2)

Stake readings were made at many stakes in the lower ablation area of Nordbogletscher of which 14 survived throughout the six summer 1978-1983. The data comprise measurements in May of winter snow accumulation together with measurements at irregular intervals throughout the summer of the lowering of the ice surface relative to the stakes. As the stakes are located in the ablation area, the measured accumulation gives the amount of snow ablation which occurs in late May to early June while the surface lowering gives the net ablation of ice.

The data used here are averages for the 14 stakes which were maintained for the full six year period (Braithwaite, 1985a). The average elevation of the 14 stakes is by coincidence identical to the elevation of Stake 53. However, the ablation in the two data sets are not the same because there is much more snow at the 14 stakes and temperatures are lower (as measured by a climate station in the middle of the glacier and at almost the same elevation as base camp).

1.3 Qamanârssûp sermia

Ablation was measured almost daily at Stake 751 at 790 m a.s.l. near the edge of Qamanârssûp sermia for the seven summers 1980-1986. Parallel climatological measurements were made for the same seven summers at the base camp which is located on bare ground a few hundred metres from Stake 751. As the elevation difference is only 30 m the climate conditions at the stake are essentially the same as measured at base camp. In addition to this manned climate station, only operated during the summers, there is also a Danish-designed automatic climate station which has operated throughout four complete years as well as for long periods in three other years. A relation between temperatures at Qamanârssûp sermia and the permanent weather station at Nuuk, at a distance of 150 km, was used to interpolate temperature data for the missing periods of record. Ablation and temperature can therefore be compared for seven complete years, i.e. for both winter and summer conditions.

The ablation readings for Stake 751 are averages of three sub-stakes and essentially refer to ablation of ice as there is normally little or no snow at the site in June-August or on the few occasions when the area was visited in the winter.

2. THE DEGREE-DAY MODEL

2.1 Theory

The basis of the degree-day model is the assumption that ablation rate at any place is proportional to the air temperature at the same place (at about 1-2 m above the surface) as long as the temperature is at or above the melting point:

$$a_t = \alpha_1 + \beta_1 . T_t \quad T_t \geqslant 0°C \tag{1}$$

where a_t and T_t are respectively the ablation rate and temperature at time t, and α_1 and β_1 are parameters. It is convenient to express the time t in days. As will be seen in Equation (6), the parameter β_1 is the factor relating ablation to degree-day total. It is therefore called the degree-day factor.

A logical variable H_t is defined such that:

$$H_t = 1.0 \quad T_t \geqslant 0°C \tag{2a}$$

$$= 0.0 \quad T_t < 0°C \tag{2b}$$

The total ablation A over a period of many days is given by:

$$A = \Sigma H_t . a_t \tag{3}$$

where Σ denotes summation from t = 1 to t = N days. Combining Equation (1) and (3) gives:

$$A = \alpha_1 . \Sigma H_t + \beta_1 . \Sigma H_t . T_t \quad (4)$$

where ΣH_t is the number of days N^* with temperatures at or above the melting point and $\Sigma H_t . T_t$ is the sum of positive temperatures for an N-day period, otherwise denoted by PDD. The mean ablation rate during the N-day period is A/N given by:

$$A/N = \alpha_1 . (N^*/N) + \beta_1 . (PDD/N) \quad (5)$$

where N^*/N is the probability of melting temperatures during the period and PDD/N is the mean of positive temperatures in the period. According to Braithwaite (1985b), mean positive temperature is non-linear with mean temperature although the two quantities converge at higher temperatures as episodes with negative temperatures become rarer.

Table I. Intercept (α_1) and slope (β_1) for regression equations linking daily ablation a_t to daily mean temperature T_t at Qamanârssûp sermia. The correlation coefficient is R and N is the number of days of record in each year.

Year	α_1 (mm d^{-1})	β_1 (mm d^{-1} deg^{-1})	R	N (days)
1980	-3	7.1	0.77	104
1981	1	7.4	0.79	101
1982	1	8.0	0.83	85
1983	6	7.7	0.77	81
1984	-2	7.9	0.76	78
1985	5	7.3	0.71	77
1986	-4	8.4	0.76	80
Mean	1	7.7		
S.D.	±4	±0.5		

2.2 Daily ablation rate

The hypotheses expressed by Equation (1) and (5) have been tested under Greenland conditions. For example, Braithwaite (1985c, p. 22) quotes regression equations for daily ablation rate for five summer 1980-1984 at Qamanârssûp sermia. These results are repeated in Table I with two extra summers of data. The results confirm a useful correlation between daily ablation rate and temperature. The correlation coefficients range from 0.71 to 0.83, indicating that 50-70% of the variance of daily ablation is explained by temperature alone. The remaining variance must be explained by other climatological elements and especially by substantial errors in the daily ablation readings, e.g. with error standard deviations of ± 13 to ± 19 mm d^{-1} (Braithwaite, 1985c, p. 21) compared with a mean daily ablation rate of 40 mm d^{-1} for the seven summers 1980-1986.

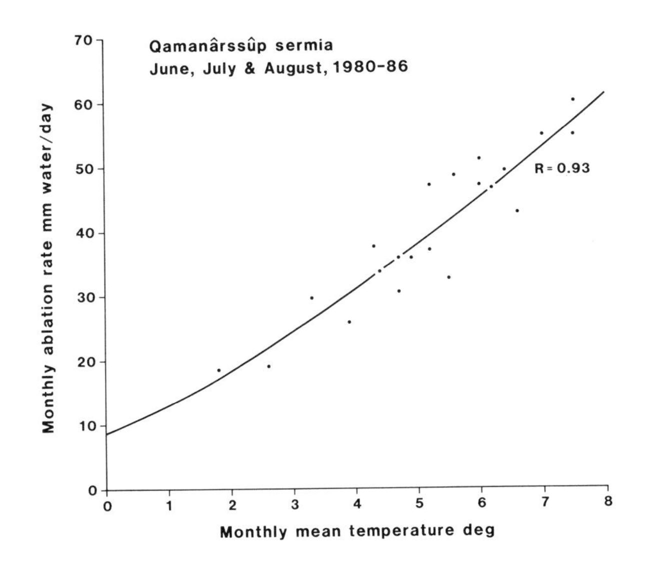

Figure 1. Correlation between monthly ablation rate and monthly mean temperature at Qamanârssûp sermia for 21 months.

2.3 Monthly ablation rate

Braithwaite & Olesen (1985) quote high correlations between monthly ablation rates and monthly positive temperatures at Nordbogletscher and Qamanârssûp sermia for five summers in each case. The results from Qamanârssûp sermia, updated with two further summers of data, are shown in Fig. 1. For convenience of the reader, the ablation data and regression line are plotted against the familiar mean temperature rather than the more exotic mean positive temperature. This is done by using a transformation based upon Braithwaite (1985b). The relation between mean ablation rate and mean positive temperature is linear according to the hypothesis in Equation (5) which means that the relation between ablation rate and mean temperature must be non-linear.

Braithwaite & Olesen (1985) lumped data from Nordbogletscher and Qamanârssûp sermia together and calculated a common equation for the two data sets. This now seems objectionable as it assumes a priori that conditions at the two sites are the same. Separate equations for the two data sets are therefore given in Table II. The results confirm useful correlations between monthly ablation rate and temperature and also that the intercept α_1 is small. The results also show a difference in degree-day factor for the two samples, i.e. 7.0 and 8.2 mm d^{-1} deg^{-1}, which may reflect a real difference in glacier-climate conditions at the two locations. However, the degree-day factor at Qamanârssûp sermia for the monthly data is not the same as the average of models using daily data shown in Table I. This shows that calculated degree-day factors depend to some extent upon the method of determination. This is hardly surprising as the degree-day factor is not a precise physical constant but is a parameter in a simple, and approximate, description of a highly complex system.

Table II. Intercept (α_1) and slope (β_1) for regression equations linking monthly ablation rate A/N to monthly mean positive temperature DDT/N at two glaciers in Greenland. R is the correlation coefficient, N is the number of days in the month, and M is the sample size in months

Data set	α_1 (mm d^{-1})	β_1 (mm d^{-1} deg^{-1})	R	M (months)
Nordbogletscher (1) (1979-1983)	-2	7.0	0.84	14
Qamanârssûp sermia (1980-1986)	-4	8.2	0.93	21

2.4 Seasonal ablation total

The above examples show that the intercept α_1 is generally small, amounting to at most a few millimetres per day. It therefore seemed reasonable to neglect the first term in Equation (4) when calculating seasonal ablation totals. The ablation is then given by:

$$A = \beta_1 . PDD \qquad (6)$$

Seasonal ablation from Nordbogletscher (1) and Qamanârssûp sermia are plotted versus degree-day totals in Fig. 2. For present purposes, winter is September-May and summer is June-August. The correlation between ablation and degree-day is extremely high with R = 0.98. The intercept is only 0.14 m water and the degree-day factor is 6.8 mm d^{-1} deg^{-1} which is lower than for daily and monthly ablation rates.

From visual inspection of the point distribution in Fig. 2, the Nordbogletscher points tend to lie on the low side of the regression line. New regression lines were therefore calculated for the two data sets separately. The resulting degree-day factors are 6.1 and 7.0 mm d^{-1} deg^{-1} for Nordbogletscher and Qamanârssûp sermia respectively. Although the four points for Nordbogletscher represent a very small sample, the results suggest a lower degree-day factor for seasonal ablation at Nordbogletscher as already suggested for monthly ablation.

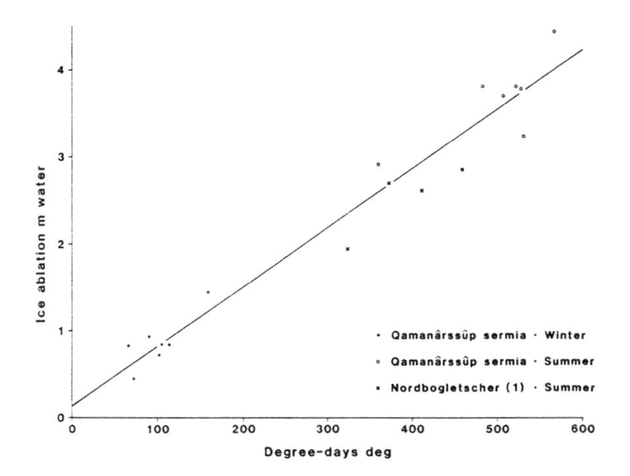

Figure 2. Comparison between ice ablation and positive degree-days totals at two glaciers in Greenland.

3. EFFECT OF RADIATION

Braithwaite & Olesen (1985) estimate that net radiation is the major source of ablation energy at both Nordbogletscher and Qamanârssûp sermia. This is in agreement with energy-balance measurements further north in Greenland (Ambach, 1963). The fact that radiation provides most of the ablation energy prompts the thought that radiation variations should be included in ablation models. This is briefly examined below.

It is often suggested that temperature acts as a proxy variable for radiation which implies that radiation data should be used instead of temperature data for calculating ablation. It is clear that incoming longwave radiation increases with temperature and probably accounts for about 10% of the degree-day factor but there is very poor correlation between ablation rate and incoming shortwave radiation (global radiation) according to Braithwaite & Olesen (1985, Table 6). Their results, updated with data for two further years at Qamanârssûp sermia, are in given in Table III. Comparing Tables II and III confirms that ablation is poorly correlated with global radiation compared to its correlation with temperature. Temperature does not therefore act as a proxy variable for global radiation.

Pollard (1980) suggests an ablation model using a linear combination of net radiation and air temperature. This is roughly equivalent to saying that α_1 varies with global radiation as the degree-day factor partly absorbs the longwave component of the net radiation. Multiple correlations of monthly ablation rate with global radiation and temperature give no improvements compared with simple correlations with

Table III. Intercept (α_2) and slope (β_2) for regression equations linking monthly ablation rate A/N to monthly mean global radiation G/N at two glaciers in Greenland. R is the correlation coefficient, N is the number of days in the month and M is the sample size in months.

Data set	α_2 (mm d^{-1})	β_2 (mm/MJ m^{-2})	R	M (months)
Nordbogletscher (1) (1979-1983)	11	1.20	0.43	14
Qamanârssûp sermia (1980-1986)	40	0.02	0.01	21

temperature alone. However, the present data refer essentially to the ablation of ice, i.e. with a restricted range of albedo, while the ablation of snow followed by ice ablation involves a large change in albedo and therefore of absorbed shortwave radiation.

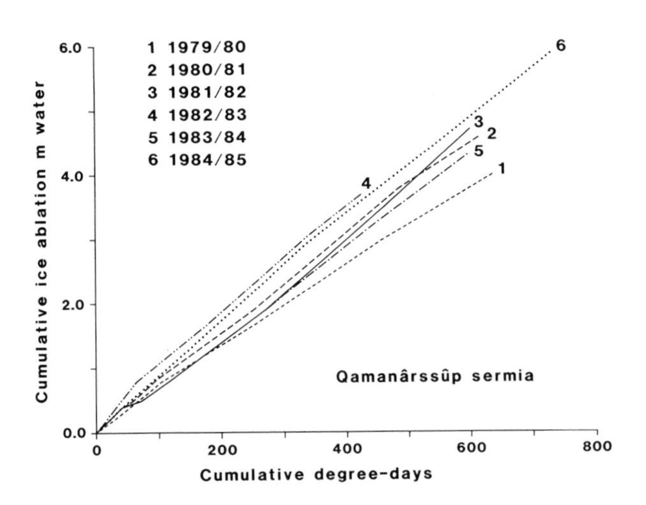

Figure 3. Cumulative ice ablation versus cumulative degree days for six balance years at Qamanârssûp sermia, West Greenland.

Another possibility is that the degree-day factor varies with the seasonal variation of shortwave radiation (Gottlieb, 1980; Lundquist, 1982). This is contradicted by the double-mass curves in Fig. 3 where ablation and degree-days are accumulated through six complete years (1 September to 31 August) at Qamanârssûp sermia. The slope at any point of such a curve is proportional to the degree-day factor. Although not perfectly straight, the curves indicate that the degree-day factor is reasonably constant throughout the year compared with obvious year-to-year fluctuations (with a coefficient of variation of about 10%). There is, for example, no marked difference between degree-day factors for summer and winter in agreement with Fig. 2.

4. EFFECT OF SNOW ACCUMULATION

The above discussion refers essentially to the ablation of ice with at most traces of snow. However, the presence of snow has a marked effect on ice ablation. This is best illustrated by considering the Nordbogletscher (2) data set. The annual ice ablation for the six years 1978-1983 is plotted against snow accumulation (left) and summer degree-days (right) in Fig. 4. Both the ablation and accumulation values refer to averages at the 14 stakes. There is a strong positive correlation between ice ablation and degree-days but there is also a strong negative correlation between ice ablation and accumulation.

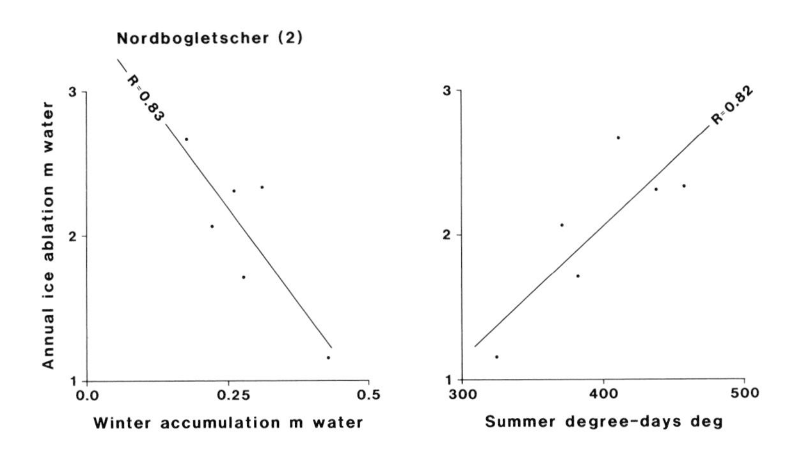

Figure 4. Annual ice ablation versus winter snow accumulation (left) and summer degree-days (right) at Nordbogletscher, South Greenland.

It is natural that snow accumulation inhibits ice ablation because the melting of the snow uses energy, parameterized here by summer degree-days, which would otherwise be used for melting ice. However, the effect of snow accumulatin on ice ablation is greatly magnified as an increase of accumulation by 0.1 m water reduces ablation by about 0.5 m water. Another interesting point is that the regression line for ablation versus degree-days in Fig. 4 has a high degree-day factor of 9.1 mm d^{-1} deg^{-1} compared to values in Tables I and II.

Both the above points can be explained by assuming that the ablation of snow follows a similar degree-day relation to the one for ice ablation but with lower degree-day facor, e.g. as implicit in the TS(3°) method of Hoinkes & Steinacker (1975). This lower degree-day factor covers two effects. First, snow melt is less than ice melt for the same temperature conditions and, second, meltwater from snow can refreeze to form superimposed ice which must be melted again.

For the present analysis a degree-day factor of 2.5 mm d^{-1} deg^{-1} is assumed for snow ablation. The number of degree-days used for snow ablation is then the amount of accumulation divided by the degree-day factor for snow. The degree-day total actually used for ice ablation is the degree-day total for the summer minus the degree-days for snow ablation. The correlation coefficient between ice ablation and this accumulation-corrected degree-day total is now 0.96 instead of 0.82 as in Fig. 4. The corresponding degree-day factor is also reduced to 7.2 mm d^{-1} deg^{-1} which is in better agreement with values in Tables I and II.

It is desirable to apply a correction to ablation data sets with snow accumulation before comparing them with data sets with little or no accumulation. This is done by transforming total snow and ice ablation into the "equivalent ice ablation". In this case the equivalent ice ablation is the ice ablation plus the snow ablation (equal to snow accumulation) multiplied by a factor of (7.2/2.5).

5. OTHER METHODS

The degree-day method for calculating ablation is not difficult to use if temperature data are available throughout the year. For example, the degree-day sum for each month is estimated quite accurately from monthly mean temperatures using the probability model of Braithwaite (1985b), monthly values are summed to give annual degree-day totals, and annual ablation is calculated using the appropriate degree-day factor. However, for some purposes it is more convenient to use even simpler methods where ablation is calculated from a single temperature statistic, e.g. from annual or summer mean temperature.

5.1 Annual temperature method

Oerlemans & Van der Veen (1984, p. 185) re-analysed Greenland data from Ambach (1972) and suggest the following non-linear equation

$$A = \gamma\ (12 + T_a)^2 \qquad T_a \geqslant -12°C \qquad (7)$$

where A is the annual ablation in m water, T_a is the annual mean temperature and γ is a parameter. The relation between annual ice ablation and T_a is shown in Fig. 5 where the points for Nordbogletscher (2) refer to equivalent ice ablation estimated by the method outlined in the previous section. The annual mean temperatures at Qamanârssûp sermia are calculated from the year-round measurements of the automatic weather station while the annual temperatures at Nordbogletscher (2) are estimated from those at Narssarssuaq weather station using a lapse rate obtained by comparing summer data at Nordbogletscher (2) and Narssarssuaq. The curves refer to linear regressions between ablation A and the variable $(12 + T_a)^2$.

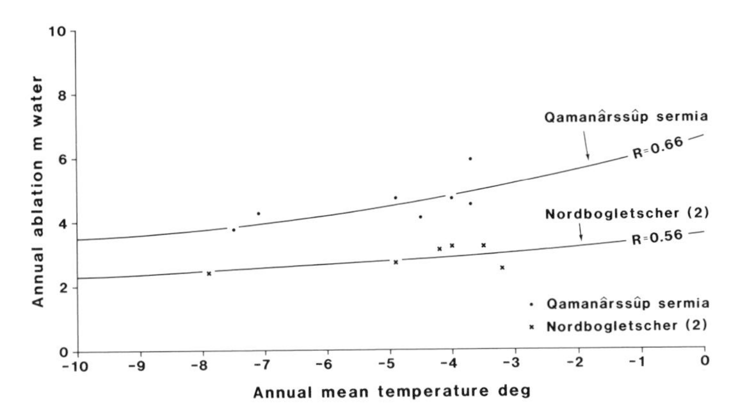

Figure 5. Annual ice ablation versus annual mean temperature at two glaciers in Greenland. Data for Nordbogletscher refer to equivalent ice ablation after correcting for effect of snow accumulation.

The relations for the two data sets are clearly different. There is higher ablation at Qamanârssûp sermia than at Nordbogletscher (2) for the same annual temperature. Moreover, both curves in Fig. 5 give much higher ablation than the equation of Oerlemans & Van der Veen (1984, p. 185) which gives values of only 1.44 and 1.27 m water respectively for the mean ablation at Nordbogletscher (2) and Qamanârssûp sermia. The main reason for these underestimations is probably that the equation of Oerlemans & Van der Veen refers to conditions near the equilibrium line, i.e. mainly involving snow ablation.

The correlation coefficients for the two curves are not especially high. For example, the degree-day curve in Fig. 1 has a correlation coefficient of 0.93 compared with 0.56 and 0.66 respectively in Fig. 5. In particular, the lower correlation for Nordbogletscher (2) probably reflects large errors in estimating annual mean temperature with lapse rates based upon summer data. A more fundamental objection to the use of annual mean temperature is the fact that ablation only occurs in about a quarter to a third of the year so that temperature variations for most of the year are irrelevant for ablation.

5.2 Summer temperature method

Krenke (1975) summarizes earlier work in Russian where he and V.G. Khodakow analysed data from many glaciers in the USSR and proposed the following non-linear equation:

$$A = (9.5 + T_s)^3/1000 \quad (8)$$

where A is again the annual ablation in m water and T_s is the summer mean temperature (for June-August).

The relation between annual ablation and the variable T_s is shown in Fig. 6. The ablation data for Nordbogletscher (2) refer once again to equivalent ice ablation while the summer mean temperature for both data sets are based upon direct measurements made at the respective base camps close to the glaciers.

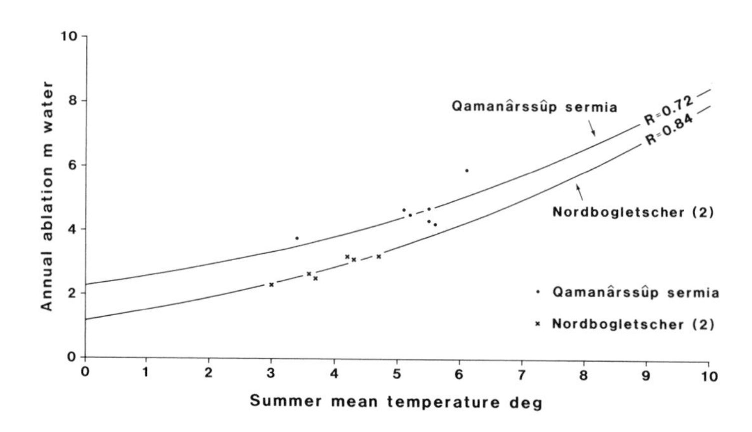

Figure 6. Annual ice ablation versus summer mean temperature at two glaciers in Greenland. Data for Nordbogletscher refer to equivalent ice ablation after correcting for effect of snow accumulation.

The relations for the two data sets run nearly parallel to each other with, once again, higher ablation at Qamanârssûp sermia than at Nordbogletscher (2). The original equation of Krenke (1975) overestimates mean ablation at Nordbogletscher by 19% but underestimates mean ablation at Qamanârssûp sermia by -30%. This represents fair agreement considering that the data come from Greenland and the equation comes from the USSR.

The correlation coefficients for the curves are better than in the previous case but are lower than for the degree-day model. This probably reflects the greater logical relation between ablation and summer temperature because 80-90% of the ablation in the present case occurs in June-August. However, it is still objectionable to compare annual ablation with mean temperature for an arbitrary period, e.g. because the ablation season at low elevations is longer than June-August while the reverse is true at high elevations. This objection can be set against the convenience of the summer temperature method.

5.3 Change of ablation with Climate

As a demonstration of the above method, ablation can be calculated as a function of temperature changes from present climate as might occur in the future due to greenhouse-effect warming of the atmosphere. The results for the two sites are given in Table IV for changes of summer temperature of 0 to +5°C from present conditions. The ± figures in the table are 95% confidence intervals which take account of the statistical uncertainties caused by small sample sizes and the less than perfect correlations between ablation and temperature. The results show that ablation will increase substantially if summer temperature increases, e.g. ablation may double if temperatures rise by 5°C.

Table IV. Calculated variations of ablation on two glaciers with change ΔT from present summer temperature. Using a recalibration of the equation of Krenke (1975).

ΔT (°C)	Nordbogletscher (%)	Qamanârssûp sermia (%)
0	100 ± 2	100 ± 4
+1	121 ± 5	114 ± 8
+3	172 ± 16	149 ± 23
+5	238 ± 32	192 ± 43

6. STATUS AND OUTLOOK

The present study confirms a relation between ice ablation in Greenland and air temperature expressed as positive temperature sums. The degree-day factor, linking ablation to temperature, shows no distinct seasonal variation although it fluctuates from year to year. There are also indications that the degree-day factor varies according to locality and that the factor for snow ablation is lower than for ice ablation.

The degree-day factor is already used for practical planning of hydropower in Greenland but a better understanding of its variability is needed. For example, even the small difference in degree-day factor between Qamanârssûp sermia and Nordbogletscher may have an important effect on cost-benefit calculations. Again, if hydropower is proposed for a hitherto unstudied area, what value should be used for the degree-day factor? This uncertainty is more important if one calculates ablation under alternative climates. For example, if variations of the degree-day factor under present climate are not perfectly understood, how does one know what value it will have under other climate conditions?

Investigations on the variability of degree-day factor in Greenland will continue. This will be done both by continuation of measurements and by more detailed analyses of data already collected. For example, multivariate statistical treatment of daily ablation readings from Nordbogletscher and Qamanârssûp sermia will be used to look for possible combinations of climatological variables which are associated with ablation variations.

ACKNOWLEDGEMENTS

This paper is published by permission of the Director, the Geological Survey of Greenland. The field station in Johan Dahl Land 1978-1983 was partly funded by the European Economic Community (EEC) and by the Danish Energy Ministry while the Qamanârssûp sermia station was wholly funded by the Geological Survey of Greenland. The work in Johan Dahl Land 1980-1983 was lead by cand. scient. Poul Clement. The automatic climate station at Qamanârssûp sermia which has given invaluable year-round data is operated by the Greenland Technical Organization.

REFERENCES

Ambach, W. 1963. Untersuchungen zum Energieumsatz in der Ablationszone des Grönländischen Inlandeises. Meddelelser om Grønland 174 (4), 311 pp.

Ambach, W. 1972. Zur Schätzung der Eis-Nettoablation im Randgebiet des Grönländischen Inlandeises. Polarforschung 42 (1) 18-23.

Braithwaite, R.J. 1980. Glacier energy balance and air temperature: comments on a paper by Dr. M. Kuhn. Journal of Glaciology 25 (93), 501-503.

Braithwaite, R.J. 1981. On glacier energy balance, ablation and temperature. Journal of Glaciology, 27 (97), 381-391.

Braithwaite, R.J. 1985a. Relations between annual runoff and climate, Johan Dahl Land, South Greenland. Grønlands Geologiske Undersøgelse Gletscher-hydrologiske Meddelelser 85/2, 25 pp.

Braithwaite, R.J. 1985b. Calculation of degree-days for glacier-climate research. Zeitschrift für Gletscherkunde und Glazialgeologie 20 (1984), 1-8.

Bratithwaite, R.J. 1985c. Glaciological investigations at Qamanârssûp sermia, West Greenland, 1983-1984. Grønlands Geologiske Undersøgelse Gletscher-hydrologiske Meddelelser 85/3, 26 pp.

Braithwaite, R.J. & Olesen, O.B. 1985. Ice ablation in West Greenland in relation to air temperature and global radiation. Zeitschrift für Gletscherkunde und Glazialgeologie 20 (1984), 155-168.

Braithwaite, R.J. & Thomsen, H.H. 1984. Runoff conditions at Paakitsup Akuliarusersua, Jakobshavn, estimated by modelling. Grønlands Geologiske Undersøgelse Gletscher-hydrologiske Meddelelser 84/3, 22 pp.

Finsterwalder, S. & Schunk, H. 1887. Der Suldenferner. Zeitschrift des Deutschen und Oesterreichischen Alpenvereins 18, 72-89.

Gottlieb, L. 1980. Development and application of a runoff model for snowcovered and glacierized basins. Nordic Hydrology 11 (1980), 255-272.

Hoinkes, H. & Steinacker, H. 1975. Hydrometeorological implications of the mass balance of Hintereisferner, 1952-53 to 1968-69. Proceedings of the Snow and Ice Symposium, Moscow 1971. IAHS Publication 104, 144-149.

Krenke, A.N. 1975. Climate conditions of present-day glaciation in Soviet Central Asia. Proceedings of the Snow and Ice Symposium, Moscow 1971. IAHS Publication 104, 30-41.

Lundquist, D. 1982. Modelling runoff from a glacierized basin. Hydrological Aspects of Alpine and High Mountain areas. Proceedings of the Exeter Symposium, July 1982. IAHS Publication 138, 131-136.

Oerlemans, J. & Van der Veen, C.J. 1984. Ice sheets and climate. 217 pp. Dordrecht: D. Reidel Publishing Company.

Olesen, O.B. & Braithwaite, R.J. this volume. Field stations for glacier-climate research, West Greenland.

Pollard, D. 1980. A simple parmeterization for ice sheet ablation rate. Tellus 32 (1980), 384-388.

HYDROMETEOROLOGICAL CONDITIONS, MASS BALANCE AND RUNOFF FROM ALPINE GLACIERS

David N. Collins
Dept. of Geography
University of Manchester
Manchester M13 9PL, U.K.

ABSTRACT

Discharge records for rivers draining from highly-glacierized basins in the upper Rhône catchment area, Switzerland, have been examined together with precipitation and air temperature data with a view to determining which variables from which stations can be used to represent interacting climatic elements that influence mass balance of and runoff from Alpine glaciers. Relationships between climate and mass balance and between climate and runoff were analysed at seasonal and annual timescales by correlation and multiple linear regression techniques. Variables were selected from summer air temperatures, fluctuations of which are similar over the catchment area, and precipitation records from valley stations with the intention of providing best levels of explanation of variances of runoff and mass balance. An attempt to estimate the relationship between precipitation and elevation was made using the catch of totalizing raingauges emptied at roughly half-yearly intervals, for which reasonable levels of correlation exist between valley and mountain gauges.

Summer air temperatures averaged over the months May through September and June through August are highly positively correlated with summer totals of runoff but strongly negatively associated with mass balance. Relationships between precipitation, best expressed by totals from October through May, May through September and May alone, with mass balance are more strongly positive than are those with runoff negative. Between 75 and 91 per cent of the variances of total summer discharge (May through September) and of net mass balance are explained by multiple regression against mean summer air temperatures and either one or two precipitation variables. Significant fluctuations of climatic conditions, glacier mass balance and runoff occurred in the period 1922-1985. Generally-declining air temperatures led to runoff reducing substantially between the 1940s and the 1970s, at the same time as glacierized areas diminished. Multiple regression models, whilst exhibiting high degrees of fit, appear to have limited predictive power on account of non-stationarity of relationships between variables used.

J. Oerlemans (ed.), Glacier Fluctuations and Climatic Change, 235–260.

1. INTRODUCTION

Climatic conditions influence both the total quantity of runoff in a year from a glacierized catchment and the net change in amount of perennial snow and ice stored within the basin. Over longer periods, variations in climatic factors lead to year-to-year changes in water yield and in mass balance. Changes in the latter through glacier dynamics determine the planimetric extent of the glacier, cumulative expansion or contraction of which in turn affects runoff. Relationships between climate and mass balance, and between climate and runoff are of interest from scientific and practical viewpoints (e.g. Meier and Roots, 1982). Since climatic measurements extend back beyond those of mass balance, and stations are more numerous and more widely distributed than are the few glaciers for which observations have been undertaken, variations in climatic characteristics may be usefully used to reconstruct changes in mass storage, and to extrapolate mass balances to unmeasured glaciers. Climate - mass balance and climate - runoff relations can also be used for forecasting purposes, for example in predicting the effects of various scenarios of impending climatic change on runoff, of particular importance with respect to assured future supplies for water resources and power systems. Precipitation and air temperature records are more widely available, over longer periods and from more stations than those of other hydrometeorological characteristics (Collins, 1985). Definition of relationships between these climatic elements and mass balance or runoff is therefore required in the absence of long term measurements of radiation.

The aims of this investigation are to identify appropriate variables to represent the climatic elements which interact to influence glacier mass balance and runoff from glacierized areas, to examine the relative importance of these elements in determining annual and seasonal variations in flow and annual changes in storage, and to quantify relationships between the climatic variables selected, net balance and discharge. Climatic, glaciological and hydrological data for highly glacierized basins in the high Alpine portion of the catchment area of the river Rhône in Kanton Wallis, Switzerland, have been analyzed for a period during which glaciers first declined in area (until the mid-1960s) before readvancing slightly (Kasser, 1981; Patzelt, 1985), a factor which has probably affected the other relationships. Climatic influences on mass balance and runoff are assessed by the use of correlation and multiple regression techniques, which linear methods have proved valuable in describing both climate - mass balance (Martin, 1975, 1977; Kasser, 1981; Günther and Widlewski, 1986) and climate - glacier-runoff linkages (Kasser, 1973, 1981; Collins, 1985, 1987).

Several questions are raised in this paper. The degree of correlation between controlling climatic variables and dependent hydroglaciological performances will depend on how representative are data from the chosen climatic stations of actual hydrometeorological conditions affecting individual glacierized basins, which variables are selected from the measurement records available to represent the climatic elements, and which variables are used to describe characteristics of the dependent series. Variables representing monthly, seasonal and annual characteristics of climatic elements, seasonal and annual behaviour of runoff and annual mass balance have been considered in this study. Length, statistical distribution, and stationarity of the data series and of the variables extracted therefrom will also influence calculated levels of association, as will any deviation from linearity in the relationships.

2. CLIMATE - MASS BALANCE AND CLIMATE - RUNOFF LINKAGES

Net annual change in mass storage of snow and ice within and total annual runoff from a basin are outcomes resulting from the different interactions amongst the same set of climatic elements (see Collins, 1985, Fig. 6). Summer energy inputs interact with precipitation to influence net balance, which is the difference in water equivalent between the amount of surplus snow, in excess of that which can be melted, held over in the accumulation area at the end of summer and the quantity of ice and firn melted during the ablation season. The proportion of radiation absorbed at the surface is reduced when the glacier is covered by snow which has a high albedo limiting utilisation of available energy. The thicker the blanket of snow in winter and the more frequently snow fall events cover the surface in summer, the slower the ascent of transient snow line and the more accumulation is favoured at higher and ablation prevented at lower elevations. Energy inputs are moderated also by cloud cover which may produce precipitation. Summer rainfall will not impart as strong a tendency to more positive (or less negative) mass balance as would the same water equivalent deposited as snow. The more energy available, precipitation remaining constant, and the more ablation of ice and firn and more melting of snow at higher altitudes, the more negative (or less positive) the net balance will be.

This glacier surface precipitation-energy interaction also influences climate-runoff relationships. The higher the average surface albedo, the lower the amount of meltwater released to runoff per unit input of heat. Summer precipitation (solid or liquid) over bare ice will contribute to runoff and offset to some extent reduced amounts of flow derived from icemelt when snow is lying or when skies are cloudy. However, a component of flow directly related to the total annual precipitation over the ice-free area of the catchment will augment runoff at times when precipitation has the inverse effect on the component of flow derived from melt on the glacier. Ice melt contributions are enhanced in dry summers when flow from rainfall over the bare-ice and ice-free areas is scarce (e.g. Fountain and Tangborn, 1985).

Whereas in a ice-free basin annual runoff will be directly related to and always less than precipitation received, that from glacierized basins can be either greater or less than the total amount of precipitation, depending on the balance between how much water is added as snow and the amount released by icemelt from glacier storage. During periods of sustained advance or retreat of glaciers, therefore, runoff will be either reduced if a fraction of precipitation is retained, or augmented if meltwater is destocked from ice storage. There will still be a relationship with summer energy inputs, but the dimensions of the ice-covered area changing through glacier dynamics will also influence the total amount of energy delivered to the glacier surface.

3. SELECTION OF DATA MEASUREMENT STATIONS

Climatic stations at which long unbroken data records are available were selected according to proximity to glaciers, elevation and the degree of correlation exhibited between hydrometeorological variables derived from those records with mass balance or runoff. Levels of correlation of each derived climatic variable between stations were also examined, together with patterns of temporal covariation of climatic variables and mass balance or runoff totals in order to ascertain whether stations respond to general climatic conditions in the upper Rhône basin or to local topographic microclimates. In the Alps, stations are generally located in valleys rather than at higher elevations, although records of precipitation measurements at gauges located at high altitude are available for assessment.

Table I. Details of hydrometeorological stations

Station	Elevation m a.s.l.	Measurements
Sion (Aérodrome)	482	air temp.; standard raingauge (ANETZ from 1978)
Sion (Couvent des Capucins)	549	air temp.; standard raingauge
Zermatt	1632	standard raingauge
Saas Almagell	1669	air temp.; standard raingauge
Trift	2625	totalizing raingauge
Gandegg	2880	totalizing raingauge
Gornergrat	3089	totalizing raingauge
Allalingletscher	3368	totalizing raingauge

Statistically significant ($p = 0.05$) Pearson's product moment correlation coefficients with values greater than 0.83 between mean summer air temperatures between 1 May and 30 September (T_{5-9}) recorded at the Couvent des Capucins and Aérodrome sites in Sion, at Saas Almagell and Zermatt (see also Collins, 1987) suggest that these stations reflect general levels of energy availability in the area. Locations of the stations in the Rhône basin are shown in Fig. 1, and details of the meteorological stations used are given in Table I. Plots of T_{5-9} for the stations in Sion are shown in Fig. 2. There is, in general, a common pattern of temporal variation, although the rank order of the warmer summers differs between stations in the overlapped period. The Aérodrome station, situated in the valley bottom, is nevertheless anomalous and relates poorly with river flow. Peak river flows in the

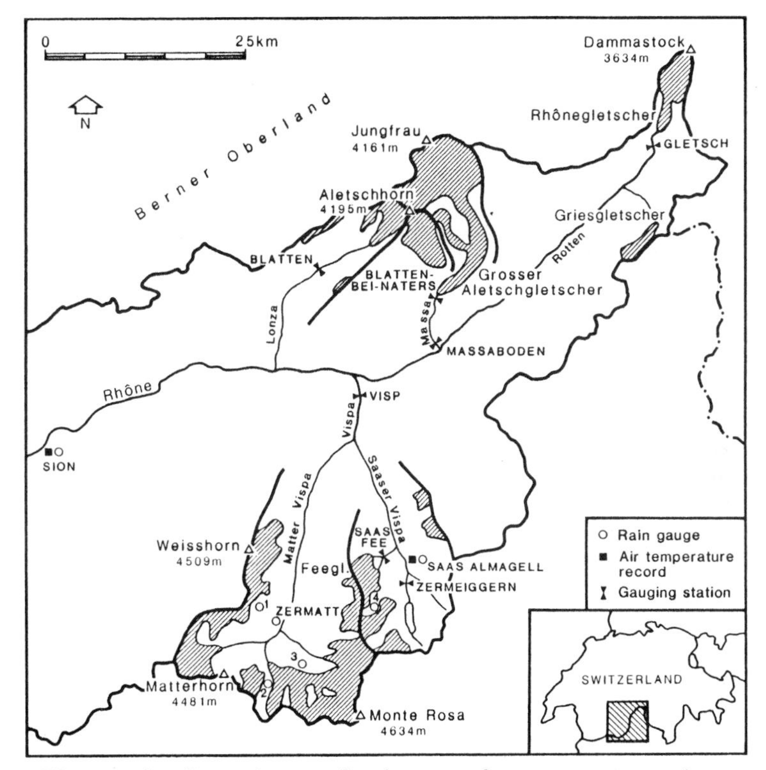

Figure 1. Locations of the catchment areas, river gauges and meteorological stations in the upper Rhône basin, Kanton Wallis, Switzerland from which records have been used in this investigation. Fee Vispa (Saas Fee), Saaser Vispa (Zermeiggern) and Vispa (Visp) drain from the Pennine Alps and the Lonza (Blatten), Massa (Blatten-bei-Naters, Massaboden) and Rhône/Rotten (Gletsch) from the Berner Oberland. Air temperature and precipitation records from stations at Sion and Saas Almagell were analysed together with those of precipitation at Zermatt and from totalising gauges at Trift (1), Gandegg (2), Gornergrat (3) and Allalingletscher (4). Glacierised areas within gauged basins are shown, together with Griesgletscher.

years after 1960 occurred in 1982 complementing the warmest summer in that period (Fig. 2), but in 1983 a higher value of T_{5-9} was recorded than in 1982 at Sion Aérodrome. Values of T_{5-9} in both these years were lower than those recorded at Couvent des Capucins in 1973 and 1976. The meteorological station at the Aérodrome was changed from conventional to an automatic (ANETZ) system in 1978.

There is less uniformity in the pattern of temporal variation of precipitation at the valley stations. Correlation between total annual precipitation (P_{11-10}) at Zermatt and Saas Almagell is 0.78 (significant at $p = 0.05$) and poorer for other pairs of stations. This may reflect localized variations in precipitation in mountain valleys, and non-linear increases of precipitation with altitude. Records from four totalizing precipitation gauges were examined in this respect. Although the actual dates on which measurements of collected precipitation were undertaken vary from gauge to gauge each year, and also from year to year, records have been standardized as if the measurement year was from 1 October to 30 September (P_{10-9}). Locations of precipitation gauges appear in Fig. 1, and elevations are listed in Table I.

Records of flows from gauges on rivers draining from basins with more than 35% glacierization were used for periods without abstraction or retention for hydropower purposes. Physiographic characteristics of the basins are presented in Table II, and the sites of the gauges are plotted in Fig. 1. Flow of the Saaser Vispa is contributed from Schwarzberggletscher and Allalingletscher, and the Massa drains Grosser Aletschgletscher and Oberaletschgletscher to a structure at Blatten-bei-Naters, which replaced a gauge downstream at Massaboden from 1965. Several small glaciers are distributed in the Lonza basin, and the Rhône/Rotten flows from Rhônegletscher. The temporal patterns of runoff are synchronized, particularly in terms of signs of first differences between total annual discharges (Q_{1-12}), if not in absolute deviations, and flows from the more highly-glacierized basins are well correlated, suggesting a broadly similar response to the prevailing hydro-meteorological conditions which, influenced by cyclones in the westerlies, affect the region as a whole (Fig. 2).

Table II. Characteristics of the glacierized basins investigated

Basin/gauge	Basin area km^2	Glacierization %/year	Gauge elevation m a.s.l.	Highest point m a.s.l.
Vispa/Visp	778.0	35.9/1950	659	4634
Lonza/Blatten	77.8	40.6/1983	1520	3897
Saaser Vispa/ Zermeiggern	65.2	44.6/1876 41.5/1960	1740	4199
Fee Vispa/ Saas Fee	16.7	53.7/1960	1761	4535
Rhône/Gletsch	38.9	56.4	1754	3634
Massa/Blatten-bei-Naters	194.7	66.0/1973	1446	4195
Massa/Massaboden	202.0	67.6/1934 64.1/1957	687	4195

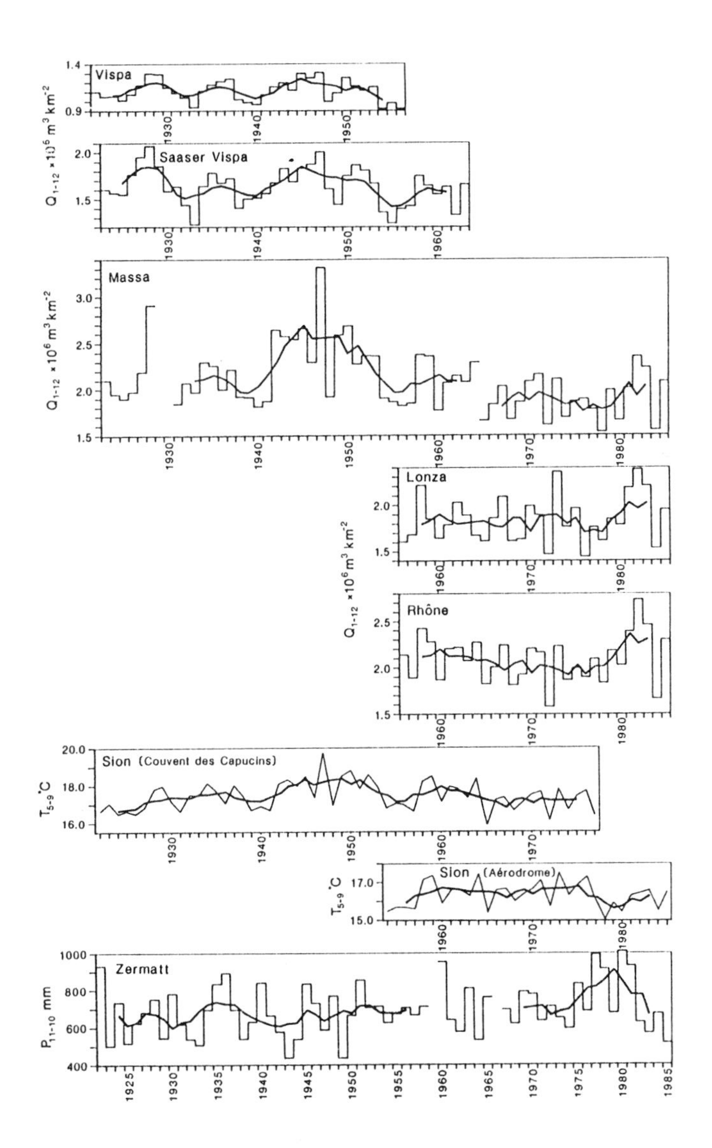

Figure 2. Annual variations and five-year running means of total annual discharge (Q_{1-12}) of the Vispa at Visp, Saaser Vispa at Zermeiggern, Massa at Massaboden (to 1964) and at Blatten-bei-Naters (from 1965), Lonza at Blatten, Rhône at Gletsch; mean summer air temperature (T_{5-9}) at Sion (Couvent des Capucins) and Sion (Aérodrome); and annual precipitation (P_{11-10}) at Zermatt.

Mass balance of Grosser Aletschgletscher (129.76 km^2 in 1957) is obtained by the hydrological method but that of Griesgletscher (6.38 km^2 in 1967, see Fig. 1) is determined by the glaciological technique. Precipitation over Aletschgletscher in the catchments of the gauges of Massaboden and Blatten-bei-Naters is estimated from measurements undertaken outside the basins at low altitude stations (IAHS, 1973). It would be inappropriate, therefore, to consider correlation of mass balance so derived with similar, external precipitation variables. The measurement year for Griesgletscher is almost fixed date, commencing in early October. The correlation between the two series of net specific mass balance measurements between 1963/64 and 1977/78 is 0.88 ($p = 0.05$) and the signs of first differences between years were the same except for one, suggesting relative similarities in response to climate from year to year. The elevation range of Griesgletscher is 2380-3260 m a.s.l.

4. SELECTION OF HYDROMETEOROLOGICAL AND RUNOFF VARIABLES

Levels of correlation amongst climatic elements and between climate and mass balance and runoff are influenced by which variables are chosen from the records of daily and monthly values to represent seasonal and annual attributes of a series. The selected variables should provide at least some conceptual and physical realism in representing the processes coupling mass balance and runoff to climate. P_{11-10} rather than P_{10-9} (where the subscripts represent the inclusive months over which precipitation is totalled) has been preferred to represent total annual precipitation over Alpine glacierized basins in relation to runoff, on the basis that precipitation in October in lower parts of a basin contributes immediately to runoff (Collins, 1985). From the viewpoint of mass balance, precipitation in October over most of the glacier area will be retained. Winter precipitation accumulation continues into March, April or May according to altitude, and can be represented by variables such as P_{10-4}, P_{10-5}, P_{11-4} or P_{11-5}. Maintenance of a high average albedo of snow and ice surfaces in spring is important in retarding melt, and single month variables (e.g. P_4, P_5) or aggregates (e.g. P_{4-6}) may be significant, even though at lower elevations this precipitation may fall as rain.

Precipitation in summer months (P_{5-9} or P_{6-9}) may suppress melt should snow fall, but some will contribute to runoff in the period identified. P_9, precipitation in September, may, if occurring as snow, bring the ablation season to an abrupt early close. Energy input in summer was represented by simple mean air temperature variables T_{5-9} and T_{6-8}. Total summer (Q_{5-9} or Q_{6-9}) and total annual (Q_{1-12}) discharges were used as runoff variables. Q_{1-12} was selected, rather than Q_{10-9}, which would have been consistent with the mass balance and precipitation variables, or Q_{11-10}, on the grounds that runoff between October and December relates to melting conditions in the preceding summer months.

5. YEAR-TO-YEAR AND LONGER-TERM VARIATIONS OF CLIMATIC CONDITIONS, MASS BALANCE AND RUNOFF

The general trends of secular variations of climatic conditions significant to hydroglaciological responses and underlying variations of runoff are illustrated by plots of five-year running means of Q_{1-12} for several basins, T_{5-9} at Sion and P_{11-10} at Zermatt in Fig. 2, and show some of the linkages that exist between climatic variables and flow. Mean quinquennial T_{5-9} at Sion (Couvent des Capucins) increased from 16.66°C (1922-1926) to 17.68°C by 1934-1938, and after a relatively cool period, reached a maximum of 18.36°C centred on 1949. High summer temperatures maintainted quinquennial averages above 18.0°C from 1942-1946 through to 1949-1953, followed by a sudden decrease in the mid-1950s (1953-1957 mean T_{5-9} of 17.05°C). After a recovery to a high of 17.9°C (1958-1962), summer temperatures at Sion declined slowly to quinquennial means of about 17.1°C. This pattern was reflected by the station at Sion Aérodrome, until anomalous behaviour commenced in 1978. A run of generally cool summers in the mid- and late-1970s was terminated by amelioration in the early-1980s as indicated by T_{5-9} at Saas Almagell (Fig. 3). With respect to neighbouring years, 1928, 1929, 1934, 1947, 1959, 1964, 1973, 1982 and 1983 were notably warm and 1948, 1957, 1965, 1972, 1978 and 1984 cool. Quinquennial means of P_{11-10} at Zermatt (Fig. 2) show wetter periods in the late 1920s, mid-1930s and early 1950s, then remaining close to the overall period mean, before increasing to the highest levels over the seventy-year period. The wettest quinquennia occurred from 1973/74-1977/78 to 1977/78-1981/82. Precipitation was reduced during warmer periods, in the 1940s and from 1982. Relatively high annual totals at Zermatt occurred in 1921/22, 1935/36, 1959/60 and 1976/77 with the highest in 1979/80, and lows in 1922/23, 1942/43, 1948/49, 1963/64 and 1984/85.

Underlying oscillations in the variation of runoff between 1922 and 1985 showed maxima in the late 1920s and mid-1930s, sustained highs from 1942 to 1953 with the exception of 1948 (and 1949 in the Vispa basin), in the early 1960s and early 1980s, closely paralleling fluctuations of pentad means of T_{5-9}. Minima in the late 1950s, and 1970s further demonstrate this directly phased relationship. Although considerable irregularity from year to year characterizes all the flow series, the pattern of fluctuations is broadly similar and synchronous. Exceptionally large annual totals of runoff were discharged from all basins which had operational gauges in 1928, 1945, 1947, 1950, 1958, 1964, 1982 and 1983, with relative lows in 1940, 1948, 1954-1957, 1965, 1972 and 1978. The overall range of variability of runoff extends from 25 per cent beneath to 30.6 per cent above the mean, for all the basins, except that unusually high flows at Massaboden in 1947, by far the warmest summer, raised the upper limit to 50.3 per cent.

About forty years of widespread glacier retreat and general reduction of the amount of snow and ice in long term storage in the Alps was terminated in 1964/65, a year with above average precipitation and the coolest summer since 1916 (Kasser, 1981). According to Kasser (1981), the glacierized area in the Rhône basin above Lac Leman was reduced by 11.3 per cent between 1934 and 1973. Year to year irregularity in sign and magnitude characterize net annual specific balance (mm or kg m^{-2}) of Griesgletscher (Fig. 3). Net storage increased in the late 1960s, but temporary gains were lost, and not replaced until the end of the 1970s.

Table III. Matrix of correlation coefficients showing interrelationships between hydrometeorological variables recorded at Saas Almagell and relationships between those variables and total annual runoff from Grosser Aletschgletscher measured at Blatten-bei-Naters, and with net annual mass balance (October-September) of Griesgletscher (M) for the periods 1968-1980 and 1967/68-1979/80, respectively.

	M	Q_{1-12}	Q_{5-9}	T_{5-9}	T_{6-8}	P_{11-5}	P_{10-5}	P_{6-9}
Q_{1-12}	-0.79							
Q_{5-9}	-0.82							
T_{5-9}	-0.85	0.86	0.90					
T_{6-8}	-0.95	0.76	0.83					
P_{11-10}	0.67	-0.29	-0.39	-0.56	-0.68			
P_{10-9}	0.70	-0.34	-0.42	-0.58	-0.71			
P_{11-5}	0.65	-0.30	-0.41	-0.59	-0.67			
P_{10-5}	0.67	-0.35	-0.43	-0.59	-0.69			
P_{6-9}	0.21	0.00	-0.01	-0.05	-0.19	0.09	0.03	
P_5	0.63	-0.19	-0.31	-0.35				0.30

Underlined values are significant at the p = 0.05 level. n = 13.

6. RELATIONSHIPS BETWEEN CLIMATIC VARIABLES AND MASS BALANCE OF GRIESGLETSCHER

Correlation coefficients computed between variables representing climatic elements recorded at Saas Almagell and net annual balance of Griesgletscher are presented in Table III. Net annual mass balance is strongly negatively correlated with mean summer air temperature variables, T_{5-9} explaining 72.3 per cent of the variance and T_{6-8} 90.3 per cent. More positive mass balance is directly associated with P_{10-9}, P_{11-10}, P_{11-5} and precipitation in May (49%, 44.9%, 42.3% and 39.7% variance explanation respectively). Strength of relationship between mass balance with summer temperature is markedly influenced by choice of variable, but less so for that with precipitation. P_{10-9} and P_{10-5} are slightly more strongly correlated with mass balance than P_{11-10} and P_{11-5} because precipitation in October over the glacier area is largely stored. Winter precipitation dominates this relationship, but much of the effect can be attributed to precipitation in May.

These relationships between climatic variables and mass balance are shown by the series plotted in Fig. 3. Positive balances of Griesgletscher are clearly associated with cool summers. Exceptionally large accumulations of snow in winters 1974/75, 1976/77, 1977/78 and

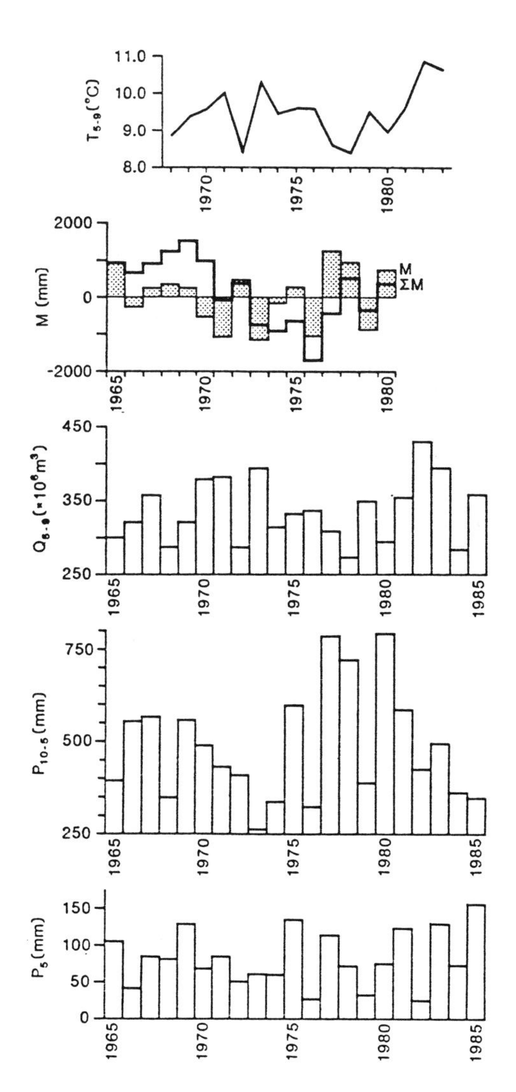

Figure 3. Year to year variations between 1965 and 1985 of mean summer air temperature (T_{5-9}) at Saas Almagell (top line); net annual changes in mass balance (M, shaded columns) and cumulative change in mass balance from October 1964 (ΣM) of Griesgletscher; total summer runoff (Q_{5-9}) of the Massa at Blatten-bei-Naters; and total precipitation at Zermatt during the winter (P_{10-5}) and (bottom line) in the month of May (P_5).

1979/80 led to positive mass balances, accentuated by the cool summers which followed in 1977, 1978 and 1980. High winter total snowfall is poorly correlated with snowfall in May. Correlation coefficients are presented only for relationships between independent climatic variables in Table III. Summer temperature is much more negatively correlated with preceding total winter precipitation than with May or total summer precipitation, notwithstanding the existence of a physical explanation for the latter but not for the former relationship.

The effects of interactions of precipitation and temperature were assessed by examining the proportions of variance in mass balance explained by hydrometeorological variables. Multiple linear regression models of the form $M = a + b_1 T + b_2 X_2$ $(+b_3 X_3)$ were fitted in order to estimate any improvement in variance explanation added by the use of one or two precipitation variables as well as a temperature variable. The temperature and precipitation variables which were tested are not strictly independent. Parameters, variables and goodness of fit of the models are listed in Table IV. As expected from the simple correlation coefficients, models involving T_{6-8} provided substantially better fit than those utilising T_{5-9}. Addition of P_5 increases explanation more than other precipitation variables (Eqs. 5 and 9) despite those variables individually having higher degrees of correlation with mass balance. The addition of a second and independent precipitation variable fails to enhance explanation (Eqs. 4, 10, 11). All the regression coefficients (b_k) are significant at the $p = 0.05$ level. A model based on T_{6-8} and P_5 would therefore be the most appropriate for description of the dependence of annual mass balance of glaciers on climatic conditions in the upper Rhône basin and for prediction. Ninety-one per cent of the variance of net specific balance of Griesgletscher is explained using records from Saas Almagell by Eq. 9, in which the variables used are conceptually sound.

Table IV. Multiple regression models of the form $M = a + b_1 T + b_2 X_2$ $(+ b_3 X_3)$ relating net annual mass balance of Griesgletscher to hydrometeorological conditions recorded at Saas Almagell for the period 1968-1980. X_2 and X_3 are precipitation variables.

Equation		X_2, X_3	a	b_1	b_2	b_3	r^2
1	T_{5-9}	P_{11-10}	8171	- 979	1.06		0.78
2	T_{5-9}	P_{11-5}	8755	-1006	0.95		0.76
3	T_{5-9}	P_5	8897	-1013	4.61		0.85
4	T_{5-9}	P_{11-5}, P_{6-9}	8241	-1009	0.89	2.27	0.79
5	T_{5-9}	P_{6-9},P_5	8724	-1020	1.04	4.32	0.85
6	T_{6-8}	P_{11-10}	12665	-1213	0.166		0.90
7	T_{6-8}	P_{10-9}	12542	-1202	0.162		0.90
8	T_{6-8}	P_{11-5}	12872	-1226	0.119		0.90
9	T_{6-8}	P_5	12327	-1179	1.15		0.91
10	T_{6-8}	P_{11-5}, P_{6-9}	12640	-1216	0.127	0.48	0.90
11	T_{6-8}	P_5,P_{6-9}	12274	-1178	1.09	0.22	0.91

Table V. Correlation coefficients showing relationships between hydrometeorological factors and runoff.

River/gauge	Meteorological station	Period	T_{5-9}	P_{11-10}	P_{11-5}	P_{6-9}	P_5	P_{5-9}
Fee Vispa/ Saas Fee	Saas Almagell	1968-1975						
Q_{1-12}			<u>0.76</u>	-0.14	-0.23	-0.03	-0.46	-0.32
Q_{5-9}			<u>0.73</u>	-0.18	-0.29	-0.01	<u>-0.58</u>	-0.38
Q_{6-9}			<u>0.70</u>	-0.20	-0.38	0.08	<u>-0.66</u>	-0.36
Rhône/Gletsch	Sion T Zermatt P	1956-1977						
Q_{1-12}			<u>0.79</u>	-0.22	-	-	-	-
Q_{5-9}			<u>0.80</u>	-0.25	-0.01	-0.28	-0.22	-
Q_{6-9}			<u>0.76</u>	-0.28	-0.03	-0.28	-0.22	-

Underlined values are significant at the $p = 0.05$ level.

7. RELATIONSHIPS BETWEEN CLIMATIC VARIABLES AND RUNOFF FROM GLACIERIZED BASINS

Curves of mean summer temperature parallel those of total annual discharge from all the basins for which records are graphed in Fig. 2. The strength of relationship between summer temperature and discharge, however, varies according to the percentage glacierization of a basin, increasing with larger ice cover (Collins, 1987). Similarly the influence of precipitation will change from a strong positive relationship in ice-free basins through to negative in heavily-glacierized basins, and the effect in relation to that of temperature will decline. In this paper, therefore, relationships are examined in detail for the basins with more than 50 per cent glacierization only. In these, runoff is strongly correlated with energy inputs. Relationships between Q_{1-12} and Q_{5-9} with both T_{5-9} and T_{6-8} are indicated by the correlation coefficients between runoff of the Massa and air temperature at Saas Almagell (Table III). T_{5-9} explains higher proportions of variance of flow than T_{6-8} (74.0 as against 57.8 per cent for Q_{1-12} and 81.0 against 69.0 per cent for Q_{5-9}). Thus, T_{5-9} is the preferred energy input variable for discharge in contrast with T_{6-8} for mass balance. Correlation coefficients for Q_{5-9} with precipitation are greater than for the equivalent variables with Q_{1-12}. Nevertheless, the most important precipitation variable P_{10-5} explains only 18.5 per cent of the variance in runoff, whereas over 40 per cent explanation of mass balance variation is accounted for by most of the precipitation variables. The relationships with runoff are confirmed by analyses of the Fee Vispa and Rhône basins (Table V). Q_{5-9} shows a marginally higher association than Q_{6-9} with T_{5-9}. P_5 at Saas Almagell explains over 34 per cent of the variance of summer runoff at the neighbouring gauge in Saas Fee. High winter precipitation in 1976/77, 1977/78 and 1979/80 with the ensuing cool summers produced lower than average summer discharges but the year-to-year variations in amount of winter precipitation occurring in May have not such a clear impact on flow (Fig. 3).

Table VI. Matrix of correlation coefficients showing interrelationships between hydrometeorological variables recorded at Saas Almagell in the period 1968-1983, and relationships between those variables and total annual (Q_{1-12}) and total summer (Q_{5-9}) flow of the Massa at Blatten-bei-Naters.

	Q_{1-12}	Q_{5-9}	T_{5-9}	P_{11-5}	P_{6-9}
T_{5-9}	$\underline{0.92}$	$\underline{0.94}$			
P_{11-10}	-0.18	-0.27	-0.38		
P_{11-5}	-0.22	-0.32	-0.41		
P_{6-9}	0.09	0.11	0.02	0.04	
P_5	-0.10	-0.22	-0.18		-0.01

Underlined values are significant at the p = 0.05 level. n = 16.

Table VII. Multiple regression models of the form $Q_{5-9} = a + b_1 T_{5-9} + b_2 X_2 \ (+ b_3 X_3)$ where X_2 and X_3 are precipitation values.

Eq.	river/ gauge	period	meteorological station	X_2	X_3	a	b_1	b_2	b_3	r^2
12	Massa/Blatten-bei-Naters	1968-1983	Saas Almagell	P_{11-5}		-250	61.2	0.0191		0.88
13				P_{6-9}	P_5	-227	58.5	0.0637	-0.0296	0.88
14				P_{11-5}	P_{6-9}	-263	61.0	0.0180	0.0614	0.89
15	Rhône/Gletsch	1956-1977	Sion (Couvent des Capucins)	P_5		-110	10.1	0.1540		0.75
16				P_{6-9}		-107	10.2	0.0113		0.75
17				P_5	P_{11-5}	-122	10.4	0.1630	0.0197	0.82
18				P_5	P_{6-9}	-118	10.7	0.1310	-0.0037	0.83

The stength of correlation between variables depends on the lengths of the series involved. The addition of data for only three years, 1981 through 1983, to those from 1968 through 1980 for the relationship between discharge of the Massa and climatic variables at Saas Almagell results in slightly higher values of the correlation coefficients of Q_{1-12} and Q_{5-9} with T_{5-9}, and P_5, and lower for relationships with P_{11-5} and P_{11-10} (see Tables III and VI).

Coefficients of determination for relationships between Q_{1-12} and Q_{5-9} with T_{5-9} expressed by bivariate linear regression show that in the range of 57.7-84.6 per cent of the variance in annual and 53.3-88.4 per cent of summer discharge totals can be explained, the actual levels depending on the pairs of stations involved. The levels of explanation can be improved on by fitting multiple linear regression models, analogous to those derived above for mass balance, which also provides a means of assessing the impact of energy and precipitation interactions on runoff. Models relating total annual discharge (Q_{1-12}) of the Rhône and Massa to climatic variables have been described by Collins (1987). Highest levels of variance explanation were obtained using either P_{11-10} alone or P_{11-5} with P_{6-9} as precipitation variables in addition to T_{5-9}. Parameters and variables of equations involving Q_{5-9} are given in Table VII with goodness of fit r^2. T_{5-9} has been used in all the multiple regression models for discharge, together with P_{11-5} where winter precipitation is incorporated. Use of T_{6-8} or P_{10-5} as variables would not improve goodness of fit with respect to either Q_{5-9} or Q_{1-12}. Multiple regression models for the Massa (for the years 1968-1983) provide trivial improvements in variance of Q_{5-9} or Q_{1-12} explained when precipitation variables are added, by comparison with an r^2 value for T_{5-9} alone of 0.88 (eqs. 12-14, and see Collins (1987, Table VII, eqs. 13 and 14)). Addition of precipitation variables improves explanation of Q_{5-9} and Q_{1-12} of the Rhône considerably, however (eqs. 15-17, and Collins (1987, Table VII, eqs. 6-8)). The value of r^2 for Q_{5-9} related with T_{5-9} alone for the Rhône is 0.64, raised by incorporation of P_5 and P_{6-9} together to 0.83. Precipitation in May again appears as the most important precipitation variable. P_5 has a positive rather than the negative regression coefficient that might be expected both from correlation analysis, and physically from albedo effects at the glacier surface.

The discharge series for the Massa is non-stationary, flows declining from maxima in the warm 1940s when glacier extent was also greater than in more recent years (see Fig. 2, Q_{1-12}) with reduction in flow continuing until the high discharges of the warm summers of 1982 and 1983 (Fig. 4, Q_{5-9}). Correlation coefficients and regression parameters will vary therefore according to which time segment of the series is used in defining relationships. Predictive ability will depend on the calibration period used (Collins, 1987). Within a period for which a multiple regression model offers a high degree of fit, however, analysis of residuals (effectively pointing to extreme years) might provide further detail of the impact of individual climatic elements on runoff in particular years. Residuals from the fitting of equation 17 for the period 1956-1977 (Table VII) are plotted in Fig. 4 together with the variables involved. In the model represented by equation 17, P_5 and P_{11-5} are the precipitation variables. Since P_5 is a subset of P_{11-5},

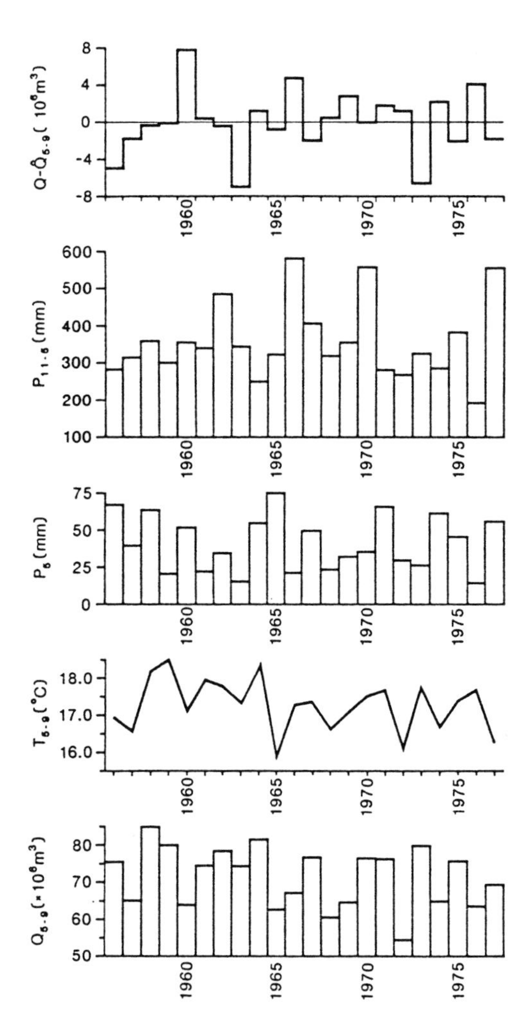

Figure 4. Year to year variations between 1956 and 1977 of deviations ($\hat{Q}_{5-9} - Q_{5-9}$) of total summer runoff of the Rhône at Gletsch predicted by the multiple regression equation $\hat{Q}_{5-9} = 10.4\ T_{5-9} + 0.0163\ P_5 + 0.0197\ P_{11-5} - 122$ from measured flows, based on summer temperature and precipitation measured at Sion (Couvent des Capucins) (top line); total precipitation measured during winter (P_{11-5}); precipitation in May (P_5) and mean summer air temperature (T_{5-9}) at Sion; and (bottom line) total measured summer discharge (Q_{5-9}) of the Rhône at Gletsch.

these are not independent variables; the spurious correlation coefficient calculated between them is 0.4. Effectively, in the multiple regression, extra weighting is given to P_5. The regression coefficients for P_5 and P_{11-5} are positive whereas respective correlation coefficients with Q_{5-9} are negative. Equation 17 underestimates Q_{5-9} most in 1956, 1963 and 1973, years with above average flows, but by no more than 9%. These years were not unusual in terms of temperature, but P_5 was below average in 1963 and 1973. Absence of excessive snow in late spring would, however, be likely to favour summer ablation. Large overestimates in 1960, 1966 and 1976 (up to 12%), in which seasons below average flows of the Rhône were registered, result from above average precipitation in winter 1959/60 and May 1960, and in the winter of 1965/66. Exceptionally low precipitation values occurred, however, in 1975/76. Analysis of residuals from regression models is inconclusive but at least suggests that the models are oversimplified, and that further factors must be involved, or that data aggregation at the monthly level is insufficiently sensitive to interactions between precipitation and energy supply.

8. DISCUSSION

Summer energy inputs, represented by the variables T_{5-9} and T_{6-8}, clearly exert a stong influence on both mass balance and runoff. The long runoff series respond both from year to year and throughout the period from the 1920s to the 1980s to ablation season temperatures. River flows declined considerably from the 1940s to the 1970s, not just as a result of generally falling temperatures, however, but also because of reduction of glacier areas. Planimetric area decrease has been small as a fraction of total glacierized area, but almost all losses have been in the ablation zone. In the Massa basin, from 1927 glacierized area reduced by 6.1% to 1957, and a further 0.7% by 1973, 7.3% in total (Kasser, 1981). Tongues of larger glaciers such as Aletschgletscher failed to respond to positive mass balances in the 1960s unlike those of many smaller ice masses which subsequently advanced. Decadal mean T_{5-9} at Sion declined from 18.08°C (1941-1950) to 17.07°C (1968-1977). In the same periods, mean discharge (Q_{5-9}) levels were 2.33 and 1.72×10^6 m^3 km^{-2} respectively, a fall of 26.2 per cent.

What fraction of the flow reduction has resulted from glacier recession is difficult to assess. From year to year, interaction between energy input and precipitation leads to wide deviations from the overall trend of the series. Inclusion of the 1930s and 1940s therefore in the period of calibration of a multiple regression model would be expected to result in flows in later years being overestimated because of glacier recession (Collins, 1987). This is not always the case, even before 1978. Kasser (1973) introduced glacier recession as an independent variable in multiple regression to predict flow of the Rhône to Porte du Scex (16.8% reducing to 13.6% glacierization) but with negligible impact on r^2. The degree of correlation between T_{5-9} or T_{6-8} and Q_{5-9} or net balance will reflect changing areas of ice exposed to melting, controls on which are the rate at which the transient snow line rises in early summer and the final height reached (a function of the interaction between energy and precipitation). Year to year changes in area of ice exposed to melting because of this effect will be much greater than those small annual incremental decreases resulting from continued retreat. Such changes probably distort the linearity of relationships with heat input, reducing goodness of fit. An expression of possible non-linearity and the influence of non-stationarity is provided by calculated correlation coefficients and parameters of multiple regression models changing with the length of series and with the portion of the series used. Collins (1987) has demonstrated the effects of choice of calibration period on the predictive ability of multiple regression models of discharge. Non-linearity of relationships with temperature will also occur because glacier recession leads to the ice which remains being at higher and cooler elevations so diminishing melting, and because of the outcome of energy-precipitation interaction in one year influencing glacier surface albedo in succeeding years. A low end of season snowline will lead to raised averaged albedo in the next year even if, in the following summer, the transient snowline reaches a high elevation, as firn from the persisting snow in the previous summer will be exposed rather than ice.

In several of the relationships, climatic elements have ambivalent influences on mass balance or runoff which are reflected in measured strengths of association. Winter precipitation totals influence runoff both by contributing the snowmelt component of flow and by reducing the amount of icemelt by maintaining high albedo. High winter snowfall therefore produces a large snowmelt component but retards the rate of ascent of the transient snowline. The net effect in these highly-glacierized basins is a negative correlation coefficient of low magnitude between winter precipitation variables and summer discharge (e.g. in the range -0.19 to -0.42 in Table III). Since melting of ice dominates runoff, with high specific yields of meltwater, the direct relationship that snow accumulation should have on runoff from the ice-free area is completely masked by the inverse effect on icemelt. With lower percentage glacierization of basins, the direct relationship will come to dominate, the transition point depending on the quantity of precipitation received. Summer precipitation will directly influence the component of runoff from the ice-free area and should contribute to runoff from the glacier ablation area. However, if precipitation occurs as snow, albedo will be raised which will decrease the predominant icemelt component. It is this effect that produces correlation coefficients between summer precipitation and runoff which are close to zero.

Precipitation at all times tends towards production of more positive outcomes of net mass balance. Winter precipitation directly influences accumulation and reduces melting in the ablation zone in the following summer. Summer precipitation in the form of snow has the same effects, and although the length of time of cover of the ablation area will be short, melting at peak rates is suppressed. Rain in summer at lower altitudes is not in itself detrimental to ablation, but whether occurring as snow or rain higher up, will be accompanied by addition to storage in the accumulation zone. These considerations are reflected in the slight positive value of the correlation coefficient between mass balance and P_{6-9}.

The correlation coefficient for mass balance with T_{6-8} is more negative than that with T_{5-9}, since in May and September, use of incoming heat energy may be reduced by snow cover. This order is reversed in that Q_{5-9} and Q_{1-12} are both more highly correlated with T_{5-9} than T_{6-8}, probably reflecting the influence of warm conditions in September boosting summer runoff totals. Seasonal distribution of variability in climatic elements and runoff will therefore affect the performance of variables. The fraction of total annual flow occurring between May and September varies independently of total annual flow (Fig. 5), in the range 85-91%. Variability of monthly flow is greatest in May and September of the summer months (Collins, 1985), and is higher in winter months than in summer, although the absolute quantities of water involved in winter are small. The proportion of P_{5-9} occurring in

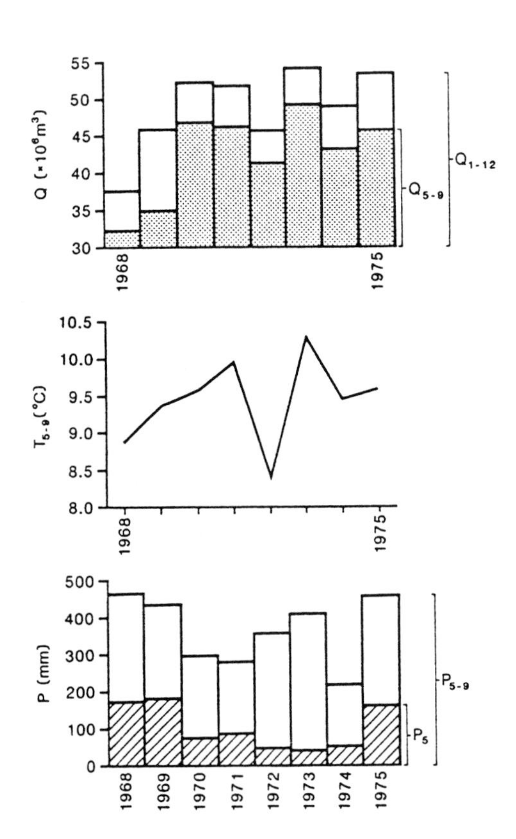

Figure 5. Proportions of total annual flow (Q_{1-12}) of the Fee Vispa at Saas Fee occurring between May and September (Q_{5-9}) (upper), mean summer air temperature (T_{5-9}) (centre) and fractions of total summer precipitation (P_{5-9}) falling in May (P_5) (lower) at Saas Almagell, in the period 1968-1975.

May is also highly variable (Fig. 5) and independent of absolute quantities, between 10 and 41%. Of P_{11-5}, P_5 accounts for between 9 and 45% (Fig. 4). The coefficient of variation for precipitation (at Saas Almagell) in May ($c_v = 0.548$) is much larger than for annual or seasonal totals (P_{11-10}: $c_v = 0.200$, P_{11-5}: $c_v = 0.311$, and P_{5-9}: $c_v = 0.238$). The occurrence of precipitation in May is therefore of considerable importance.

High levels of explanation of variance at around 90 per cent by multiple regression of net mass balance against T_{6-8} and whichever precipitation variables (Table IV), are matched by similar levels against Q_{5-9} with respect to hydrometeorological variables at Saas Almagell (Tabel VII). Temperature and precipitation at Sion (Couvent), a station located at a considerably lower elevation than and at a great distance from Rhônegletscher provide nevertheless up to 83% explanation of variance of runoff at Gletsch. Actual year to year deviations of summer temperature from long term means are consistent over the Alpine

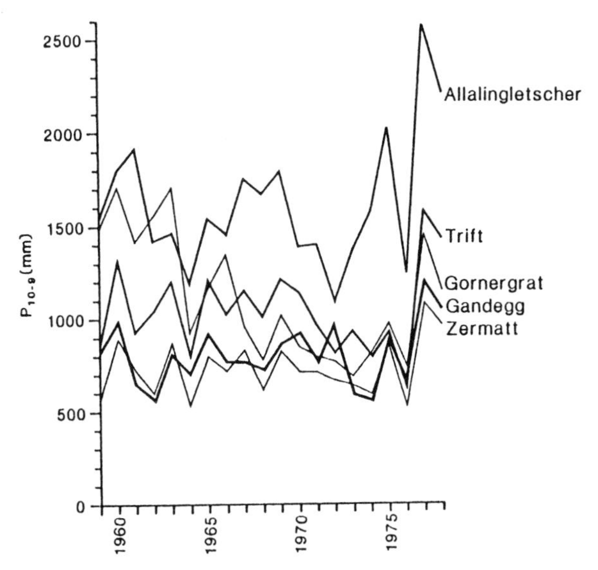

Figure 6. Total annual precipitation (P_{10-9}) recorded at the totalising gauges at Allalingletscher, Gandegg, Gornergrat and Trift, and at the meteorological station at Zermatt between 1959 and 1978.

area, whereas precipitation is much more variable making selection of stations critical. Conversion of temperatures using a standard lapse rate to those experienced at the equilibrium lines of glaciers is feasible (and was undertaken by Günther and Widlewski (1986)), but estimating precipitation variation with elevation is more difficult. There are few accurate measurements of either snowfall or summer rainfall at high altitude, where precipitation measurement errors are great because of turbulence caused by wind and because thick snowpack necessitates the installation of raingauges several metres above the (summer) ground surface. Measurement of snow accumulation in precipitation gauges is unreliable, and sites are unlikely to be representative of the spatial variability expected over glacier accumulation basins. Whilst estimation of precipitation in high mountain areas remains an unsolved problem, some indication of the relationship between precipitation and altitude in the Rhône basin is offered by records from the totalizing raingauges. The four totalising gauges were selected from those available (see Schweizerische Meteorologische Anstalt, 1985) on grounds of representing the upper height ranges of the Vispa basin, continuity of record and a high degree of consistency in behaviour from year to year. These variations are shown together with measurements from the conventional precipitation gauge at Zermatt for the period 1959-1978 in Fig. 6, and correlation coefficients between the records are presented in Table VIII. Taking first differences of the series improves the calculated levels of association (Table IX). A first difference represents the deviation, smaller or greater, of a measurement in one year by comparison with that in the previous year. Broadly parallel behaviour in sign and quantity of deviation from year to year is indicated by the correlation coefficients of Table IX. Use of first differences of annual mass balance and discharge data would also have led to improved levels of correlation. The variation of precipitation with elevation as indicated by these measurements is shown for several years in Fig. 7. The relationship is irregular, non-linear and non-superimposable. Estimation of precipitation levels over glacier and glacierized basins from valley or mountain gauges remains subject to large errors.

Separation of the runoff component arising from snowmelt and rainfall over the ice-free area from that produced by the glacier would improve the relationships between runoff and climatic elements, but is rendered impractical by the difficulty of estimating precipitation increase with altitude. This problem also arises in the development of conceptual models of mass balance and runoff which are physically-based rather than founded on linear regression. Development of such models is necessary to overcome the non-stationarity and non-linearity of climate - runoff relationships, to allow for the ambivalent effect of precipitation inputs, and to permit the form of precipitation (rain or snow) to be determined from prevailing air temperatures. The time basis for data input will need to be shorter than the monthly periods employed in this investigation.

Consistent deviations of summer temperatures over this area of the Alps have led to parallel variations in discharge of glacier-fed streams. Ablation characteristics of glaciers in summer can therefore be assumed to have varied consistently over the area. As precipitation, which varies both spatially and non-linearly with elevation, has a stronger influence on annual mass balances, these might be expected to be less constrained. However, cumulative mass balance over several years shows the impact of the years with exceptionally high winter or May inputs of precipitation. Extreme inputs have tended to be widespread in particular years throughout this region. There is a surprisingly strong negative correlation between quantity of winter snowfall and air temperature of the following summer (e.g. 1976/77; 1977/78; 1979/80 - high snowfall totals and low summer temperatures, and 1972/73 - reduced snowfall and high summer temperatures) for which teleconnection there is no simple physical explanation. Only summer temperatures have been considered in this paper, but winter temperatures control the cold content of snow. In spring, heat energy has to be supplied to raise the temperature of the snowpack to 0°C before melting can take place.

Table VIII. Matrix of correlation coefficients between measurements of total annual precipitation recorded at the gauge at Zermatt and at the four totalising precipitation gauges.

	Zermatt	T	Ga	Go
Trift	0.89			
Gandegg	0.79	0.72		
Gornergrat	0.42	0.50	0.27	
Allalingletscher	0.80	0.71	0.56	0.36

Underlined values are significant at the p = 0.05 level. n = 20.

Table IX. Matrix of correlation coefficients calculated between first differences of total annual precipitation records from the gauge at Zermatt and the four totalising precipitation gauges.

	Zermatt	T	Ga	Go
Trift	0.90			
Gandegg	0.81	0.75		
Gornergrat	0.78	0.79	0.65	
Allalingletscher	0.82	0.75	0.58	0.61

n = 19.

The results of linear regression analysis in this paper compare favourably with those of Martin (1975, 1977) for the mass balance of Glacier de Sarennes with meteorological elements at Lyon Bron, 130 km distant. A value of r^2 of 0.77 was obtained using P_{10-5}, P_6 and the sum of mean maximum July and August temperatures, and 0.84 when May and June temperatures were added. Precipitation in spring was also found by Martin (1977) to affect mass balance positively although precipitation in June was used rather than May. Günther and Widlewski (1986) obtained weaker correlation coefficients between the mass balance of Griesgletscher and respective temperature variables at Guetsch (30 km to the north-east) for the period 1963-1980 than those reported here for Saas Almagell from 1968 to 1980. Correlation coefficients with hybrid precipitation variables based on records from Zermatt and Airolo (20 km east) for P_{6-9} and P_{10-9} in that period were greater than those presented in this paper.

The relationship between climatic variables and mass balance is probably non-stationary. In the short series examined, within which glacierized area was relatively stable, a regression model calibrated on the earlier part of the series might be expected to produce reasonable predictions of mass balance in later years. Fitting the model described by equation 11 for the years 1967/68-1976/77 gives:

$$M = 13779 - 1285\ T_{6-8} + 0.29\ P_5 - 0.70\ P_{6-9}$$

using climatic variables from Saas Almagell. This model estimates substantially more positive (or less negative) net mass balances in the years 1978 to 1980 than were measured by between 51 and 491 kg m^{-2}. A similar model for runoff from Rhônegletscher underestimates Q_{5-9} in the same years. That there exist groups of years in which poor fits of predicted to observed values arise probably results from glacier surface albedo conditions at the end of the ablation season in one year influencing those of the next. It is unfortunate that climatic data at Sion (Couvent des Capucins) and Saas Almagell are not available for later and earlier years respectively.

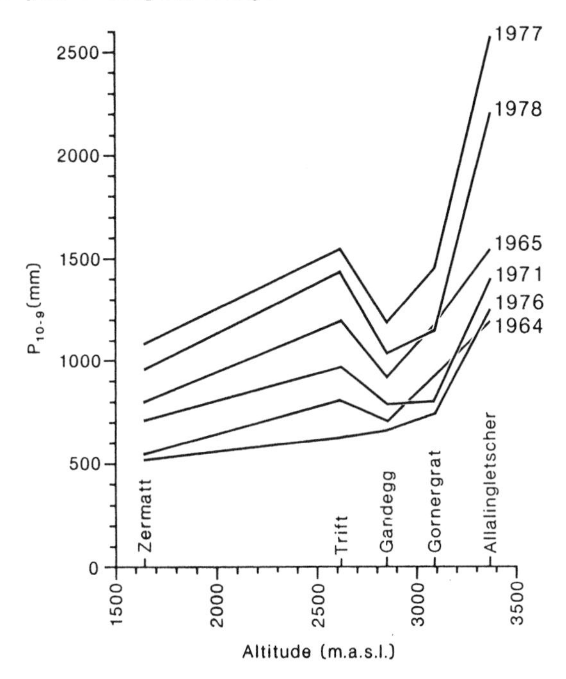

Figure 7. Relationships between precipitation (P_{10-9}) and altitude in the higher areas of the Vispa basin indicated by measurements at Zermatt and at the totalising gauges at Allalingletscher, Gandegg, Gornergrat and Trift for certain years between 1958/59 to 1977/78.

9. CONCLUSIONS

Marked climatic fluctuations have influenced both mass balance of glaciers and runoff from heavily glacierized basins in the Rhône catchment area in the period 1922-1985. Generally-declining mean summer air temperatures have occurred at the same time as larger glaciers continued to recede, and precipitation tended to increase. Annual net balances have tended towards negative with and summer total river flows have responded directly to summer energy inputs. Precipitation has had a lesser impact on runoff than on net annual mass balance.

This investigation has attempted to identify the climatic and runoff variables which provide the best relationships between hydrometeorological elements, mass balance and runoff from glaciers, and has sought the best combinations of those variables to explain the variances of net balance and summer runoff. Mean air temperature in the months June through August emerged as the most important individual variable for mass balance and temperatures in the months May through September for total summer runoff between May and September. Winter precipitation (October-May) and total annual precipitation (October-September) are the most important precipitation variables,when treated independently of temperature, although precipitation in May alone is almost of equal status with respect to mass balance.

Multiple regression models relating total summer runoff and net annual mass balance to mean summer air temperature and precipitation variables provide a high degree of fit, between 75 and 91 per cent of the variances of Q_{5-9} and net balance being explained by the best combinations of variables from the optimum stations. Precipitation in May and total summer precipitation or both are the best individual and collective precipitation variables for explanation of variation in both mass balance and runoff. Regression coefficients for these variables have positive signs when Q_{5-9} is regressed against precipitation and temperature together although simple regression coefficients with precipitation are negative. The magnitudes of both sets of coefficients are small. Ambivalent relationships between precipitation and runoff necessitate the use of physically-based models for predictive purposes. A better account of the interaction or albedo effect of precipitation on incoming energy supply should be offered by such models, which would also allow the phase (solid or liquid) of precipitation over glacier surfaces in May and September to be determined using ambient air temperatures. These factors will be of importance in models devised to predict future impacts of climatic variations on storage of snow and ice in and on runoff from Alpine catchments. However, estimation of the variation of precipitation with altitude in Alpine basins, essential for such modelling, remains an outstanding problem in mountain glaciology and hydrology.

Net mass balance and summer runoff totals are moderately negatively correlated, as a result of different interactions amongst the set of climatic elements. Multiple regression models are consistent in underestimating runoff in those years when predicted mass balance is more positive than the actual measured value. Other variables, such as cold content of the snowpack and radiation, may improve the fit of regression models, but physically-based conceptual models, which incorporate varying average surface albedo by allowing the transient snowline to rise, will probably be required for predictive purposes.

ACKNOWLEDGEMENTS

The assistance of Landeshydrologie und -geologie, Bundesamt für Umweltschutz, Bern and Schweizerische Meteorologische Anstalt, Zürich in making available discharge and climatic records respectively is gratefully acknowledged.

REFERENCES

Collins, D.N. 1985. Climatic variation and runoff from Alpine glaciers. Zeitschrift für Gletscherkunde und Glazialgeologie, 20, 127-145.

Collins, D.N. 1987. Climatic fluctuations and runoff from glacierised Alpine basins. In: "The influence of climate change and climatic variability on the hydrologic regime and water resources". International Association of Hydrological Sciences Publication 168, 77-89.

Fountain, A.G. and Tangborn, W.V. 1985. The effect of glaciers on streamflow variations. Water Resources Research 21 (4), 579-586.

Günther, R. and Widlewski, D. 1986. Die Korrelation verschiedener Klimaelemente mit dem Massenhaushalt alpiner und skandinavischer Gletscher. Zeitschrift für Gletscherkunde und Glazialgeologie 22, 125-147.

I.A.H.S. 1973. "Flucuations of glaciers 1965-1970". International Association of Hydrological Sciences/UNESCO. Paris, 357 pp.

Kasser, P. 1973. Influences of changes in the glacierised area on summer runoff in the Porte du Scex drainage basin of the Rhone. In: Symposium on the Hydrology of Glaciers. International Association of Scientific Hydrology Publication 95, 221-225.

Kasser, P. 1981. Rezente Gletscherveränderungen in den Schweizer Alpen. In: Jahrbuch der Schweizerischen Naturforschenden Gesellschaft, wissenschaftlichter Teil, 1978. Birkhauser, Basel, 106-138.

Martin, S. 1975. Corrélation bilans de masse annuels - facteurs météorologiques dans les Grandes Rousses. Zeitschrift für Gletscherkunde und Glazialgeologie 10, 89-100.

Martin, S. 1977. Analyse et reconstitution de la série des bilans annuels du glaciers de Sarennes, sa relation avec les fluctuations du niveau de trois glaciers du Massif du Mont Blanc (Bossons, Argentière, Mer de Glace). Zeitschrift für Gletscherkunde und Glazialgeologie 13, 127-153.

Meier, M.F. and Roots, E.F. 1982. Glaciers as a water resource. Nature and Resources, UNESCO 18, 7-14.

Patzelt, G. 1985. The period of glacier advances in the Alps, 1965 to 1980. Zeitschrif für Gletscherkunde und Glazialgeologie 21, 403-407.

Schweizerische Meteorologische Anstalt 1985. Annalen der Schweizerischen Meteorologischen Anstalt 1984. Zürich.

ON THE BLOWING SNOW IN ADELIE LAND, EASTERN ANTARCTICA

A contribution to I.A.G.O. (Interaction - Atmosphere - Glace - Ocean)

Gerd Wendler
Geophysical Institute
University of Alaska
Fairbanks, AK 99701

ABSTRACT

Measurements of blowing snow were carried out in Adelie Land, Eastern Antarctica, with a blowing snow measuring device developed after Schmidt (1977). The instrument photo-electrically measures the number and size of the snow particles. The logarithm of the mass fluxes were found to vary with the wind speed, a result previously reported.

We know the annual wind speed from an automatic weather station (AWS), which reports over satellite year round. Hence we were able to calculate the annual mass flux of snow; a value of 6.3×10^6 kg m^{-1} a^{-1} was found, which is a credible value for a windy area (mean annual wind speed 12.8 m/s).

1. INTRODUCTION

Adelie Land, Eastern Antarctica, is one of the windiest places in the world. This was first pointed out by Mawson (1915), who wintered over at Cape Denison, and was astounded by the strength of the wind. He expressed this in the title of his book, "Home of the Blizzard". The winds are of katabatic nature. A shallow layer of air at the surface is cooled for most of the year due to negative radiation balance, and this layer, having a density higher than the air aloft at the same height further down the slope, flows down the ice slopes of the Antarctic Continent. This is a very widespread phenomenon in Antarctica. Indeed, there is not a single meteorological parameter on any continent, which influences the surface climate of the whole continent as much as the katabatic wind does for Antarctica (Schwerdfeger, 1984). Hence, it is not surprising that this wind has been studied widely (e.g. Ball, 1960; Streten, 1968; Weller, 1969; Loewe, 1970; Radok, 1973). It can reach exceptionally high speeds if it is reinforced by a synoptic weather pattern, or if it occurs at a preferred area of the Antarctic surface where funnelling occurs (Parish, 1984). For example, Poggi (1979) measured a maximum of 96 m/s at Dumont d'Urville, one of the strongest winds measured anywhere in the world.

J. Oerlemans (ed.), Glacier Fluctuations and Climatic Change, 261–279.

Strong winds over a snow surface transport large amounts of snow. At moderate wind speeds this transport takes place close to the surface, and is called drifting snow. At higher wind speeds, normally values around 10 m/s, the snow is elevated to greater heights, and one speaks of blowing snow. A manifestation of this transport of snow is to be found in sastrugi, a kind of snow dune, which cover large parts of Antarctica. They are normally directed along the resultant wind vector. Mather and Miller (1967) used the direction of these sastrugi to deduce the mean wind field over Antarctica. Their results are in good agreement with modelling studies which were carried out more than a decade later (Parish, 1984). Such sastrugi are very pronounced in Adelie Land, and can reach heights exceeding one meter.

2. AREA, TIME OF MEASUREMENT AND CLIMATOLOGY.

Some eights years ago, a joint U.S. - French experiment was started to investigate the katabatic wind regime in Eastern Antarctica (Wendler and Poggi, 1980; Poggi et al., 1982). Blowing snow is of importance for the understanding of the katabatic wind. It has at least two effects on the wind speed (Kodama and Wendler, 1986):

1) Snow suspended in the air increases the density of the air, and thus increases the katabatic force;
2) Air moving down the Antarctic slope is adiabatically warmed, which results in unsaturated air. When blowing snow is present, the surface area of the snow is substantially increased, and the air will stay close to saturation pressure all the time. This sublimation of snow particles cools the air.

Both processes represent a positive feedback on the wind speed. This explains the observation that after blowing snow has started, an increase in the intensity occurs.

Besides its effect on the wind velocity, blowing snow deserves interest by itself. Large amounts of snow are redistributed, hence having great importance for glaciological studies (Radok, 1970). It has been described as a low level avalanche moving slowly to the edge of the Antarctic continent and then out to the ocean. Therefore, it is of great importance for the mass balance of Antarctica (Loewe, 1970). The magnitude of this mass flux is poorly known (Meier, 1986).

As part of the joint U.S. - French experiment, measurements of the blowing snow were carried out. This paper reports on detailed measurements made at D 47 (67°23'S, 138°43'E) in Adelie Land at an altitude of 1560 m (Fig. 1). D 47 is located some 110 km south of the year-round station Dumont d'Urville. The slope angle is of the order of 6.5×10^{-3}. Further, D 47 is also the site of an automatic weather station (AWS), which reports over satellite on a year-round basis (Renard & Salinas, 1977; Stearns & Savage, 1981). Hence, a large amount of climatological data is available (Wendler & Kodama, 1985). These data give not only seasonal and day-to-day observations, but also the diurnal course. The mean annual wind speed for D 47 is 12.8 m/s, the maximum is

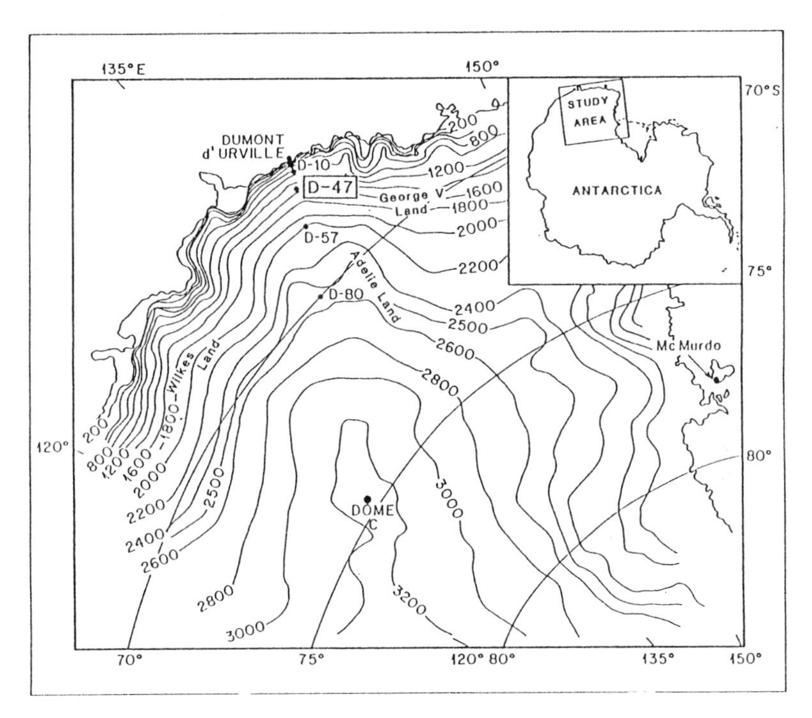

Figure 1. Location map of our measuring site in Antarctica. Altitude lines are given in meters.

observed in winter (13.5 m/s seasonal mean), while a minimum occurs in summer with 11.3 m/s. The directional constancy, which is the resultant wind vector divided by the sum of all wind speeds has a mean annual value of 0.94. This is a very large value and means that winds which do not blow down the slope coming from a southerly direction are a very rare occurrence. The surface of the whole area is covered with sastrugi, which have the same direction as the resultant wind. This agreement was already used by Mather, who deduced the surface wind field from the sastrugi direction. Sometimes two directions of sastrugi can be observed. This happens when a storm with a somewhat crosswise direction to the mean resultant wind occurs. Fig. 3 displays such an occasion. Normally, they disappear after a while, and only one direction is displayed (Fig. 2).

The mean annual temperature is, at -25.7°C, relatively mild. The seasonal minimum is, of course, in winter (-30.1°C), while the summer has a mean temperature of -17.2°C. Positive temperatures were never measured, as the station is located just above the dry snow line, which is of importance for the intensity of the blowing snow.

Our snow blowing measurements were carried out in November and December of 1985. During the time period the temperature varied between -13°C and -22°C.

Figure 2. Photo of sastrugi field at our measuring site, D 47, eastern Antarctica.

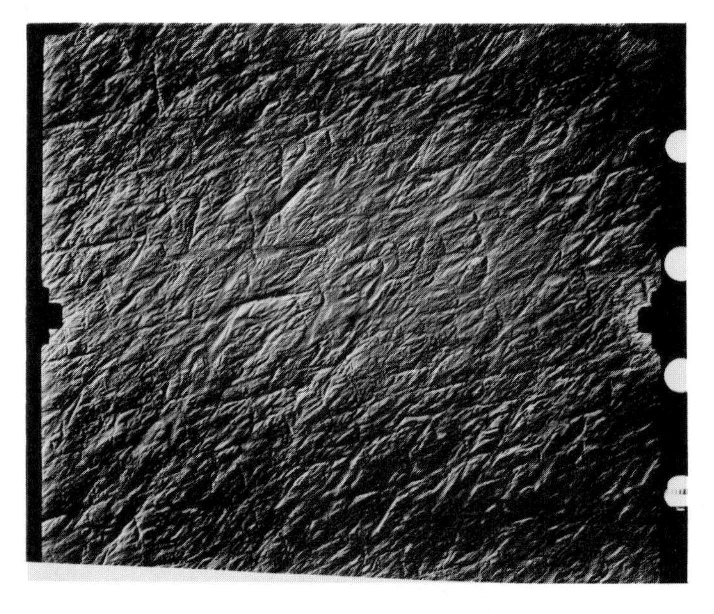

Figure 3. Aerial photo taken from an LC 130 over Adelie Land. Note the cross direction of the sastrugi to the right. Photo: Dr. J. Kelley.

3. INSTRUMENTATION

Measurements of blowing snow date back a long time. In Antarctica, the Byrd snow drift project was one of the larger efforts to obtain a better understanding (Budd et al., 1966). Altogether, this topic has received substantial attention in Antarctica, e.g. Mellor and Radok (1960), Budd (1966), Loewe (1970), Naruse (1970), Kobayashi (1972), Shillians (1975) and Kobayashi (1978).

All devices which catch snow particles directly have a certain collection efficiency, which can be fairly low and is often not known very well. Hence, some twenty years ago the concept of detecting snow particles photo-electrically was first applied. Such instrumentation was first tested in Antarctica by the Australian Antarctic Division (Landon-Smith & Woodberry, 1963). Hollund et al. (1966) developed the instrument further. Their work indicated the possibility of detecting individual snow particles by their shadows on photosensitive semi-conductors. Not only the number of particles, but also information on size and speed could be obtained. The instrument was in the following years further refined, mainly to avoid obstruction of the wind flow and to increase the signal to noise ratio (Rogers & Sommerfield, 1968; Schmidt & Holub, 1971; Schmidt, 1971). Finally, Schmidt (1977) wrote a detailed paper on the snow particle counter including design criteria. Following his design, two instruments were built by Mimken and Hill, Fairbanks, Alaska. A description of the design criteria which differed from those of the original design by Schmidt are given in the appendix authored by G. Mimken.

While we had problems with one of these instruments, the other one performed well and was applied in the present study in Adelie Land. A photograph of the instrument at the measurement site is given in Fig. 4

Figure 4. The photo-electric blowing snow measuring instrument during a good weather period, installed at D 47, eastern Antarctica.

4. CALIBRATION

The calibration of the snow blowing sensor of Schmidt - Mimken is not easy (Schmidt, 1977), especially as far as the mean sizes of the particles are concerned. Our sensor was built in such a way that the output voltage for frequency is a linear function of the number of particles. We verified this calibration curve in the laboratory with an electronic impulse generator, and good agreement was found.

More difficult was the calibration of the particle size. Also Schmidt (1977) showed more scatter for size than for frequency. Hence we carried out elaborate calibration procedures in the laboratory with rotating wires of different sizes after a method suggested by Schmidt (1977). After initial difficulties, correct procedures, e.g. rotation speed, were established. In. Fig. 5 the calibration curve is given. In contrast to Schmidt's original design, our instrument was designed in such a way that a linear relationship between size and output voltage existed. Further, our curve goes through (0,0).

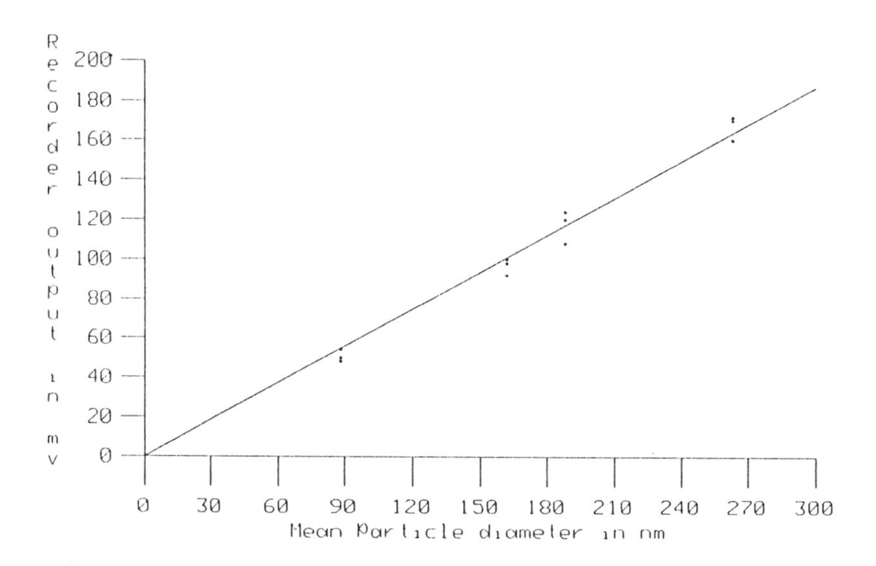

Figure 5. Calibration curve of the Schmidt-Mimken photo-electric snow blowing measuring device, derived in the laboratory for size of the particle. Note the good linear relationship between output voltage and size.

Field measurements in Adelie Land gave particle sizes between 100 and 300 µm. This is a relatively small size, however, as blowing snow is a very frequent occurrence, snow flakes are broken up rapidly, and small particles are found. The values we found are of the same magnitude as found by Schmidt (1977). Further, a visual inspection of the sizes was carried out in the field with the help of a fine grid and a magnifying glass. Generally, diameters between 100 and 300 μm were observed, agreeing well with our calibration.

We wanted to check our snow blowing device in the field under more realistic conditions. Therefore, a mechanical snow collecting device was built. It has a round opening with a diameter of 16 mm. In the first 80 mm, the wind was decelerated three to four times, then in the vertical direction by another factor of nearly two. The snow fell into an inner box, which was surrounded by an outer box. The boxes were buried in the snow so that the disturbances in the wind field were minimal.

On 18 December 1985, at 9.55 h, a comparison of the two was carried out for a one hour period at a height of 38 cm. The mean wind speed during the period was 17.1 m/s at a 4 m altitude, with gusts exceeding 20 m/s. Moderate to strong blowing snow was observed during the period. 78.8 g of snow were found in the box, which results in a flux of 30.8 g cm^{-2} h^{-1} or 85.5 g m^{-2} s^{-1}.

Our photo-electric device gave the mean diameter of the particles as 185 μm and a mean frequency of 1480 particles/second. However, we made our calculations for shorter time periods, so that a higher accuracy could be obtained. As the flux is proportional to the third power of the diameter, mean values normally underestimate the total flux. The photo-electric device gave a flux of 35.6 g cm^{-2} h^{-1} or 98.9 g m^{-2} s^{-1}. Assuming there to be no error in the photo-electric device, these values would give a collection efficiency of 86% for the mechanical device.

Further, a comparison with snow drift data measured at Mizuho, the Japanese Antarctic station, was carried out (Kobayashi, 1978). If adaptation for our measuring heights and our wind speeds was performed, good agreement was found. Hence, it was concluded from laboratory calibration, field intercomparison and by a comparison with results from other investigations under similar conditions that the photo-electric snow blowing device gave reliable data.

5. WIND SPEED

A meteorological tower four meters high was equipped with four wind sensors located at 0.5, 1.0, 2.0 and 4.0 m respectively, above the snow surface. The anemometers measured the revolutions electronically. These counts were given for ten second intervals on a Campbell Data Logger CR 7, which was programmed to integrate the values for ten minute intervals. Some profiles of the wind speed for periods of neutral stability conditions (temperatures between 0.5 and 4.0 m within 0.1°C) are given in Fig. 6. The typical logarithmic dependency with height is observed, which is to be expected for neutral conditions and relatively strong wind velocities with a large amount of mixing. During periods with low wind speeds and a negative radiation budget (stable conditions) these profiles change, but this is of little interest here, as we are concerned about blowing snow. Roughness parameters of the order of z_0 = 0.1 mm were observed, which is in good agreement with the literature

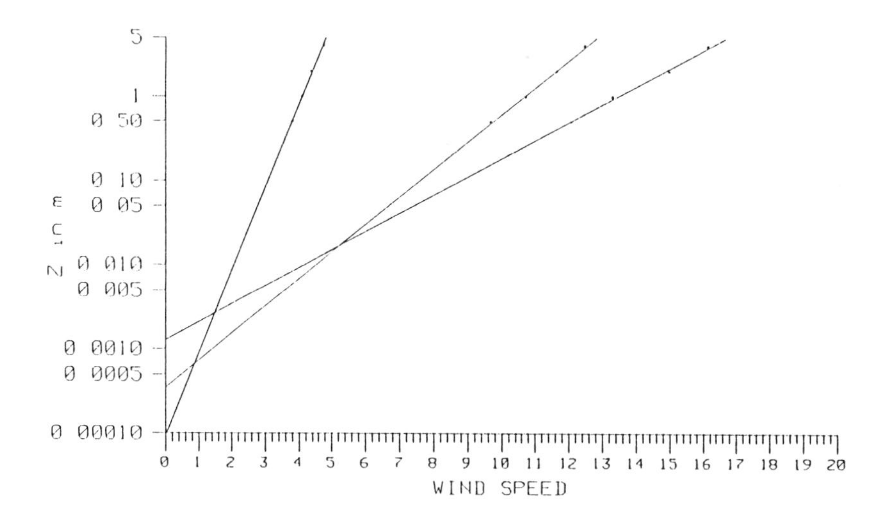

Figure 6. Wind speed against the logarithm of the height for neutral atmospheric conditions at D 47, eastern Antarctica. Note that with increasing wind speed (blowing snow) the surface characteristic changes, and the roughness parameter increases.

(e.g. Liljequist, 1957). Temperature profiles were, as expected for windy periods, very close to neutral.

For windy periods (with wind speeds at the 4 m level above 10 m/s) the roughness parameter increases. This result, that blowing snow changes the surface characteristics, has been previously observed, e.g. König (1985), who carried out his measurement on the Ekström Ice Shelf of Antarctica.

6. BLOWING SNOW AND WIND SPEED

We tried to establish relationships between the ten minute averaged wind speeds and the amount of blowing snow, as measured by the particle counter. Neither for the mean wind speed, nor with the maxima of wind speed within the ten minute period and the particle count could satisfactory results be obtained. Tabler as cited by Schmidt (1977) had already expressed the opinion that a ten minute time interval might be too long. Schmidt (1977) verified this with his measurements in the Rocky Mountains and we reconfirmed it for Antarctica.

A cup anemometer with a very short response time (type TS-2A) was therefore installed at one height (110 cm). It was intercalibrated with the anemometers of the towers, and a mean deviation of 2% in speed was found, which was acceptable. After this, the wind speeds were integrated for four second intervals, which were compared to the output of the blowing snow measuring device. It could be seen that maxima in wind speed are correlated to maxima in frequency as well as in size. In other words, stronger winds do not only pick up more snow particles, but also larger ones. In Figure 7 the frequency and mean size of the blowing snow particles is plotted against the wind speed for one specific height (30 cm) and one specific date (1 December 1985). It can be seen that there

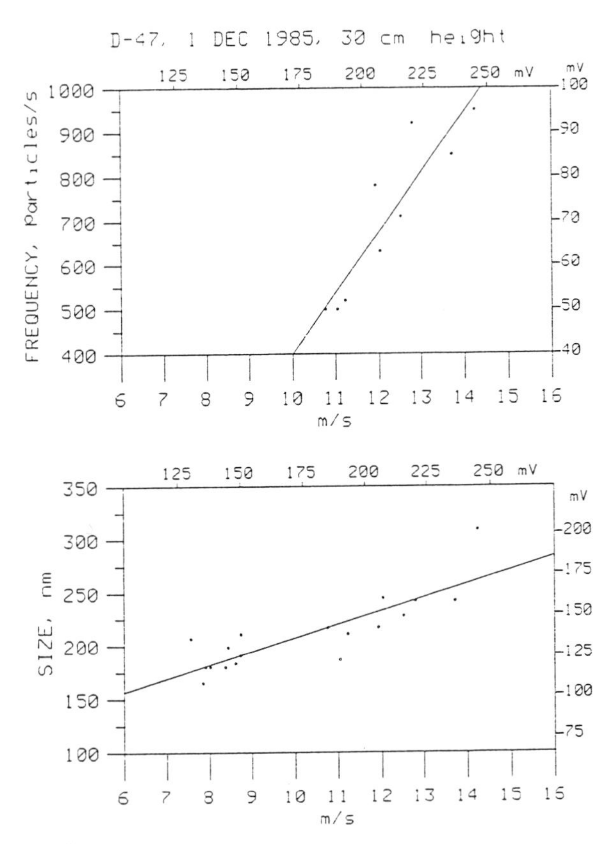

Figure 7. Number of particles per second (frequency) and mean size of the particles against wind speed for one specific event (1 December 1985) and one specific height (30 cm) at D 47, eastern Antarctica.

is a linear relationship between frequency and wind speed and size and wind speed, with the number of particles being much more sensitive to wind speed changes that the sizes. Hence it was possible to calculate the flux at a specific height (F_z) and for a specific wind speed:

$$F_z = \rho \; cf \, \frac{\pi}{4} \, d^3$$

ρ = density of the snow (we assumed 0.9 g cm^{-3})
c = instrument constant (depends on the distance between the two slits and the distance between the sender and receiver)
f = frequency of the particles (a function of wind speed and the height of the sensor)
d = mean diameter of the particles (a function of wind speed and the height of the sensor)

Measurements were carried out for other events (Table I), however, the calculations presented here are based on the measurements of 1 December 1985, as all measurements are not completely evaluated. The flux at a

Table I. Blowing snow measurements

Date	Time	Height (cm)	Recording frequency	Wind speed (m/s)	Comments
25 Nov	16.15 to 18.20	10	minute and hourly	10.3 to 12.6	No special wind instrument
27 Nov	11.20 to 16.00	10	minute and hourly	9.8 to 11.4	fast responding wind sensor installed
28 Nov	8.00 to 15.00	5,10 and 20	minute and hourly	7.1 to 11.0	
1 Dec	9.00 to 12.00	5,10, 30 and 60	minute and hourly	11.0 to 14.0	
6 Dec	10.00 to 15.00	5	hourly	12.8 to 15.0	winds up to 17.6 m/s
9 Dec	10.30 to 13.30	5,10, 30 and 60	minute and hourly	8.2 to 10.4	
13 Dec	9.30 to 15.00	5,10, 30 and 60	minute and hourly	10.0 to 16.0	snow fall possible visibility <20 m
19 Dec	9.50 to 11.30	38	hourly	15.0 to 19.0	intercomparison with other snow gauge

specific height could be plotted against wind speed. As Kobayashi observed at the Mizuho Station, a linear relationship between the logarithm of the snow transport and the wind speed was found. In Fig. 8, this relationship is presented for three heights (10 cm, 30 cm and 60 cm) for the above mentioned blowing snow event of 1 December 1985. While the amount of blowing snow is a strong function of height, the gradient of the amount of the blowing snow as a function of height is presented for different wind speeds, as obtained from the 1 December 1985 event.

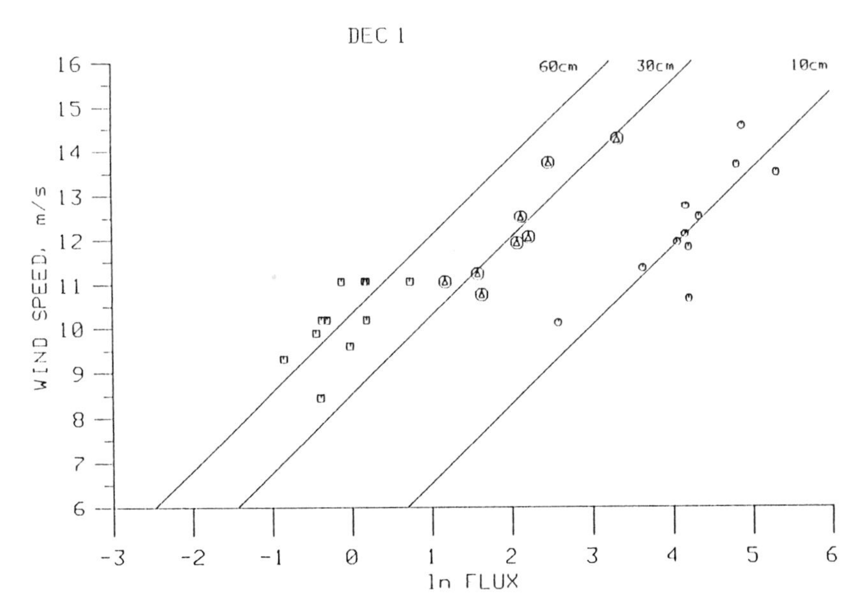

Figure 8. The logarithm of the flux against wind speed for three specific heights (10 cm, 30 cm, 60 cm) on 1 December 1985 at D 47, eastern Antarctica.

We can now integrate with respect to height to obtain the total flux (F):

$$F = \int_{z_0}^{\infty} F_z \, dz$$

As the amount of blowing snow mathematically goes to infinity at the surface, a minimum height of 1 mm was used for z_0 . The following relation holds for the amount of blowing snow:

$$\ln(F) = 0.21U - 0.45$$

with F expressed in g s^{-1} m^{-1}, and U is wind speed in m s^{-1} at 3 m altitude. This formula gives lower values than the one found by Budd et al. (1966); however, the slope is very similar.

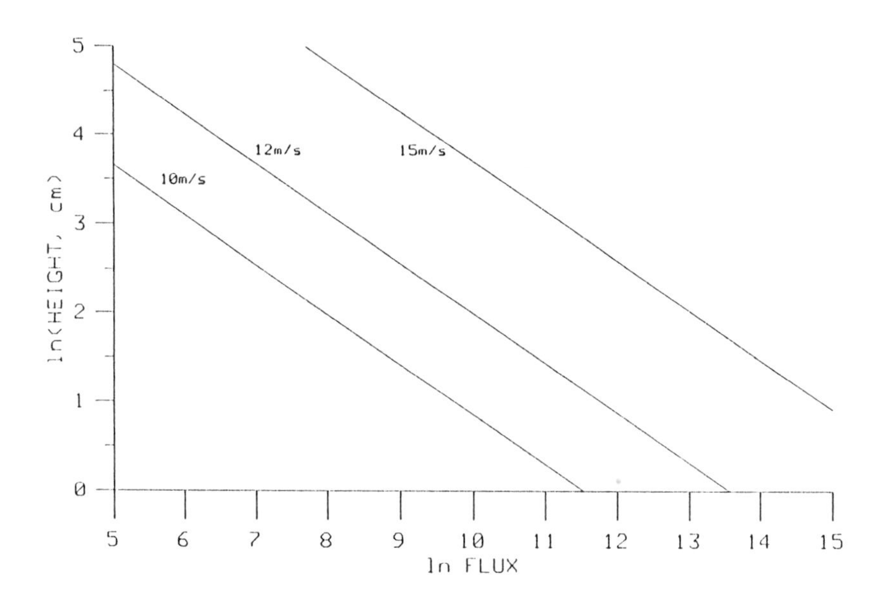

Figure 9. Flux against height (both logarithmic) for different wind speeds for D 47, eastern Antarctica. The curves were derived from the measurements of 1 December 1985.

7. TOTAL ANNUAL TRANSPORT

The automatic weather station (AWS) (Stearns & Weidner, 1986) reports meteorological data of D 47 year-round: including the wind speed at the nominal height of 3 m. In Fig. 10 the annual course of the wind speed is presented. Absolute maxima of 55 m s^{-1} were observed. We note a winter maximum, and a slight minimum in summer. To calculate the flux, the mean wind speed is of lesser importance than its distribution, as strong blizzards transport most of the snow. Hence, in Fig. 11 the mean monthly frequency distributions of the wind speed are presented.

From hourly wind speeds, the total amount of drifting snow was calculated. A value of 6.3×10^6 kg m^{-1} a^{-1} was found: a very large value. Kobayashi offered a value of nearly an order of magnitude less for Mizuho Camp, but Loewe (1970) estimated values up to 70×10^6 kg m^{-1} a^{-1} as possible for windy areas. As our site is in one of the windiest areas of Antarctica, this value is quite reasonable.

However, one must be somewhat careful with this value, as it is based only on the summer observations. For example, Kobayashi (1972) showed that variations of up to an order of magnitude are possible for the same wind speed. This agrees with our observation, that the density of the blowing snow decreases with time, even if the wind speed stays constant. After one or two days of blizzard, there is hardly any loose snow available, and the snow becomes so hard that footprints are no longer visible. This is in contrast to the findings of Schmidt (1986), who made his measurements in the Rocky Mountains, and found an increased flux over hard surfaces. Our observations agree with those of Loewe (1970) who also worked in Adelie Land, and states:

> "Occasionally, the density of the drift can become very small, even with strong winds. After a prolonged period of gale from the same direction, the snow surface will become firmly compressed and so perfectly streamlined that no more snow can be removed. Such a case was experienced at Port-Martin on 8 March, 1951, when the drift almost ceased although the wind remained of the order of 40 m s^{-1} with unchanging direction."

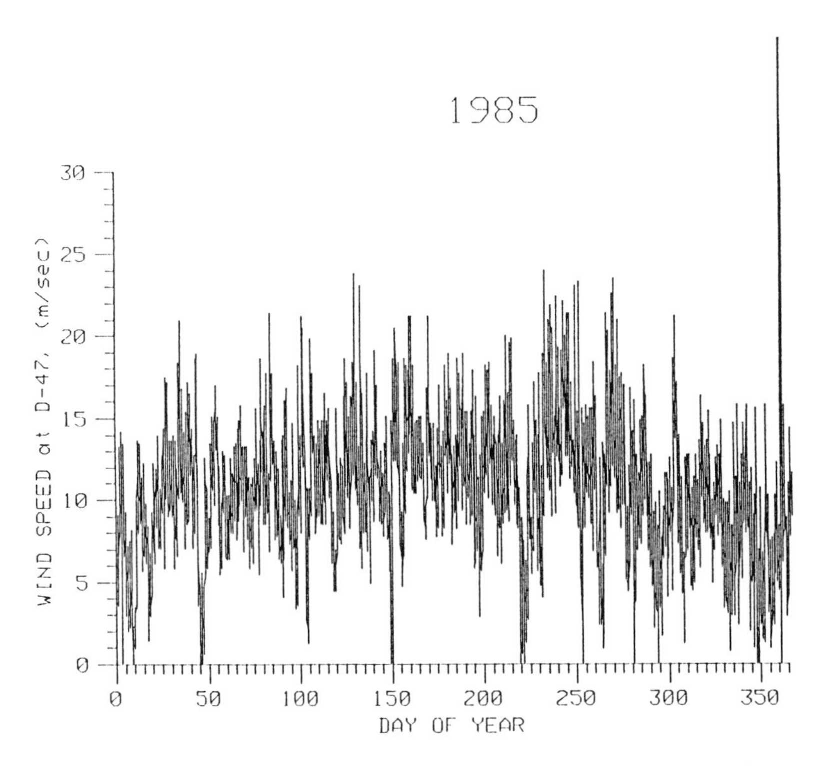

Figure 10. Annual wind speed distribution at D 47, eastern Antarctica for 1985. Winter data for 1985 were not available; 1984 data were substituted.

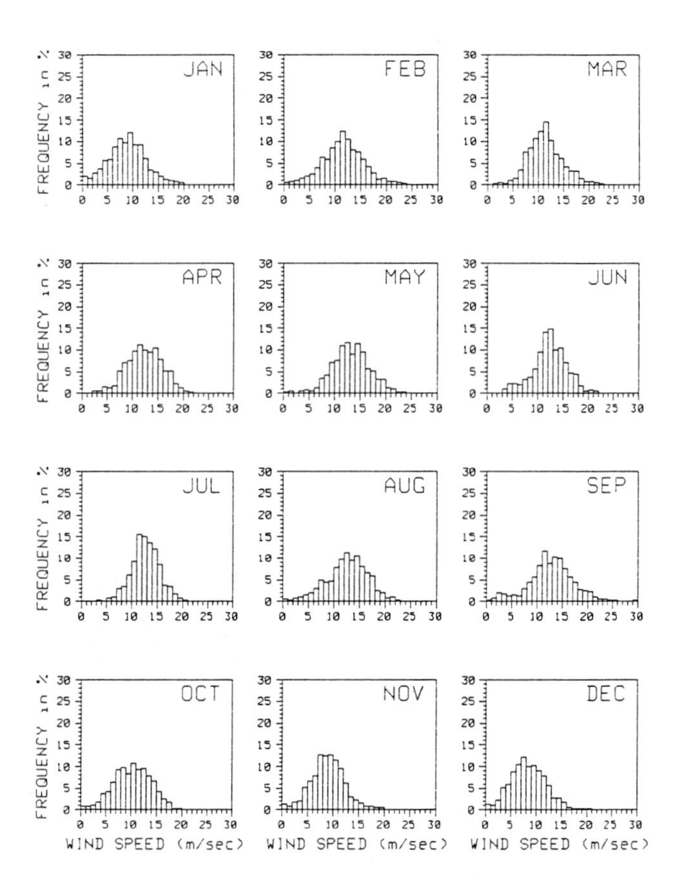

Figure 11. Frequency distribution of the wind speed each month at D 47, eastern Antarctica.

8. CONCLUSION

A photo-electric blowing snow measuring device has been introduced, which has two advantages over the traditional snow traps:

1) Little or no disturbance of the wind field.
2) High time resolution, about two orders of magnitude better than the traditional ones. This makes it much easier to relate wind speed variation to blowing snow flux.

An estimate of the annual wind flux gave a high but reasonable value (6.3×10^6 kg m^{-1} a^{-1}), and shows the importance of the blowing snow for the mass balance of Antarctica. Maybe as convincing as a scientific argument is Fig. 12, which shows the tail of a downed LC 130 in Adelie Land nearly covered by snow.

ACKNOWLEDGEMENTS

This study was supported by NSF Grant DPP 8100161. My thanks go to many people from the U.S. Antarctic Research Program and Expeditions Polaires Françaises, without whose help this study could not have been carried out. Furthermore, my thanks go to F. Brill for the data handling, computer work and drafting, which he carried out with great dedication.

Figure 12. Photograph of the tail of a downed LC 130 in eastern Antarctica, nearly covered by snow.

APPENDIX

Blowing snow device, constructed by Mimken and Hill
by G. Mimken

In the design of the instrument, we closely followed the work of Schmidt (1977). Because of the large effort required to calibrate this type of instrument it was decided to follow not only the rationale of the original design, but also its essential electronic and mechanical characteristics.

The instrument was built in stainless steel to the original design dimensions. This gave an instrument that was durable in a variety of environments, and was capable of being deployed both on the coast where it was subject to salty atmosphere, and inland.

There were three major differences in the electronic system that could be considered worthy improvements. They are listed as follows:

A) The power source for the lamp uses the ±15 VDC system power with the provision for monitoring lamp status to indicate a failure of the light source.

B) Use of modules to provide signal conversion functions instead of discrete components as in the original design. System scaling remained the same in all respects. This was a reflection of advancements in semiconductor technology.

C) The addition of a module to provide output giving the log of the particle frequency in addition to the linear particle frequency output of the original design.

Additionally, we did not require the use of the visual range computer and so used a high speed, two channel chart recorder for the size and frequency outputs.

The table below gives our size calibration results:

	Wire Size	Output
A)	90 μm	48 mV
B)	164 μm	90 mV
C)	188 μm	115 mV
D)	260 μm	164 mV

Frequency was calculated using a signal generator, the results are listed below:

	Freq	Output
A)	0 Hz	0 mV
B)	1000 Hz	105 mV
C)	2000 Hz	210 mV
D)	3000 Hz	315 mV

	Freq	Output
E)	4000 Hz	420 mV
F)	5000 Hz	524 mV
G)	6000 Hz	627 mV
H)	7000 Hz	730 mV

The results of these calibrations demonstrate the functionality of the design.

REFERENCES

Ball, F.K. 1960. Winds on the ice slopes of Antarctica. Antarctic Meteorology, Oxford, Pergamon Press, 9-16.

Budd, W.W. 1966. The drifting of non-uniform snow particles. In: M. Rubin (ed.), Studies in Antarctic Meteorology. American Geophys. Union, Antarctic Res. Ser., 9, 59-70.

Budd, W.F., Dingle, W.R.J., and Radok, U. 1966. The Byrd snowdrift project, outline and basic results. In: M. Rubin (ed.), Studies in Antarctic Meteorology. American Geophys. Union, Antarctic Res. Ser., 9.

Hollung, O., Rogers, W.E. and Businger, J.A. 1966. Development of a system to measure the density of drifting snow. Joint Tech. Rep. Elec. Eng., Dep. Atmos. Sci., University of Washington, Seattle, 54 pp.

Kobayashi, D. 1972. Studies of snow transport in low-level drifting snow. Contributions from the Institute of Low Temperature Science, Series A, 24, 1-58.

Kobayashi, S. 1978. Snow transport by katabatic winds in Mizuho Camp Area, East Antarctica. Journ. Met. Soc. of Japan, 56, no. 2, 130-139.

Kodama, Y. and Wendler, G. 1986. The wind and temperature regime along the slope of Adelie Land, Antarctica. Journ. Geophys. Res. 91 (D6), 6735-6741.

König, G. 1985. Roughness length of an Antarctic ice shelf. Polarforschung 55 (1), 27-32.

Landon-Smith, I.H. and Woodberry, B. 1965. The photoelectric metering of windblown snow. Interim Report Series A (IV), Glaciol. Publ. 79, 18.

Liljequist, G. 1957. Energy exchange of an Antarctic snowfield, Norwegian-British-Swedish Antarctic expeditions, 1949-1952. Scientific Results, 2, part 1C.

Loewe, F. 1970. The transport of snow on ice sheets by the wind. Studies on drifting snow, Met. Dept., University of Melbourne, 13, 1-69.

Mather, K.B. and Miller, G.S. 1967. Notes on topographic feature affecting the surface winds in Antarctica, with special reference to katabatic winds. Tech. Rep. UAG-R-189, Univ. Alaska, Fairbanks, 125 pp.

Mawson, D. 1915. The home of blizzard, being the story of the Australian Antarctic expedition 1911-1914. Heinemann, London, 338 pp.

Meier, M. 1986. Glaciers, ice sheets and sea level: Effects of a CO_2-induced climatic change. Rep. of Workshop, Ad Hoc Comm. on the relation between land ice and sea level, Comm on Glac., Polar Res. Bd., 330 pp.

Mellor, M. and Radok, U. 1960. Some properties of drifting snow. In: Antarctic Meteorology, Pergamon Press, 333-346.

Naruse, R. 1970. Measurement of drifting snow on the coastal region of Antarctica, near Syowa Station (in Japanese with English abstract). Contributions from the Institute of Low Temperature Science, Series A, no. 28, 147-154.

Parish, T. 1984. A numerical study of strong katabatic winds over Antarctica. MWR 112, 545-554.

Poggi, A. 1979. Personal Communication

Poggi, A., Delunay, D., Mallot, H. and Wendler, G. 1982. Interactions Atmosphere-Glace-Ocean en Antarctique de l'Est. Proceedings of the Argos Users Conference, Paris, 1982, 103-111.

Radok, U. 1970. Boundary processes of drifting snow. Studies of Drifting Snow, Met. Dept., University of Melbourne, 13, 1-20.

Radok, U. 1973. On the energetics of surface winds over the Antarctic Ice Cap. Energy fluxes over polar surfaces. Proceedings of IAMAP, IAPSO/SCAR/WMO Symposium, Moscow, WMO Tech. Note 179, 2-5.

Renard, R.J. and Salinas, M.G. 1977. The history, operation, and performance of an experimental automatic weather station in Antarctica. Naval Postgraduate School, Monterey, CA, NPS-63Rd7710, 57.

Rogers, W.E. and Sommerfield, R.A. 1968. A photoelectric snow particle counter. Am. Geophys. Union Trans, no. 49, 690 pp.

Schmidt, R.A. 1984. Transport rate of drifting snow and the mean wind wind speed profile. Boundary Layer Met. 34, 213-241.

Schmidt, R.A. 1977. A system that measures blowing snow. USDA Forest Service Research Paper, RM-194, 80.

Schmidt, R.A. 1971. Processing size, frequency and speed data from snow particle counters. USDA Forest Service Research Note RM-196, 4.

Schmidt, R.A. and Holub, E.W. 1971. Calibrating the snow particle counter for particle size and speed. USDA Forest Service Research Note RM-189, 8.

Schwerdtfeger, W. 1984. Weather and Climate of the Antarctic. Elsevier Science Publishers B.V., 261 pp.

Shillians, I.M. 1975. Effect of inversion winds on topographic detail and mass balance on inland ice sheet. J. Glaciology 14, 85-90.

Stearns, C. and Savage, M. 1981. Automatic weather station, Austral summer 1982-1983. Antarctic Journal, 1985 Review, 189-191.

Stearns, C. and Weidner, G. 1986. Antarctic Automatic weather stations, Austral summer 1984-1985. Ant. J. of the U.S. Annual Review 1985, 189-191.

Streten, N.A. 1968. Some characteristics of strong wind periods in coastal East Antarctica. J. Appl. Met. 7, 46-52.

Weller, G. 1969. A meridional surface wind speed profile in Mac Robertson Land, Antarctica. Pure and Appl. Geophy. 77, no.6, 193-200.

Wendler, G. and Poggi, A. 1980. Measurements of the katabatic wind in

Wendler, G. and Poggi, A. 1980. Measurements of the katabatic wind in Antarctica. Ant. J. of the U.S., 1980 Review, 193-195.
Wendler, G. and Kodama, Y. 1985. Some results of climatic investigations of Adelie Land, eastern Antarctica. Zeitschrift für Gletscherkunde und Glazialgeologie 21, 319-327.

ALTITUDINAL SHIFT OF THE EQUILIBRIUM LINE IN GREENLAND CALCULATED FROM HEAT BALANCE CHARACTERISTICS

W. Ambach and M. Kuhn
Universität Innsbruck
Innsbruck
Austria

ABSTRACT

Applying a degree-day method, the altitudinal shift of the equilibrium line is calculated for the Greenland Ice Sheet by assuming changes in temperature and in cumulative accumulation. For a disturbance of $\delta T_a = 1K$, the altitudinal shift of the equilibrium line is obtained between $96\ m < \Delta h < 168\ m$, depending on the gradient of cumulative accumulation. For a disturbance $\delta c = +100\ kgm^{-2}$ the altitudinal shift of the equilibrium line amounts to $-78\ m < \Delta h < -48\ m$. Regions with negative gradient of cumulative accumulation react essentially more sensitive as regions with positive gradient.

1. INTRODUCTION

The mass balance of the Greenland Ice Sheet is essentially determined by the position of the equilibrium line. In order to calculate the ltitudinal shift of the equilibrium line induced by climatic warming or by a change in cumulative accumulation, a method is applied, which is based on earlier work by Kuhn (1980, 1987b). The parameters necessary are deduced from heat balance measurements carried out at Camp IV-EGIG and at Carrefour station (Fig. 1) at the western EGIG-profile (Ambach, 1963, 1977).

2. DEGREE-DAY METHOD

The heat balance equation at the equilibrium line reads:

$$\tau H = Lkc \qquad (1)$$

where τ means the number of days with ablation, i.e. days with averaged temperature $\geqslant 0°C$. Note that an air temperature of 0°C is not necessarily the threshold of ablation (Kuhn, 1987a), but it is a practical approximation. Equation (1) is true independent of the definition of τ since it is the product τH that determines the energy

J. Oerlemans (ed.), Glacier Fluctuations and Climatic Change, 281–288.

sum available for melting. H is the time average of the heat flux density available for melting, L the latent heat of fusion and c the cumulative accumulation. The factor k takes into account the formation of superimposed ice (Ambach, 1985).

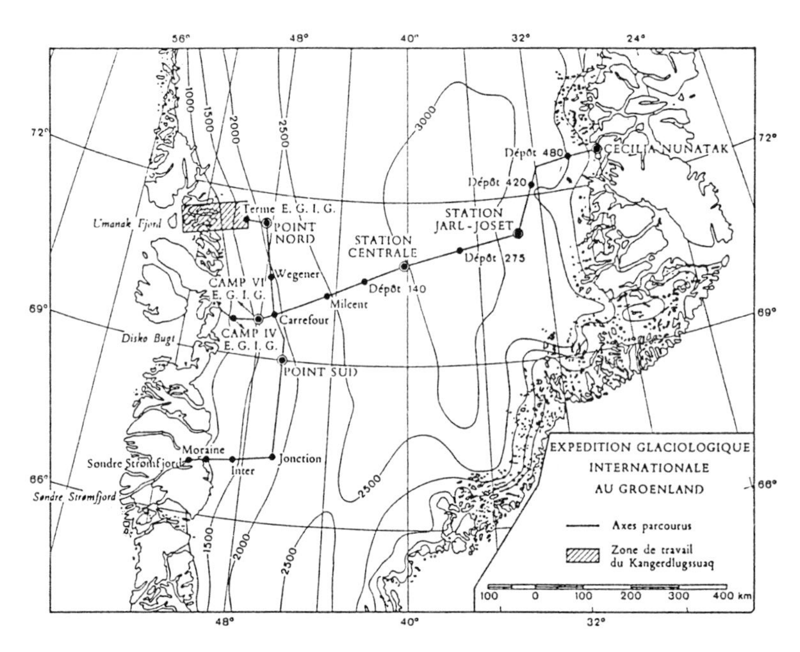

Figure 1. Routes of the International Glaciological Greenland Expedition (EGIG). Map drawn by Expéditions Polaires Françaises.

If the equilibrium

$$\tau_0 H_0 = Lkc_0 \tag{2}$$

at the altitude h_0 is disturbed by $\delta\tau$, δH or δc, equilibrium will be reestablished at a new altitude $h' = h_0 + \Delta h$ with

$$(\tau_0 + \delta\tau + (\partial\tau/\partial z)\Delta h)\ (H_0 + \delta H + (\partial H/\partial z)\Delta h) = Lk(c_0 + \delta c + (\partial c/\partial z)\ \Delta h) \tag{3}$$

For simplicity, δH and $\delta\tau$ are discussed as functions of air temperature δT_a only, and surface temperature $T_s = 0°C$. The conversion holds:

$$\delta H = \mu \delta T_a \tag{4}$$

$$\delta\tau = \gamma \delta T_a \tag{5}$$

The numerical values were evaluated from heat balance measurements at Camp IV-EGIC (Ambach, 1985):

$$\tau_0 = 35 \text{ d}$$

$$c_0 = 450 \text{ kg m}^{-2}$$

$$H_0 = 7.15 \text{ MJ m}^{-2} \text{ d}^{-1}$$

$$L = 0.33 \text{ MJ kg}^{-1}$$

$$k = 5/3$$

$$\mu = 1.1 \text{ MJ m}^{-2} \text{ d}^{-1} \text{ K}^{-1}$$

$$\gamma = 10.4 - 0.011\, \Delta h \quad \text{dK}^{-1} \quad \text{for } \Delta h > -200 \text{ m}$$
$$= 13.5 \text{ dK}^{-1} \quad \text{for } \Delta h < -200 \text{ m}$$

$$\partial T_a/\partial z = -0.0073 \text{ K m}^{-1}$$

The range of values of $\partial c/\partial z$ accepted here for the Greenland Ice Sheet is

$$-0.32 \text{ kg m}^{-2}/\text{m} < (\partial c/\partial z) < 0.55 \text{ kg m}^{-2}/\text{m} \qquad (6)$$

Negative values were measured in profiles at 77°N and 70°N (Benson, 1962) between 1000 and 2500 m a.s.l. (77°N) and 2000 and 3000 m a.s.l. (70°N). Positive values were obtained in the western part of the EGIG-profile (Ambach, 1985) up to 1300 m a.s.l.

3. ALTITUDINAL SHIFT OF THE EQUILIBRIUM LINE

3.1 Altitudinal shift of the equilibrium line following a temperature change

The solutions of eq. (3) are shown in Fig. 2 for changes in temperature δT_a keeping $\delta c = 0$. Regions with negative values of $\partial c/\partial z$ react more sensitively to a given δT_a than regions with positive values (Ambach & Kuhn, 1985). In the range of $\partial c/\partial z$, given by eq. (6), which is indicated by the dashed lines in Fig. 2, a raise of the equilibrium line of $96 \text{ m} < \Delta h < 168 \text{ m}$ is predicted for $\delta T_a = 1\text{K}$.

3.2 Altitudinal shift of the equilibrium line by change in cumulative accumulation

In a similar way Fig. 3 presents the consequences of changes in cumulative accumulation δc as function of $\partial c/\partial z$ keeping $\delta T_a = 0$. As before, regions with negative values of $\partial c/\partial z$ react more sensitively than regions with positive values. For $\delta c = 100 \text{ kgm}^{-2}$ and for the limiting values of $\partial c/\partial z$ given by eq. (6), the altitudinal shift of the equilibrium line amounts to $-78 \text{ m} < \Delta h < -48 \text{ m}$.

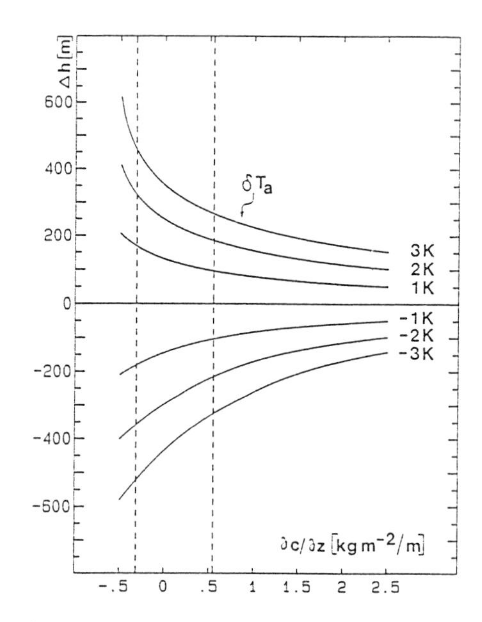

Figure 2. Altitudinal shift of the equilibrium line versus the gradient of cumulative accumulation depending on disturbances in temperature.

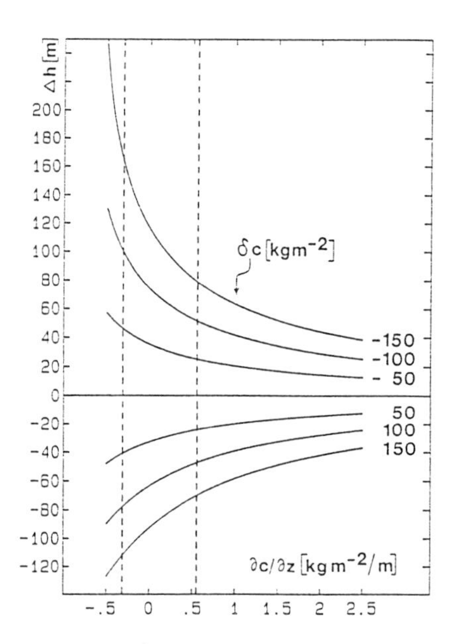

Figure 3. Altitudinal shift of the equilibrium line versus the gradient of cumulative accumulation depending on disturbances in cumulative accumulation.

5. CONSEQUENCES RELATED TO MASS BALANCE

The total net ablation and total net accumulation for the entire Greenland Ice Sheet are calculated for disturbed conditions by

$$A' = A_0\ \overline{a}'/\overline{a}_0 + \overline{a}'\ \Delta S_a \qquad (7)$$

$$C' = C_0\ \overline{c}'/\overline{c}_0 + \overline{c}'\ \Delta S_c \qquad (8)$$

with $\Delta S_a = -\Delta S_c$. A, C mean total net ablation and the total net accumulation, $\overline{a}$, $\overline{c}$ the mean specific net ablation and mean specific net accumulation, and ΔS_a, ΔS_c the change in ablation and accumulation area by the altitudinal shift of the equilibrium line. The subscript zero corresponds to the undisturbed condition, the primed values to disturbed conditions. Furthermore, it is evident from Fig. 4 that

$$\overline{a}'/\overline{a}_0 = h'/h_0 \qquad (9)$$

where h_0 and h' are elevations above the ice margin which is kept at its present altitude (Fig. 4). For the numerical evaluation the following values are assumed: C_0 = 500 km^3 w.e., A_0 = 300 km^3 w.e., $\overline{a}_0$ = 1.1 m w.e., $\overline{c}_0$ = 0.31 m w.e. (Weidick, 1978), and h_0 = 900 m. The difference $C_0 - A_0$ is due to calving. The area between 1000 and 2000 m a.s.l. is determined as 550 10^3 km^2 from a map. For small altitudinal shifts of the equilibrium line the corresponding change in area is calculated by linear interpolation with $\Delta S/\Delta z$ = 550 10^3 km^2/km. The average position of the equilibrium line is assumed at z = 1500 m a.s.l. for the undisturbed conditions.

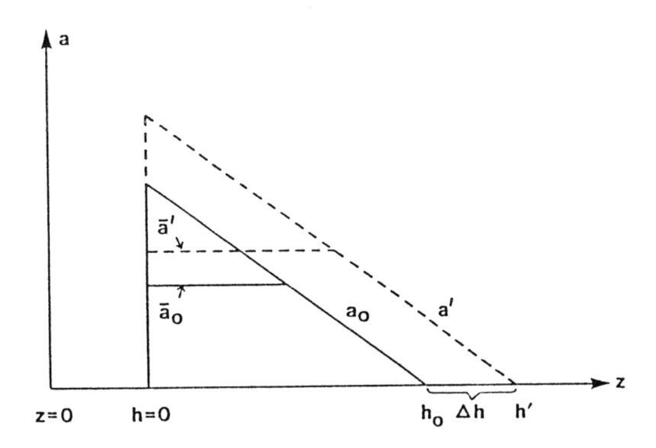

Figure 4. Dependence of the specific net ablation a_0, a' on the altitude for undisturbed and disturbed conditions. z = 0 sea level, h = 0 ice margin, h_0, h' undisturbed and disturbed position of the equilibrium line above the ice margin.

5.1 Mass balance for $\delta T_a > 0$ and $\delta c = 0$

Under the condition that the disturbance in cumulative accumulation $\delta c = 0$ and with the simplification that in the accumulation area of the Greenland Ice Sheet the mean specific net accumulation of the accumulation area is essentially equal to mean specific cumulative accumulation, it follows that $\bar{c}' = \bar{c}_0$, thus eqs. (7) and (8) read:

$$A' = A_0 (h_0 + \Delta h)/h_0 + \bar{a}_0 \Delta S_a (h_0 + \Delta h)/h_0 \qquad (10)$$

$$C' = C_0 + \bar{c}_0 \Delta S_c \qquad (11)$$

Results of the evaluation are shown for the limiting cases $\partial c/\partial z = 0.55$ kgm^{-2}/m (Fig. 5a) and $\partial c/\partial z = -0.32$ kgm^{-2}/m (Fig. 5b). It is essential to note that condition $A' + C' = 0$ is already fulfilled for small changes in temperature $1K < \delta T_a < 2K$ with $\delta c = 0$ (Fig. 5a,b).

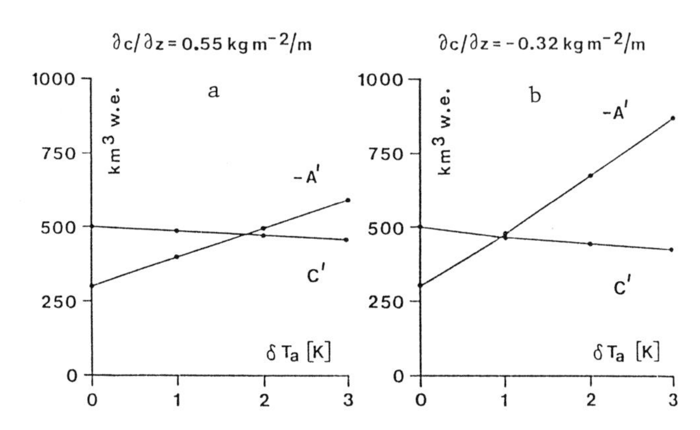

Figure 5. Total net ablation A' and total net accumulation C' of the Greenland Ice Sheet depending on disturbances in temperature for the limiting cases.
a) $\partial c/\partial z = 0.55$ kgm^{-2}/m,
b) $\partial c/\partial z = -0.32$ kgm^{-2}/m

5.2 Mass balance for $\delta c > 0$ and $\delta T_a = 0$

In a similar way as in section 5.1, the values A' and C' are calculated by eq. (7)-(9) and further

$$B' = C' - A' \qquad (12)$$

$$\Delta B = B' - B_0 \qquad (13)$$

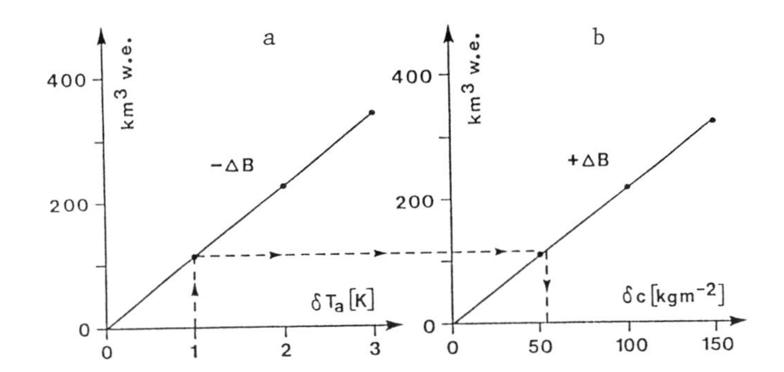

Figure 6. Changes in the total mass balance of the Greenland Ice Sheet depending on disturbances a) in temperature, b) in cumulative accumulation. For example, dashed line indicates equivalent changes with contrary signs.

ΔB versus δT_a and δc is shown in Figs. 6a,b for $\partial c/\partial z = 0.55$ kgm^{-2}/m. Assuming a feedback of positive values of δT_a with positive values of δc, the signs of ΔB are contrary. Figs. 6a,b show at which condition the change in mass balance is compensated for simultaneous disturbances of $+\delta T_a$ and $+\delta c$. For example, in Figs. 6a,b it is shown, that for $\delta T_a = +1$K and $+\delta c = +50$ kgm^{-2} the total mass balance remains unchanged.

ACKNOWLEDGEMENTS

The manuscript is part of a project which is financially supported by the Österreichische Akademie der Wissenschaften, Wien. The authors thank Mr. J. Huber for the cooperation in the evaluation of the data.

REFERENCES

Ambach, W. 1963. Untersuchungen zum Energieumsatz in der Ablationszone des grönländische Inlandeises (Camp IV-EGIG, 69°40'05"N, 49°37'58"W). Meddelelser om Gronland, 174,4.

Ambach, W. 1977. Untersuchungen zum Energieumsatz in der Albationszone des grönländischen Inlandeises. Nachtrag. Meddelelser om Gronland, 187, 5.

Ambach, W. 1985. Characteristics of the heat balance of the Greenland Ice Sheet for modelling. J. Glaciology 31 (107), 3-12.

Ambach, W. and M. Kuhn 1985. Accumulation gradients in Greenland and mass balance response to climatic changes. Zeitschr. für Gletscherkunde und Glazialgeologie 21, 311-317.

Benson, C.S. 1962. Stratigraphic studies in the snow and firn of the Greenland Ice Sheet. SIPRE Research Report 70, 38 pp.

Kuhn, M. 1980. Climate and glaciers, sea level, ice and climatic change. In: Allison, I. (Ed.), proceedings of the Canberra Symposium, Dec. 1979, IAHS Publ. 131, 3-20.
Kuhn, M. 1987a. Micro-meteorological conditions for snow melt. J. Glaciology, in press.
Kuhn, M. 1987b. The response of equilibrium line altitude and mass balance profiles to climatic fluctuations. This volume.
Weidick, A. 1978. Glacial history of Greenland: A review. International Polartagung, October 4-7, 1978, Berlin. Oral presentation.

THE EVOLUTION OF THE ENGLACIAL TEMPERATURE DISTRIBUTION IN THE SUPERIMPOSED ICE ZONE OF A POLAR ICE CAP DURING A SUMMER SEASON

Wouter Greuell and Johannes Oerlemans
Institute of Meteorology and Physical Oceanography
University of Utrecht
The Netherlands

ABSTRACT

The aim of the present investigation was to provide more insight into the processes affecting the evolution of the englacial temperature distribution at a non-temperate location on a glacier. Measurements were made in the top 10 m of the ice at the summit of Laika Ice Cap (Canadian Arctic) during the summer 1975 (by Blatter et al.). This location is in the superimposed ice zone. The model simulation includes calculation of the surface energy fluxes, of radiation penetration, of the englacial temperature and density distribution, and of the formation, penetration and refreezing of melt water.

In the first kind of experiments the energy fluxes from the atmosphere were tuned in such a way as to obtain the right amount of ablation. With these energy fluxes as a boundary condition the consequences of melt water penetration and refreezing for the englacial temperature distribution were proofed to be considerable. In the second kind of experiments the measured temperature at the interannual surface was used as boundary condition, and to start with the temperature below the interannual surface could only be affected by conduction. The measured and the calculated temperatures match until melt water penetrates to the interannual surface. Thereafter, calculations give too low temperatures. Most of this energy deficiency will probably be due to radiation penetration, whereas a minor part of it may be caused by melt water penetration into open veins or an error in the assumed interface temperature.

1. INTRODUCTION

Two kinds of parameters determine the ablation at a glacier surface. On one hand there are the meteorological elements like incoming radiation, temperature, humidity and wind velocity. On the other hand temperature, density and structure of the upper snow and ice layers affect the mass balance. In fact, most of the parameters of both groups are interrelated

J. Oerlemans (ed.), Glacier Fluctuations and Climatic Change, 289–303.

and affected by the melting process. If a computer model is constructed for the simulation of mass balance the number of variables and of their relations has to be restricted, of course. In most cases the energy fluxes between atmosphere and glacier are calculated from standard meteorological quantities and additional assumptions concerning the state of the snow or glacier ice are made. Albedo and the roughness lengths are generally prescribed and the temperature of the snow or ice is assumed to be 0°C over the entire depth of the glacier at the location to which the calculation should apply. The validity of this last assumption for alpine glaciers was tested in previous work (Greuell and Oerlemans, 1986). It appeared that in the Alps the "zero degree assumption" hardly affects the calculated mass balance at lower elevations, say below the equilibrium line. However, larger errors are made at higher elevations with the error increasing with elevation.

The englacial temperature and density distributions affect the mass balance in two ways. Firstly, the long wave outgoing radiation and turbulent fluxes depend on the surface temperature. Secondly, melt water formed at the surface may penetrate and refreeze at lower depths depending on temperature and density distribution. In that case the melt water does not run off and therefore does not contribute to the ablation. The amplitude of the cold wave penetrating into the snow of alpine glaciers during the winter season or during nighttime in the summer season increases with elevation. More melt water is then needed to eliminate the cold wave.

The purpose of the present study is to obtain more insight into the evolution of the englacial temperature profile. This is done by means of a data set and a computer simulation. As location for the simulation the summit of Laika Ice Cap (Canadian Arctic, 75°53' N, 79°10' W, 530 m a.s.l.) was chosen. This location is in the superimposed ice zone. Here the temperature profile down to a depth of 10 m was measured during the summer season 1975 (Blatter, 1985). To start with the computer simulation will show to what extent we are able to simulate the evolution of the temperature profile with the mere knowledge of some standard meteorological variables. Then a sensitivity experiment will be done to estimate the effects of penetrating and refreezing melt water.

Finally, instead of the calculated energy fluxes at the atmosphere-glacier interface, the measured temperatures at the interannual surface will be used as a boundary condition. The comparison of the calculated and the measured temperature distribution below the interannual surface leads to some considerations about the roles of penetrating radiation and of melt water penetration into veins in the ice below this surface. Another purpose of the experiments presented here is to test the computer model that we primarily developed for mass balance studies (Greuell and Oerlemans, 1986), but which is also suitable for the present study as it includes a calculation of the temperature and density profile.

2. THEORY

In the general case the evolution of the temperature distribution in a glacier is described by the following equation:

$$\rho c \frac{dT}{dt} = \nabla \cdot (K \nabla T) + W \tag{1}$$

where ρ, c and K are the density, the specific heat and the conductivity of snow/ice, t is time, T is temperature and W the generation of energy per unit volume and per unit time. W includes the following processes:

- The energy fluxes between atmosphere and glacier. Nearly all of this energy will be absorbed in or emitted from the uppermost centimeters of the snow or ice. Only the short wave radiation will partly penetrate and be absorbed deeper.
- The formation and refreezing of melt water are sink and source of energy, respectively.
- Deformational heat
- The energy flux between glacier bed and glacier.
- Cooling and/or freezing of rain water adds energy to the glacier.

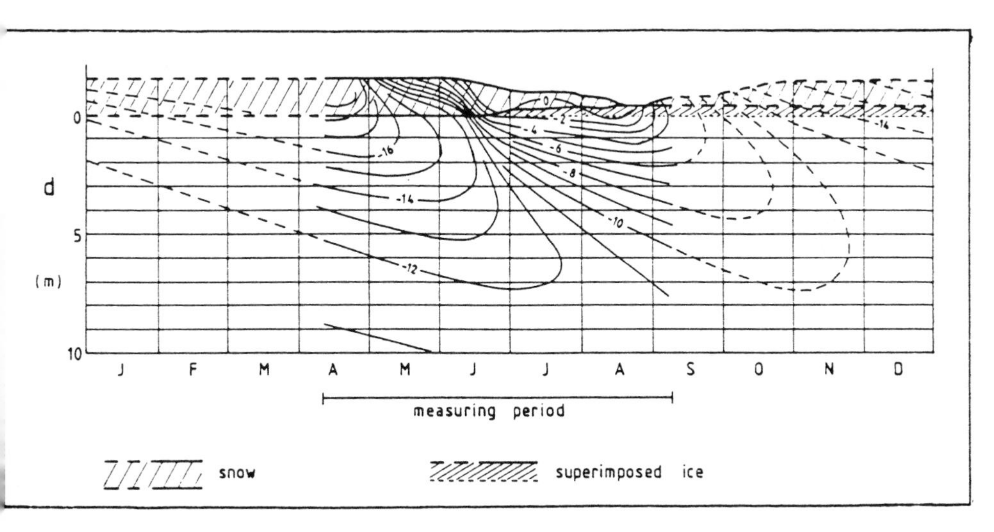

Figure 1. Evolution of the englacial temperature distribution in the top 10 m at the summit of Laika Ice Cap during 1975 as based on measured data (from Blatter, 1985).

3. MEASUREMENTS

In the summer season of 1975 the evolution of the temperature profile on the summit of Laika Ice Cap, situated on Coburg Island in the Canadian Arctic was measured (Blatter, 1985). This was done as part of a glaciological program carried out by members of the North Water Project team. The measurements covered the uppermost 10 m of the vertical column at the summit of the ice cap. A hole of this depth was drilled by hand with the aid of a SIPRE core drill in early April 1975. A cable with thermistors (accuracy ± 0.2°C) was inserted and the hole filled with snow and water. The temperature stabilized some 2 weeks later. Measurements were taken 17 times with the first measurement on April 16 and the last one on September 5.

The evolution of the temperature profile is shown in Figure 1. In the ice under the interannual surface the penetration of the winter cold wave has a time lag roughly proportional to depth, and the amplitude of the variation decreases with depth. The evolution of the temperature distribution in the layer above the interannual surface is more complicated. Schematically the following periods can be distinguished (dates are calculated with the computer model, see next sections):

a. At the beginning of the measurements the layer is about 1.5 m deep and has a mean density of about 400 kg m^{-3}.
b. Until June 9: gradual warming of the surface layers.
c. June 9 - June 15: melt water is formed at the surface, penetrates and refreezes completely in the "cold" snow of the layer above the interannual surface thereby raising temperature and density of this layer. On June 15, for the first time, the layer is warmed up to 0°C over its full depth.
d. June 15 - August 17: this is the ablation season proper. The layer continually looses energy, by conduction to the ice underneath, and thus tends to cool down. During a few short periods this cooling down is reinforced by a negative energy flux from the atmosphere. However, this flux is positive on most days and melt water is formed, part of which is used to keep the layer at 0°C. The rest runs off. At the bottom of the layer the density attains the density of ice after a while. The ice formed that way is called superimposed ice and the layer formed by it gradually thickens. On many days during this period snowfall takes place.
e. August 17 - September 5: melt water formation stops. In fact, the accumulation season starts.

Mass balance measurements at the same site give a winter accumulation for 1974-1975 of 69 cm w.e. (=water equivalent). Summer ablation amounted to 37 cm w.e. Thus, the net mass balance was 32 cm w.e. (Müller, 1977), which is close to the annual mean of 30 cm w.e. (Blatter, 1985). Aerial photographs taken in 1959 and 1971 show that the total mass balance of Laika Ice Cap is strongly negative (Blatter, 1985).

For the simulations in the present study the following meteorological data were used:

a. Temperature (2 m), humidity (2 m), wind velocity (10 m), sunshine duration and amount and kind of precipitation. All these variables were measured with different frequencies during the whole period at Base Camp station (4 m.a.s.l.), situated near the terminus of the glacier. Although with interruptions the temperature was also measured at the site of the englacial temperature measurements (530 m.a.s.l.) and during the whole period on nearby Marina Mountain (700 m.a.s.l.).

b. No radiation measurements were available for 1975. Thus, they had to be calculated from the other parameters and suitable formalae. For the global radiation under clear skies a formula given by Meyers and Dale (1983) was used, for the incoming long wave radiation a formula from Kimball et al. (1982).

For more information about the frequency of the measurements, the assumptions used to calculate the values of the variables at the summit of the glacier and the adaption of the radiation formulae to local conditions the reader is referred to Greuell and Oerlemans (1987).

For the simulations only daily mean values of the meteorological variables were used. The amplitude of the diurnal temperature variation is about ±2°C in April and May and decreases to ±0.5 - ±1.0°C during the melt season. These variations are so small that they are obscured by larger non-periodic variations (Blatter, pers. comm.). The humidity variations do not at all coincide with the 24 hr cycle (Blatter, pers. comm.). Only the global radiation is subject to a daily cycle in dependence of the zenith angle variation.

4. THE MODEL

The model used for the calculations is described in more detail in Greuell and Oerlemans (1986). It was originally developed for the simulation of mass balance. Essentially, the surface fluxes are computed from standard meteorological variables (temperature, humidity and wind velocity at one level, and cloudiness) and from the surface temperature, the albedo and roughness lengths. In a sub-model the evolution of the englacial temperature and density distribution are calculated.

Global and incoming long wave radiation are calculated by means of the before mentioned formulae. The turbulent fluxes are calculated by means of the Monin-Obukhov similarity theory (see e.g. Businger, 1973). The values of the albedo and the roughness lengths were reconsidered, because conditions on polar ice caps are different from conditions in the Alps, the region for which the model was originally designed. For the albedo (α) we used:

$$\alpha(n) = \alpha_{n=0} + 0.06\ n \qquad (2)$$

with $\alpha_{n=0} = 0.8$ in the absence of melt water at the surface, $\alpha_{n=0} = 0.54$ in the presence of melt water at the surface and n is cloudiness.

The aerodynamic roughness length z_o varies between 0.1 mm and 1.3 mm. It increases in the presence of melt water at the surface. It also increases with density and decreases with the thickness of the snow layer. The roughness lengths for temperature and humidity are equal to 0.01 mm. The values of the albedo and the roughness lengths and the conditions for their variation were detained from the work by Holmgren (1971) about the Devon Island Expedition. It is remarkable that he found that the albedo of the frozen superimposed ice or firn below the spring snow pack at his station in the superimposed ice zone did not drastically differ from that of pure fine snow. This means that the albedo is hardly affected by a fresh snowfall event.

Penetration of short wave radiation into deeper layers may have serious consequences for the energy budget if the snow or ice is not at the melting point, so an attempt was made to include this in the model. If penetration is important, the surface layer receives less energy, the underlying layers more. This may even cause melt water production below the surface, whereas the surface itself remains frozen (see Holmgren, 1985). In snow the consequences of radiation penetration are of lesser importance, because the extinction coefficient is much larger, and penetrating and refreezing melt water may undo the different partitioning of the energy. The second effect of radiation penetration concerns the turbulent fluxes and the outgoing long wave radiation, leading to an enhanced total energy flux from the atmosphere. It should be borne in mind that this effect is only present if the temperature of the very surface is below the melting point.

Absorption and scattering of radiation in snow or ice depends on wave length. Infra-red radiation hardly penetrates; in the visible part of the spectrum the extinction coefficient is much lower. Following Holmgren (1985) short wave radiation penetration was modelled as follows: the global radiation was divided into 2 parts. One part (36%), with wavelengths larger than 0.8 μm, is completely absorbed in and reflected from the model layer at the surface (6 cm thick). The rest partly penetrates into deeper layers, according to

$$Q_s = I_s\ (1 - \alpha_s)\ e^{-\beta_s z}, \tag{3}$$

where Q is the net radiation at depth z below the surface, I the global radiation at the surface, α the albedo and β the extinction coefficient. The subscript s refers to radiation with $\lambda < 0.8$ μm. In the layer above the interannual surface β_s is linear in ρ with $\beta_s = 10\ m^{-1}$ if $\rho = 400\ kg\ m^{-3}$ and $\beta_s = 4\ m^{-1}$ if $\rho = 910\ kg\ m^{-3}$. In the underlying ice: $\beta_s = 1.3\ m^{-1}$. Radiation penetration will be discussed in some more detail in Section 5.2.

The processes affecting the evolution of the temperature profile were already mentioned in Section 2. In the present investigation some of these processes were not taken into account, namely:

- horizontal advection and deformation since there is hardly any movement in the top 25 m (the thickness of the model) at the summit of the ice cap.
- the flux at the bed.
- cooling and freezing of rain since rain fall events were rare in the summer season of 1975 (5.3 mm according to our calculations).

Thus, the change of temperature is determined by conduction, the energy flux from the atmosphere, and the formation, refreezing and run off of melt water.

A few words should be said about how the model deals with the latter. The melt water formed in the top layer moves downwards. Melt water formed in one or more of the other layers is added "en route". If it reaches a model layer with a temperature below the melting point, this layer is warmed up by refreezing the melt water. If the resulting temperature exceeds 0°C, the temperature is put equal to 0°C and the remaining amount of melt water is computed. This melt water penetrates downwards into the next model layer, etc. After penetration through the lowest model layer above the interannual surface the remaining melt water runs off. In the model the density increases by this refreezing process only. Processes like compaction and metamorphosis are not considered. The initial density of snow layers accumulating during the summer season is assumed to be 400 kg m^{-3}.

Conductivity is obtained from a formula given by Paterson (1981):

$$K = 2.1 \times 10^{-2} + 4.2 \times 10^{-4} \rho + 2.2 \times 10^{-9} \rho 3 \qquad (4)$$

Thus, K = 2.1 W m^{-1} K^{-1} for ice.

The sub-model simulating the evolution of the englacial temperature and density distribution consists of 42 layers with increasing thickness from top (6 cm) to bottom (about 3 m at a depth of 25 m).
Equation 1 is solved numerically by an implicit scheme. The temperature in the lowest grid point is constant. The effect on the temperature profile calculations of this assumption was proved to be negligible. The time step for the calculations is 30 minutes and the initial conditions are the temperature and density profile obtained from the first measurements. At that time the ice was covered by 1.5 m of snow with a mean density of 400 kg m^{-3}. This value was used for the whole layer. Initial temperature conditions below 10 m were specified by values obtained from a linear interpolation between the 9.5 m-value of April 16 and the 16- and 23 m-values of September 5, when temperatures were measured in a deeper hole nearby. Once a day the grid and the englacial profiles were adjusted to the new surface level as it is determined by accumulation, melting and evaporation.

5. EXPERIMENTS

Two kinds of experiments were carried out. In the first an attempt was made to calculate the right amount of ablation for the summer season 1975 (37 cm w.e.). After tuning in such a way that this value was indeed obtained, the resulting error in the temperature profile calculations and the effect of the melt water penetration and refreezing on the temperature profile were studied. The second kind of experiments solely dealt with the evolution of the temperature profile below the interannual surface. Here the boundary condition at the top of the model was given by the measured temperature at the interannnual surface.

5.1 Experiments with surface energy flux calculations

The initial run was done with the meteorological data as described in Section 3. The ablation thus calculated exceeded the measured value significantly. The calculated net balance for 1974-1975 even became negative. Thus, for matching the measured and the calculated amount of ablation one or more of the meteorological variables, glaciological parameters or properties of the model had to be adjusted. The effects of the following were investigated: reduction of the global and the incoming long wave radiation, higher albedo, lower temperatures, higher conductivity and a smaller aerodynamic roughness length. As long as the adjustment remained within reasonable limits, with none of them the desired result could be obtained with the exception of the incoming long wave radiation. This flux had to be multiplied by a factor 0.88. Such a reduction of 12% appears to be a large amount. It seems improbable that a 12% error in the mean incoming long wave radiation is due to the calculations. While the calculation of the incoming long wave radiation from temperature, vapour pressure and sunshine duration certainly causes large errors in individual dayly means, the mean value for the whole period must have an error much smaller than 12% as the formula used was established by regression analysis of the previous year's data. The regression analysis was done with data from Base Camp station. Then, the formula was applied on the ice cap. An order of magnitude estimate shows that the error possibly caused by this displacement will certainly be much smaller than 12%. Some support for a systematic error in the measurements comes from long wave radiation measurements on another ice cap in the Canadian Arctic (Bradley, 1985) made under comparable conditions. These data have a mean value 14% lower than the mean of the measured values on Coburg Island. Indeed, the accuracy of long wave radiation measurements is generally low. So the 12%-reduction was maintained, although no direct evidence for an error in the measurements on Coburg Island could be obtained. For the simulation of the 1974-ablation season exactly the same multiplication factor was needed to match the calculated and the measured amount of ablation (18 cm w.e.).

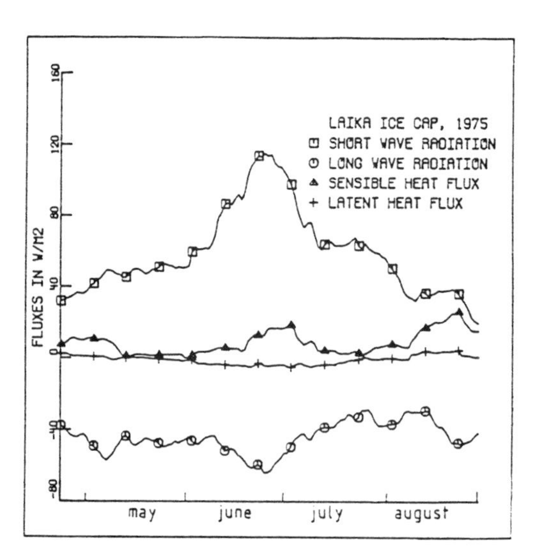

Figure 2. Running means (15 days) of the calculated energy fluxes.

Figure 2 shows the course of the energy fluxes at the surface during this summer (15 days running means). The radiation fluxes are of opposite sign and they dominate the turbulent fluxes. There are rather abrupt changes in the net short wave radiation. This is caused by a positive feedback mechanism. A melting surface has a lower albedo so that the total energy flux towards the glacier is enhanced and that melting conditions can be sustained more easily.

A similar run was made without radiation penetration. The calculated amount of ablation hardly changed (41 cm w.e. against 37 cm w.e. for the original run). Primarily one would expect that penetrating radiation enlarges the amount of ablation, because of the increased total energy flux from the atmosphere during non-melting conditions at the surface. In fact, this effect would enlarge ablation by 9 cm w.e. However, melt water formation at the surface is suppressed, so that the albedo will be high during longer periods and ablation is reduced.

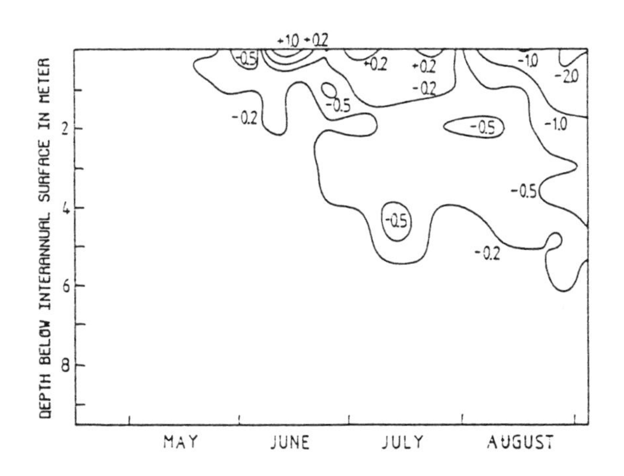

Figure 3. Error in the temperature calculation after run without interface temperature control. Calculations are based on the energy balance calculations at the surface. Contour labels give degrees centigrade. Positive values correspond to calculated temperatures being too high.

An evaluation of the evolution of the calculated temperature profile can be obtained from Figure 3. It gives the difference with the measured temperature profiles. The layer above the interannual surface is omitted from these comparisons as the fluctuations in this layer are dominated by surface energy fluxes with high frequencies, which are not represented in the input data (only daily means of the meteorological variables were used). The model calculations generally give too low temperatures. The error grows during the summer and larger discrepancies occur near the surface. Apparently the downward energy flux through the interannual surface is too small. In Section 5.2 ways of eliminating the discrepancy will be discussed.

In order to assess the effect of the penetrating melt water on the englacial temperature an experiment in which all of the melt water immediately runs off was carried out. Melt water keeps the temperature of the layer above the interannual surface close to the melting point during the ablation season. If conduction would be the only energy transport mechanism, temperatures would be substantially lower (see Figure 4). It is evident that the simulation of melt water penetration and refreezing thereof is essential at the site of these measurements.

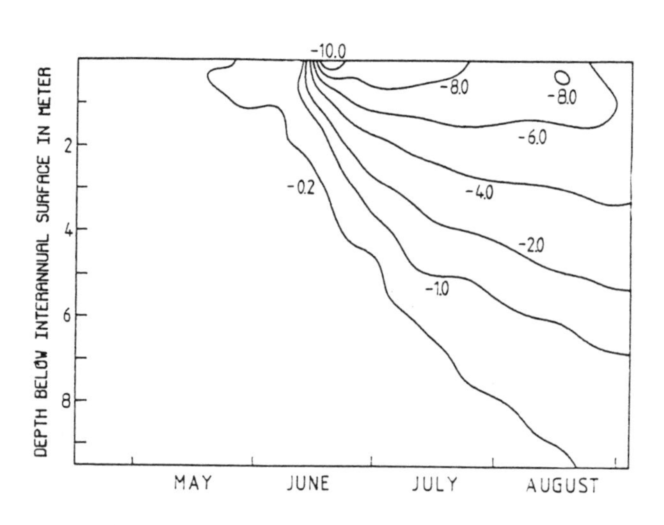

Figure 4. See Figure 3. However, in this case melt water penetration was inhibited.

5.2 Experiments with interface temperature control

The second kind of experiments solely dealt with the temperature distribution below the interannual surface. The temperature at this surface was taken from the measurements and in a first experiment radiation penetration was neglected. The calculated temperature distribution then is a solution of Equation 1 with $W = 0$ and $k = K/\rho c$ independent of time and depth and equal to 1.15×10^{-6} m^2 s^{-1} . The error made under these conditions can be evaluated from Figure 5. Three periods can be distinguished.

Until the first decade of June the error is within the accuracy of the observations. During this period no melt water penetrates to the interannual surface and the thick snow cover will almost completely inhibit radiation penetration into the ice. So, energy will only be redistributed by conduction. These circumstances were used to test the value of the diffusivity of ice as used in this study. By varying k in a systematic way we found that the optimal value is within 10% equal to 1.15×10^{-6} m^2 s^{-1} .

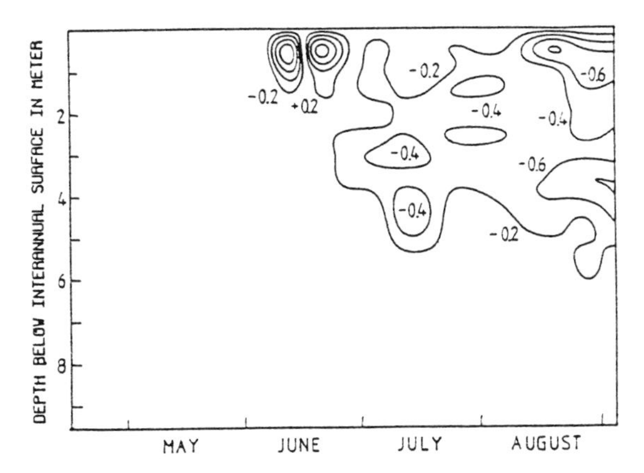

Figure 5. Error after run with temperature at the interannual surface prescribed by the measured data. Radiation penetration was neglected. Values give degrees centrigrade and positive values correspond to calculated temperatures being too high.

In June the calculation close to the interannual surface becomes more inaccurate. Both positive and negative deviations are obtained, with maxima about plus and minus 0.8°C. Two possible explanations may be given:

a. The interface temperature at this time of the year rises rapidly from about -11°C on June 14 to close to the melting point on June 18, caused by the refreezing of penetrating melt water. As the boundary condition was determined by an interpolation of the measured values (only 4 observations in June), the error in the interface temperature on some days during this time of the year will be large.
b. Melt water penetrates and refreezes below the interannual surface, possibly only along the thermistor string. However, this should only provoke positive deviations.

Like in the run without interface temperature control (Figure 3), it seems that the downward energy flux through the interface is underestimated in the period thereafter. This may be caused by:

a. Melt water penetration and refreezing. However, in that case the deviation in the uppermost measurement point at 0.5 m depth should be large at the onset of the ablation season and decrease afterwards, because the available space in the veins would be occupied by the refreezing melt water later in the season. This seasonal trend is not found. The source of the error seems to persist during the whole period. In order to match the calculated and the measured temperature distribution at the end of the measurements about 14 mm w.e. of melt water would have to penetrate and refreeze.

b. Radiation penetration. Again about 4.6 MJ m^{-2}, that is the energy released when a layer of 14 mm water freezes, would be needed to match the calculated and the measured temperature profiles at the end of the measurements. It is difficult to estimate whether this amount of radiation penetration is feasible. The bulk extinction coefficient for pure ice at some depth is known within reasonable limits of accuracy, β = 1.1 - 1.5 m^{-1} (Grenfell and Maykut, 1977). However, things become complicated in snow, because the extinction coefficient is very dependent on structure and density. Moreover, it varies with wave length, so that even in a homogeneous snow or ice cover the bulk extinction coefficient changes with depth. As the spectral composition of the incoming radiation depends on the state of the atmosphere in general, and especially on cloud conditions, the problem is even more complicated. In view of all these difficulties, and in view of the limited suitability of the present model to simulate the detailed structure of the snow cover, only a rough estimate of the effects of radiation penetration could be obtained. It appeared that with the parameterization as proposed in Section 4 the mismatch could not be eliminated. More radiation had to penetrate and therefore another parameterization was adopted. Again radiation with wavelengths greater than 0.8 µm (36%) was completely absorbed in or reflected from the uppermost model layer. The rest of the radiation was extinguished according to Equation 2, with in the layer above the interannual surface β = 20 m^{-1} for a mean density less than 500 kg m^{-3} and $\beta = \beta_c$ for a greater mean density. This "critical" density was reached on June 24. In the underlying ice β = 1.3 m^{-1}. Then, by trial and error β_c was found, so that the calculated and the measured profile for September 5 matched. A value of 1.8 m^{-1} was found. This is a very low value relative to the values given by Holmgren (1985) for fine grained snow during the early melt periods (β = 10 m^{-1}) and for loose weathered superimposed ice (β = 4 m^{-1}). It should be mentioned that the artificial difference between the extinction coefficients of the ice on both sides of the interannual surface might not be justified. The ice on both sides seems to be about the same (Blatter, pers. comm.). However, the total absorption below the interannual surface is not affected by putting the extinction coefficient below the interannual surface equal to its value above this surface, since it is only determined by the extinction coefficient above the interannual surface and the albedo. The albedo below the interannual surface might be higher than assumed in the calculations. Blatter (pers. comm.) reported dirt layers in the ice due to blowing sand. Figure 6 shows the error after the run with β_c = 1.8 m^{-1}.

c. Interface temperatures are too low. A run was made with the interface temperature 0.5°C higher from June 24 on. This interface temperature increase cannot completely eliminate the discrepancy.

Hooke (1983) simulated the evolution of the temperature profile during a summer season on Storglaciären (Swedish Lappland). To match calculations and measurements, he also had to introduce an additional energy flux at the interannual surface. Since his measurement site became snow free rather early in summer, he could estimate the amount of radiation penetration more accurately than it could be done in this study. He concluded that the "energy deficiency" in his calculations could completely be eliminated by radiation penetration.

Since a systematic error in the interface temperature measurements of more than 0.2°C seems unlikely, such an error could only explain a minor part of the mismatch. For warming by the refreezing of melt water open veins in the ice layer from the previous year as a consequence of weathering of the superimposed ice should exist. Such veins were reported by Hooke (1983) on Storglaciären, but according to Blatter (pers. comm.) superimposed ice was not exposed very long in summer 1974 and seemed to be rather compact. Furthermore, as argued before, the effectiveness of this process should decrease with time. Thus, it seems that refreezing of melt water is unimportant so that radiation penetration should play a major role in explaining the extra energy flux.

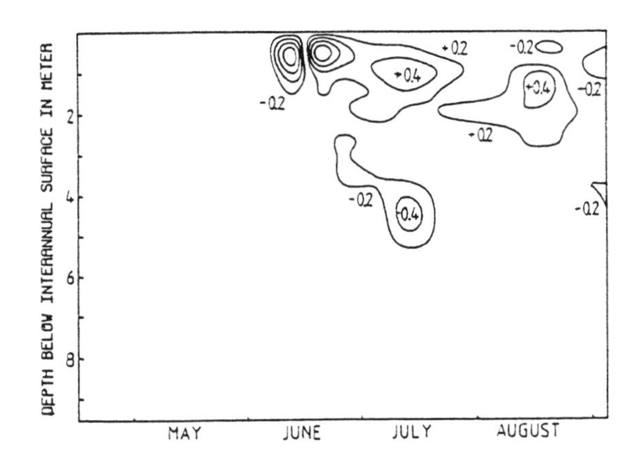

Figure 6. See Figure 5. However, radiation penetration was allowed in such a way as to match the calculated and the measured profile of September 5.

6. CONCLUSIONS

With the present investigation an effort was made to obtain insight into the processes affecting the temperature distribution in the uppermost layers of a non-temperate location on a glacier or ice cap. Besides, the suitability of the model used for this investigation was tested. The simulation of the temperature distribution in the layer above the inter-annual surface was not considered in the comparison between the measured and the calculated data, since it strongly depends on energy inputs with high frequencies which were not represented in the input data.

It seems that the temperature distribution in a glacier can be calculated very accurately provided the glacier consists of a homogeneous ice body, the temperature at the boundary of the body is known and no other energy fluxes than conduction interfere. If melt water penetration and refreezing thereof is not considered, large errors will occur in the case melt water is formed (see Figure 4). Inclusion of this process in the model leads to much better results (see Figure 3), but a minor discrepancy remains and its source seems to coincide with the ablation season. One of the possible error sources is penetration and refreezing of melt water in the ice below the interannual surface. This will depend on the structure of the ice. Moreover, as soon as the snow cover becomes thin, radiation penetration should be included in the model. The present model can only give rough estimates, but the results indicate that radiation penetration accounts for the major differences between the observations and the calculations. Hooke's (1983) calculations for a site in the ablation zone on Storglaciären lead to a similar conclusion. If accurate estimates of the intensity of the penetrating radiation are desired, a model specially designed for this purpose should be used. Such a model should take account of the spectral dependence of the extinction coefficient, of the spectral distribution of the incoming radiation and of the detailed structure of the snow/ice cover.

In many studies the annual mean air temperature is compared with the temperature at some depth in the glacier. This depth should be great enough for the annual temperature variations to be damped largely. A depth of 20 m seems to be enough, and in order to take the energy flux from below into account Hooke (1983) proposed to extrapolate the 20 m-value to the surface by means of the 20 m-temperature gradient. In the case of Laika Ice Cap the annual mean air temperature (about -14.8°C) is a couple of degrees centigrade lower than the 20 m-value extrapolated to the surface (-11.7°C). This difference will largely be due to warming by refreezing melt water.

The model failed to calculate the thickness of the layer of superimposed ice. The grid used in the model is built up from the glacier surface downwards. Since melting, evaporation and snow fall change the position of the surface, and since numerical problems caused us to stick to the grid point distance, especially at the surface, the grid has to be adjusted regularly. So the grid moves relative to material points and mass diffuses. Only mass transport through the interannual surface could be inhibited. It is recommanded that future models use a grid in which the individual grid points stick to material points. However, special attention should be paid to the grid point distances close to the surface in view of numerical problems.

For more details the reader is referred to an extended version of this paper (Greuell and Oerlemans, 1987)

ACKNOWLEDGEMENTS

We kindly thank Dr. Heinz Blatter, Prof. Dr. A. Ohmura and Giovanni Kappenberger from the Geographical Institute of the ETH, Zürich for the collection and the evaluation of the meteorological and glaciological data. Dr. Heinz Blatter also helped us to find the relevant data and reviewed the preceding versions of this paper.

REFERENCES

Blatter, H., 1985. On the Thermal Regime of Arctic Glaciers. Zürcher Geographische Schriften, Heft 22.

Businger, J.A., 1973. Turbulent transfer in the atmospheric surface layer. In Workshop on micrometeorology, D.A. Haugen, Editor, Published by the American Meteorological Society.

Grenfell, T.C. and G.A. Maykut, 1977. The Optical Properties of Ice and Snow in the Arctic Basin. Journal of Glaciology, Vol. 18, No. 80., p. 445-463.

Greuell, W. and J. Oerlemans, 1986. Sensitivity Studies with a Mass Balance Model including Temperature Profile Calculations inside the Glacier. Pater submitted to the Zeitschrift für Gletscherkunde and Glazialgeologie.

Greuell, W. and J. Oerlemans, 1987. The evolution of the englacial temperature distribution in the superimposed ice zone of a polar ice cap during a summer season. Internal Report, R-87-1.

Holmgren, B., 1971. Climate and Energy Exchange on a Sub-Polar Ice Cap in Summer (6 parts). Meteorologiska Institutionen Uppsala Universitetet, Meddelande Nr. 107.

Holmgren, B., 1985. The Interaction of Energy Fluxes at the Surface of an Arctic Ice Field (in press).

Hooke, R.L., J.E. Gould and J. Brzozowski, 1983. Near surface temperatures near and below the equilibrium line on polar and subpolar glaciers. Zeitschrift für Gletscherkunde and Glazialgeologie, Bd. 19, S.1-25.

Kimball, B.A., S.B. Idso and J.K. Aase, 1982. A model of thermal radiation from partly cloudy and overcast skies. Water resources research, vol. 18, No. 4, 931-936.

Meyers, T.P. and R.F. Dale, 1983. Predicting daily insolation with hourly cloud height and coverage. Journal of climate and applied meteorology, Vol. 22, No. 4, p. 537-545.

Müller, F., 1977. Fluctuations of glaciers 1970-1975 (Vol. III). IAHS-UNESCO, ISBN 92-3-101462-5.

Paterson, W.S.B., 1981. The physics of glaciers, Oxford, Pergamon Press.

ENERGY BALANCE CALCULATIONS ON AND NEAR HINTEREISFERNER (AUSTRIA) AND AN ESTIMATE OF THE EFFECT OF GREENHOUSE WARMING ON ABLATION

Wouter Greuell and Johannes Oerlemans
Institute of Meteorology and Oceanography
University of Utrecht
The Netherlands

ABSTRACT

During some 10 days in July 1986, measurements of meteorological variables and ablation were made at two sites on and near Hintereisferner (Austria), namely at GLACSTAT (2500 m), located on the glacier tongue, and at ROCKSTAT (2440 m), just in front of the tongue. In this paper energy balance calculations for both stations are presented. About 90% of the ablation energy at GLACSTAT came from radiation. This large value is a consequence of the low albedo (0.16) and the long periods of sunny, relatively cold weather. A detailed comparison is made of the surface energy budget at the two sites. While melting is important at GLACSTAT, the loss of energy by the turbulent fluxes and the outgoing long wave radiation is larger at ROCKSTAT.

Finally, the consequences of the rising concentrations of atmospheric trace gases on ablation are discussed. The data from GLACSTAT will be adapted to estimated past and projected future concentrations. If all relevant trace gases are considered and under the specific conditions of the experiment total ablation for the 10-days period will increase by 11% in the 50 years after 1986.

1. INTRODUCTION

It is a well known feature that glaciers are generally retreating since the middle or the end of the last century (e.g. Patzelt (1970), Reynaud (1983) and Meier (1984)). As most glaciers in any part of the world show this trend we may assume that the retreat is caused by climatic change. In principle in this case climatic change may be a change in any or a combination of the variables which determine the energy fluxes between atmosphere and glacier. Thus, one of the necessary steps in establishing the relation between climate and glacier length is a better understanding of the relation between climate and these energy fluxes.

From 12 till 22 July 1986 a field programme was carried out on the Hintereisferner (46°48'N, 10°56'E), a valley glacier in the Austrian Alps. Emphasis was laid on measurements of the meteorological variables

J. Oerlemans (ed.), Glacier Fluctuations and Climatic Change, 305–323.

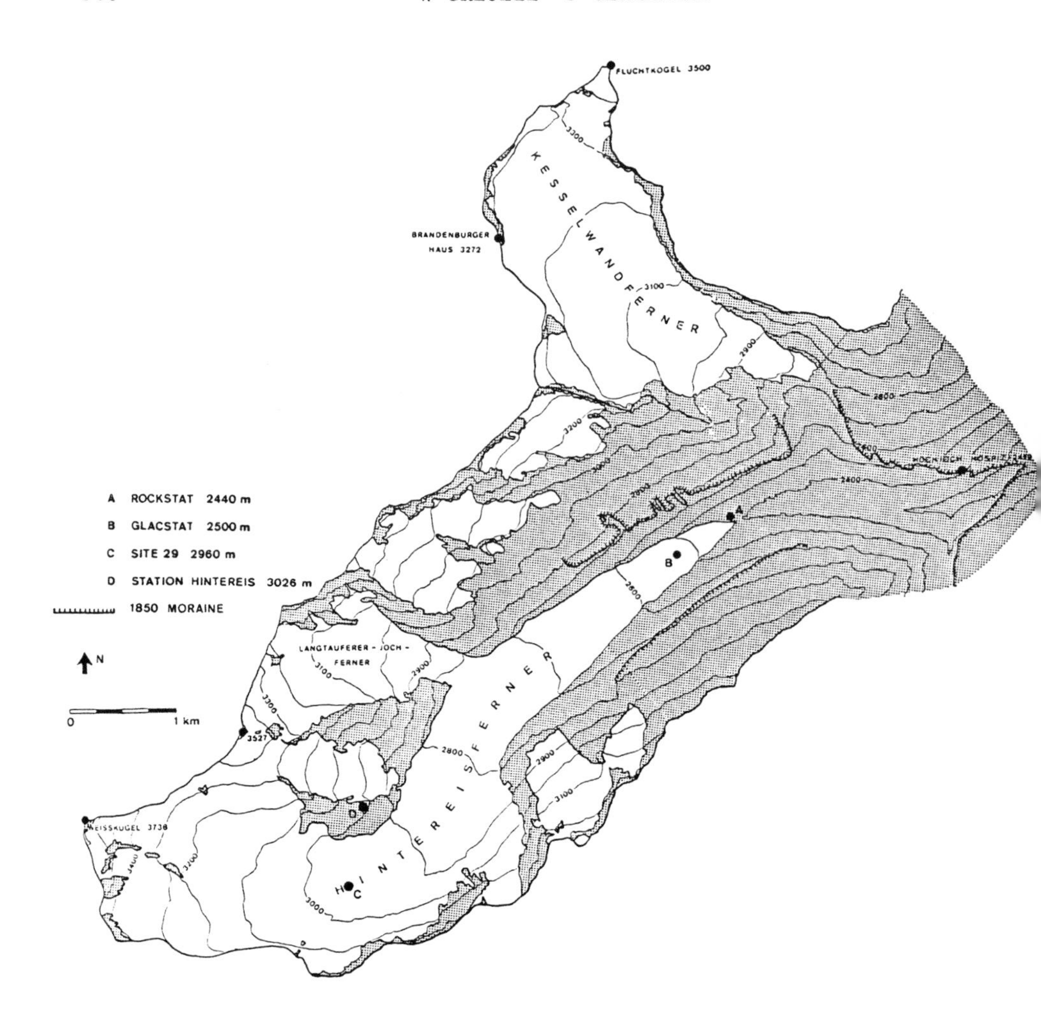

Figure 1. Catchment area of the Hintereisferner and the Kesselwandferner. White surfaces represent the 1969 glacierized area. Since 1969 the Hintereisferner has retreated so that the ice edge was about midway between GLACSTAT and ROCKSTAT in 1986, while the altitude at the site of GLACSTAT decreased to about 2500 m a.s.l.

from which the energy fluxes can be determined. Moreover, the ablation rate was measured. One of the purposes of the experiment was to obtain a better understanding of the energy balance of the glacier surface and its surroundings at different sites in exactly the same weather conditions.

Four institutes took part in the experiment: the Institut für Meteorologie und Geophysik (University of Innsbruck), the Hydrological Institute (Wallingford, U.K.), the Kommission für Glaziologie (Munich) and the Instituut voor Meteorologie en Fysische Oceanografie (University of Utrecht). The latter operated two stations: one on the glacier tongue (GLACSTAT, 2500 m a.s.l.) and one just in front of it (ROCKSTAT, 2460 m a.s.l.). The other institutes carried out their measurements higher up in the firn area. Measurements and results from these stations will be reported in Harding et al. (1987). Figure 1 gives an impression of the setting.

GLACSTAT and ROCKSTAT essentially consisted of 1 mast each. Both masts supported 11 sensors (see Figure 2). The signals from each of the sensors were automatically recorded at 2 minute intervals. Moreover, twice a day ablation was measured along 11 stakes distributed across the glacier tongue and all at the elevation of GLACSTAT (2500 m).

The weather was fine with almost cloudless skies during 6 days out of 10. During 3 days the weather was variable and one day was predominantly foggy.

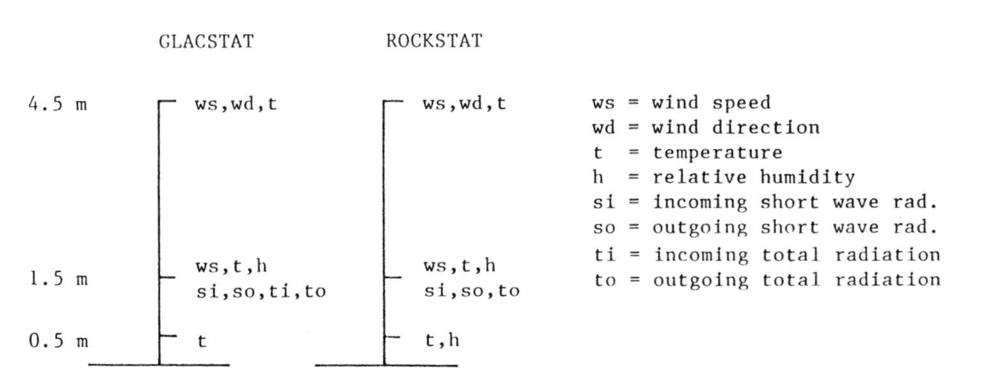

Figure 2. GLACSTAT and ROCKSTAT masts with positions of sensors. Temperature and humidity sensors are not ventilated.

In Section 2 of this paper the computer model used for the energy calculations will be discussed. Then, in Section 3 the calculations itself will be presented for the 2 stations near the terminus. Some model sensitivity experiments mainly concerning the turbulent exchange will be performed. It will also be shown that the energy gained by incoming radiation is lost in completely different ways at the 2 stations. In Section 4 the impact of the increasing concentrations of atmospheric CO_2 and other trace gases is studied. Probable changes in temperature and incoming long wave radiation due to the increasing trace gas concentrations are inferred from literature. Then, our field data are adapted accordingly and the ablation is calculated for the 1880 pre-industrial level of trace gas concentrations and for the predicted 2036 level of trace gas concentrations.

2. CALCULATION OF THE ENERGY FLUXES

If the energy transport by rain drops is neglected, the total energy flux between the atmosphere and the glacier or the soil is the sum of the radiative and the turbulent fluxes. Fluxes towards the surface will be called positive hereafter.

2.1 The radiative fluxes

At the two stations considered here the total radiative fluxes could be obtained from the measurements of incoming and outgoing short wave and incoming and outgoing total radiation through horizontal surfaces. At GLACSTAT, where the surface sloped about 10° to the north-east, the data were corrected in order to obtain the fluxes through a surface parallel to the atmosphere-glacier interface (see Mannstein, 1985). The incoming long wave radiation was not measured at ROCKSTAT. Therefore the values from GLACSTAT were used. The measurements of the outgoing long wave radiation were rejected. It appeared to be more accurate to calculate the outgoing long wave radiation from the surface temperature, which itself was also calculated (see Section 2.3).

2.2. The turbulent fluxes

Essentially, the turbulent fluxes can be calculated from profile measurements of wind velocity, temperature and humidity. In this study, starting from pairs of values of each of these variables two theories will be considered for the calculations. In both cases the fluxes are equal to the product of an exchange coefficient and the temperature or vapour pressure difference between the two levels. In the first case logarithmic profiles of the variables concerned are assumed, e.g.

$$u(z) = \frac{u_*}{k} \ln \frac{z+z_{ou}}{z_{ou}} \qquad (1)$$

with u the wind velocity, z the height, u_* the friction velocity, k von Karman's constant and z_{ou} the roughness parameter of the wind velocity profile. This leads to an exchange coefficient proportional to the wind velocity difference and the height, and independent of the stratification (hereafter s.i.-theory). The second method used here is based on the Monin-Obukhov similarity theory (hereafter M.O.-theory) see e.g. Businger (1973). It takes into account the effect of the stratification on the exchange coefficient. The errors in the measurement of the temperature and relative humidity were estimated to be 0.3°C and 5%, respectively. Calculations are only reliable if the differences in temperature and vapour pressure between the levels of the measurements are much larger than the errors in the individual measurements. Often this was not the case. However, the temperature and vapour pressure differences between the levels of the measurements and the surface were much larger, even if the measurements from the lowest level were used. So, pairs of values of temperature, vapour pressure and wind velocity were obtained from the measurements at 1.5 m in the atmosphere and from calculated surface values (see next section). The surface roughness parameters for GLACSTAT were taken from literature, while they served as tuning parameters for the calculations at ROCKSTAT.

Another problem is posed by the structure of the glacier wind. This wind is characterized by a low level wind velocity maximum (0.5-3 m above the glacier) and a thermocline at about the same level (see Holmgren, 1971). Turbulent exchange theories, like the two used in this study, are not valid close to such a wind velocity maximum. They should be applied only to the layer clearly below the maximum (Holmgren, 1971). Nevertheless, the values from 1.5 m were used for the calculations. This level is likely to be below the wind maximum and the associated thermocline when turbulent exchanges are large. As indicated by more detailed measurements of the wind profiles on the glacier high temperatures tend to cause large wind velocities, which in turn are associated with a higher wind velocity maximum. Even in cases where the wind velocity maximum and thermocline are situated below 1.5 m, the error in the calculation of the sensible heat flux introduced by this phenomenon will be restricted. The effects of taking a smaller wind velocity and a higher temperature oppose each other. However, the stability is overestimated so that calculations with the M.O.-theory are affected more than the calculations with the s.i.-theory.

The vapour pressure over the ice is simply calculated by taking the saturation vapour pressure corresponding to the surface temperature. A similar procedure does not work for ROCKSTAT, since the vapour pressure might be well below its saturation value. The availability of water in the soil is not calculated. It depends on factors like precipitation, evaporation, horizontal advection of water, water holding capacity and permeability of the soil. Thus, another method had to be used for the calculations of the latent heat flux. The values of temperature and humidity at the middle and the lower level were used to calculate the sensible and the latent heat flux and the Bowen ratio (ratio of the sensible and the latent heat flux). Then, the sensible heat flux was calculated once again from the temperatures at 1.5 m and the surface, and the latent heat flux was obtained by dividing this last calculated sensible heat flux by the Bowen ratio.

2.3 The surface temperature

In fact, the modelling of the englacial temperature was derived from an energy balance model including the calculation of englacial temperatures. This model is described by Greuell and Oerlemans (1986). The temperature profile in the uppermost layers of the glacier and the soil was calculated from:

$$\rho c \frac{\partial T}{\partial t} = K \frac{\partial^2 T}{\partial z^2} + W \qquad (2)$$

with ρ the density, c specific heat and K the conductivity of the ice or the soil, T the temperature, t the time, z the depth below the interface and W the production or consumption of energy. W includes the formation and refreezing of melt water (only at GLACSTAT) and the total energy flux from the atmosphere. Melt water does not penetrate into the ice. However, the ice close to the surface contained some water, so that a water holding capacity (C) was assumed. Equation 2 was solved numerically on a grid of 50 points extending downwards to about 4 m and with a downwards increasing grid point distance. The timestep was the same as used for the energy flux calculations, namely 2 minutes.

3. RESULTS

3.1 Results from GLACSTAT

Firstly, the total ablation calculated was compared with the amount of ablation measured along the stakes, which was 510 ± 20 mm w.e. for the whole period (see Table I).

Table I: Calculated ablation for the whole period in mm w.e.

"Standard run": z_{ou} = 1.33 mm; z_{ot} = z_{oe} = 0.01 mm; s.i.-theory; C = 10 mm; temperature, vapour pressure and wind velocity from 1.5 m.

Measured ablation	510 ± 20
"Standard run"	550
C = 0 mm	573
No turbulent fluxes	505
Roughness lengths × 5	569
Temperature from 50 cm	545
M.O.-theory	539

In a first run, called "standard run" hereafter, the roughness parameter for the wind profile, z_{ou} , was taken to be 1.33 mm and the roughness parameters for the temperature (z_{ot}) and the humidity (z_{oe}) profiles were 0.01 mm. These values are taken from Hogg et al. (1982) and should be valid for ice in the ablation zone. The value for z_{ou} agrees well with the value obtained from detailed wind velocity profile measurements (z_{ou} = 1 mm), which we performed at GLACSTAT. In this "standard run" the s.i.-theory was applied and the water holding capacity of the ice was assumed to be 10 mm. Calculated ablation amounted to 550 mm w.e., which is evidently too much.

In the next run the effect of the water holding capacity was studied. With some water present the surface temperature decline associated with negative energy fluxes is delayed until the water in the ice is frozen. The higher surface temperature reduces the total energy flux through the outgoing long wave radiation and the turbulent fluxes. Comparing experiments with C = 0 mm and C = 10 mm in Table I we see that the water holding capacity of 10 mm reduces the total ablation by 23 mm.

In fact the surface temperature never drops below the melting point if C = 10 mm. According to our field observations a water holding capacity of at least 10 mm seems to be realistic.

In the next runs some sensitivity experiments regarding the turbulent fluxes were performed. Firstly, the turbulent fluxes were put equal to zero during the whole period. The resulting amount of ablation then equals the measured amount of ablation.

If the values of the 3 roughness parameters are each multiplied by 5, the turbulent fluxes are multiplied by a factor 1.5. An inspection of values for the roughness parameters in literature (e.g. Holmgren (1971) and Kuhn (1979a)) and estimates from our field data indicate that such a multiplication by 5 is certainly more than the upper limit in the uncertainty in the roughness parameters.

Temperature was also measured at a lower level (50 cm). If the values from this level are taken instead of the values from 1.5 m, total ablation decreases by 5 mm w.e. This decrease is caused by the existence of the thermocline.

In a further experiment the M.O.-theory was used. In this case the turbulent fluxes cause about 29 mm w.e. of ablation. For the s.i.-theory this was 40 mm w.e. The quotient of these values, 0.725, is a measure of the damping of the turbulent fluxes by the stability of the stratification. However, this damping is certainly overestimated by the use of wind velocity and temperature data from above the wind velocity maximum and its associated thermocline (see Section 2.2).

All but one of the runs mentioned in Table I give too high values for the total ablation. The exception is the run during which the turbulent fluxes were zero. However, this is certainly an unrealistic situation. Thus, it seems that changing model assumptions like the ones mentioned in Table I cannot produce a calculated amount of ablation equal to the measured amount of ablation. Therefore, the discrepancy will probably be due to instrumental error. An analysis of the effects of the possible errors in the data on the calculation of the total ablation was performed. It appeared that an error in the incoming total radiation measurements seems by far the most likely cause of the discrepancy.

Average values of the fluxes and some of the meteorological variables for the whole period along with maximum and minimum hourly means can be found in Table II. About 90% of the calculated ablation is due to radiation. This is an amazingly high value. According to Hoinkes and Steinacker (1975) this value varies between 50% and 80% in the Alps. The relatively large contribution of the radiation must be attributed to: a. the abnormally high percentage of cloudless skies. The sun impinged on the instruments during about 63% of the maximum possible time; b. the extremely low albedo, 0.16 (a similar value was found by Dirmhirn and Trojer (1955) at the very end of the Hintereisferner); c. an average temperature equal to 4.0°C. In spite of the cloudless skies this is almost equal to the long year average (see Kuhn et al., 1979b).

The daily courses of the temperature, the vapour pressure, the wind velocity and the fluxes are shown in Figures 3 and 4. Figure 3 is an average for 5 almost cloudless days; Figure 4 is an average for 3 days with poor weather. In this case this is a designation for days with

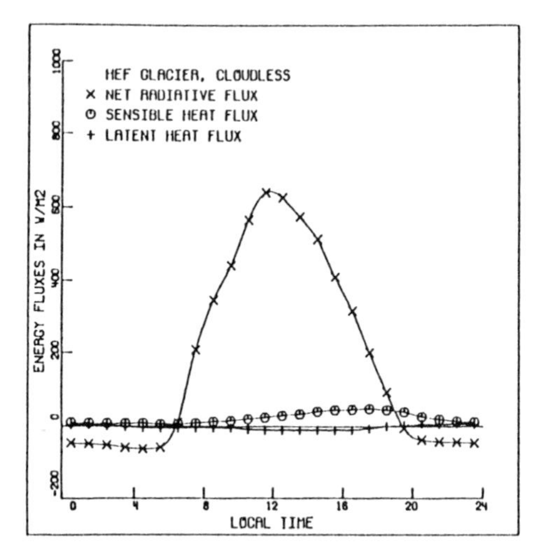

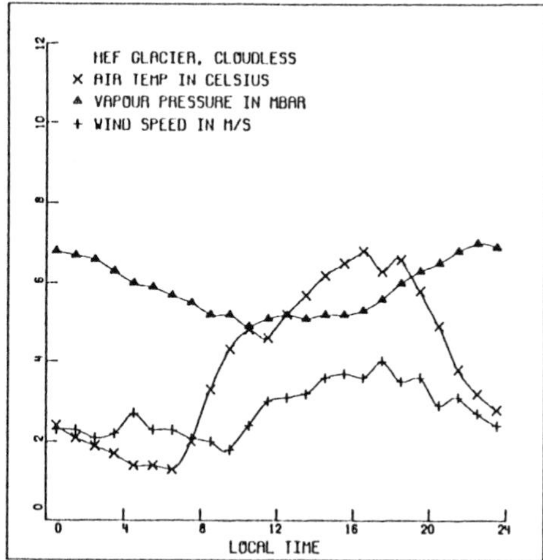

Figure 3. Daily variations of the fluxes calculated during the "standard run", and daily variations of temperature, vapour pressure and wind velocity at the middle level of the measurements (1.5 m). These are average values for 5 almost cloudless days (14, 15, 16, 17 and 21 July 1986) at GLACSTAT.

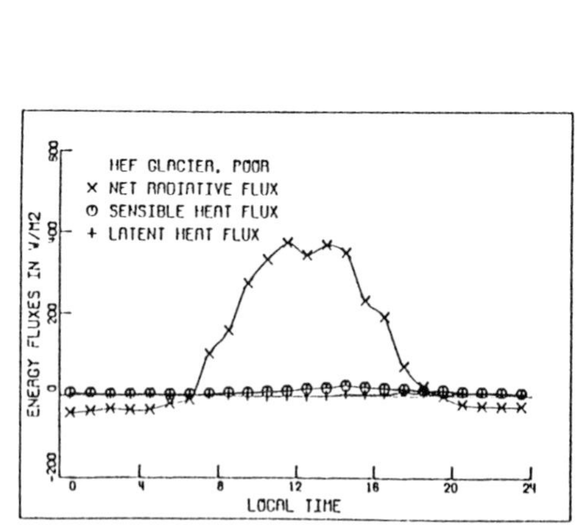

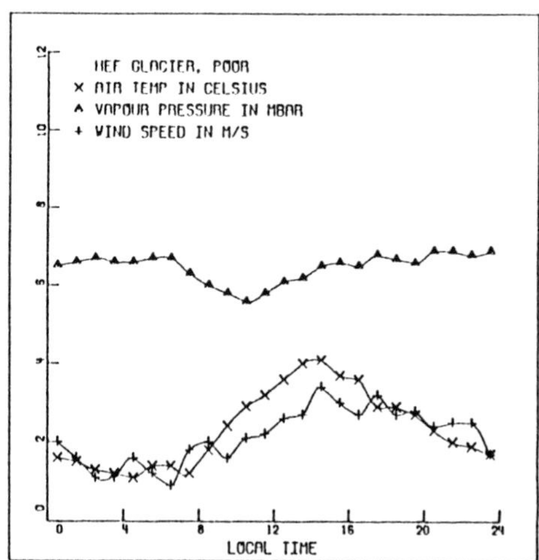

Figure 4. See Figure 3. However, these are average values for 3 "poor weather days" (13, 19 and 20 July 1986).

overcast skies and rain fall and fog during much of the time. To point out some characteristics of the daily variations we turn to the cloudless days. We find that the wind velocity and the temperature are continually rising during day-time. The rise of the temperature is interrupted at about 11 a.m. This can be understood by the movement of the thermocline associated with the glacier wind. During the early morning the thermocline is generally below 1.5 m, the height of the temperature sensor. Then, along with the increasing wind velocity the thermocline is rising. During 4 out of the 5 cloudless days considered here the thermocline passed the sensor at about 11 a.m. and thus caused a temporary decline of the temperature. The vapour pressure has a daily cycle with minimum values around noon.

Table II: Means for the whole period (12-22 July 1986) and maximum and minimum hourly means for the whole period. Variables marked with an asterisk are calculated.

	GLACSTAT			ROCKSTAT		
	mean	max	min	mean	max	min
global radiation in W m^{-2}	281	1053	0	268	1072	0
albedo	0.16	-	-	0.12	-	-
inc. long wave r. in W m^{-2}	271	316	210	271	316	210
surface temperature in °C*	0.0	0.0	0.0	10.1	29.8	-1.8
net radiation in W m^{-2}*	191	905	-106	138	769	-118
sensible heat f. in W m^{-2}*	22	118	-6	-40	47	-211
latent heat f. in W m^{-2}*	-4	36	-40	-88	12	-570
temperature in °C	4.0	12.2	-1.1	6.1	16.0	-0.3
vapour pressure in mbar	5.9	8.8	2.4	6.9	8.9	4.5
wind velocity in m s^{-1}	2.8	6.8	0.4	2.5	5.6	0.7
total energy flux in W m^{-2}*	209	946	-130	10	345	-184
ablation in mm w.e.*	550	10.2	0.0	-	-	-

The daily cycles of the turbulent heat fluxes are directly related to the daily cycles of temperature, vapour pressure and wind velocity, as the surface temperature always remains at the melting point. The dominance of the net radiative flux is evident, irrespective of the type of weather.

3.2 Results from ROCKSTAT

The method used for the calculations of ROCKSTAT has been explained in Section 2.2. The calculations were tuned by comparing calculated surface temperatures with a sequence of surface temperatures measured with a thermo-eye. Two kinds of tuning parameters were available: a. variables describing the thermal properties of the soil, notably ρ , c en K (Equation 2) and b. the surface roughness parameters. By varying the thermal properties (essentially $(\rho c K)^{\frac{1}{2}}$) the amplitude of the daily variation of the surface temperature can be tuned. By varying the roughness parameters (essentially z_{ou} as the quotient z_{ou}/z_{ot} and $z_{ot} = z_{oe}$ are prescribed) the magnitude of the turbulent fluxes change. In that way the daily mean of the surface temperature can be tuned. The following values were found to give a good result: $\rho = 2000$ kg m^{-3}, $c = 927$ J kg^{-1} K^{-1}, $K = 0.966$ J m^{-1} K^{-1}, $z_{ou} = 10$ mm, $z_{ot} = z_{oe} = 0.1$ mm. The values of the thermal parameters agree quite well with the values given by Carslaw and Jäger (1947) for an average soil ($\rho = 2500$ kg m^{-3}, $c = 840$ J kg^{-1} K^{-1} and $K = 0.966$ J m^{-1} K^{-1}).

Averages for the whole period can be found in Table II. It seems to be of special interest to compare these values with those for GLACSTAT. The average temperature ($\Delta T = 2.1$°C) and the vapour pressure ($\Delta e = 1.0$ mbar) are somewhat higher at ROCKSTAT due to the large negative turbulent fluxes during day-time. The wind velocity is slightly reduced (about 10%) at ROCKSTAT.

In Table III it is demonstrated how the incoming total radiation is lost at the 2 stations. The available amounts are almost equal and put

Table III: Loss of the available incoming radiative energy. Averages for the whole period of the measurements.

		GLACSTAT	ROCKSTAT
Gain by:	incoming total radiation	100% (552 Wm^{-2})	100% (539 Wm^{-2})
	sensible heat flux	4%	—
Loss by:	outgoing short wave radiat.	-8%	-6%
	outgoing long wave radiat.	-57%	-68%
	sensible heat flux	-	-7%
	latent heat flux	-1%	-16%
	melting	-38%	-
	storage	-	-2%

to be 100%. At both stations most of the available energy is lost by the outgoing long wave radiation. However, the percentage is definitely larger at ROCKSTAT (-68%) than at GLACSTAT (-57%) due to the higher surface temperature. For the same reason the turbulent fluxes are clearly negative at ROCKSTAT (-23%), while their sum is slightly positive at GLACSTAT (+3%). Thus, at ROCKSTAT much more of the incoming energy is returned to the atmosphere. At GLACSTAT a large part of the available energy is lost by melting (-38%). The remaining energy is lost by reflection in the short wave part of the spectrum and stored in the soil at ROCKSTAT. It should be remarked that this storage is determined by the assumed initial temperature distribution of the soil.

Like for GLACSTAT the daily variations of the fluxes are averaged over the cloudless days and the "poor weather days" (Figures 5 and 6). Turning to the cloudless days we see that the turbulent fluxes are very small during night-time, so that the energy lost by radiation is nearly balanced by the positive soil heat flux. During day-time about half of the energy gained by radiation is lost by evaporation, while the rest is in about equal parts lost by the sensible heat flux and stored in the soil.

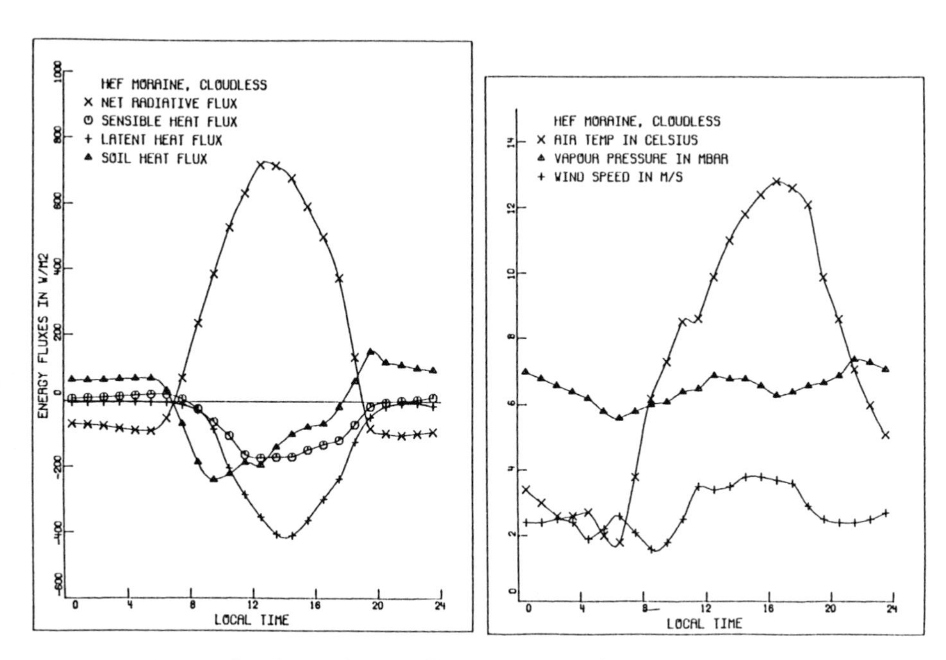

Figure 5. See Figure 3. However, these are average values for ROCKSTAT.

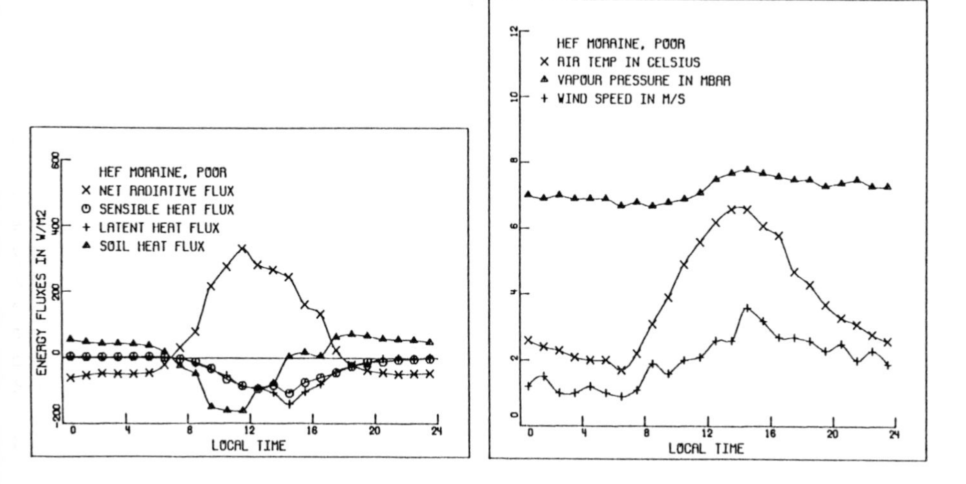

Figure 6. See Figure 5. However, these are average values for the "poor weather days".

While the method used for the calculations gives some confidence in the sum of the turbulent fluxes, the division between the sensible and the latent heat flux should be interpreted carefully. The Bowen ratio used for this division has been calculated from profile values of temperature and vapour pressure as explained in Section 2.2. The error in the measurements of these variables causes large errors in the Bowen ratio. However, the results agree quite well with those obtained by Rott (1979) in a nearby valley. Large evaporation fluxes as found in this study are only possible if the soil is wet. Observations of the soil wetness were not made.

4. INCREASING CONCENTRATIONS OF CO_2 AND OTHER TRACE GASES

4.1 Qualitative analysis of the effect on ablation

The possible climatic effects of the increasing concentrations of CO_2 only (e.g. Ramanathan, 1981 and Hansen et al., 1981) and of all relevant trace gases (Ramanathan et al., 1985) are treated in the cited literature. For the calculations 'radiative-convective' models are used. In this sub-section the consequences for the individual components of the energy balance and the resulting ablation of a glacier will be considered in a qualitative way. Then, in the next sub-section a calculation will be presented to arrive at a quantitative estimate of the 'CO_2-effect' on the energy budget and its components.

Table IV gives the changes of the variables and the fluxes affected by the increasing trace gas concentrations. Six locations are considered. In the horizontal we distinct a glacier (G) and a location (N) beyond the thermal influence of glaciers (let us say the Netherlands). In the vertical we have the surface itself (0), a level (1) about 1.5 m above the surface and a level (2) where the air is beyond the thermal influence of glaciers (let us say 2 km above the surface). Let ΔT_a be the temperature change at level 2. The corresponding change of the vapour pressure (Δe_a) is calculated with the assumption of fixed relative humidities. According to Hansen et al. (1981) this is more realistic than fixed absolute humidities. By definition ΔT_a is the same at G_2 and N_2. Let us now turn to turn to N_0 and N_1. The 'radiative-convective' models cited above provide vertical profiles of ΔT. It appears that the ΔT's at N_0 and N_1 are almost equal to ΔT_a. The same is valid for Δe and Δe_a. So temperature and vapour pressure difference between levels N_0 and N_1, and consequently the turbulent fluxes, hardly change (see Hansen et al., 1981).

The incoming long-wave radiation increases (ΔL_i) due to the higher emissivity of the atmosphere directly related to the increasing trace gas concentrations. The positive feedback mechanisms $\Delta T_a \leftrightarrow \Delta L_i$ and $\Delta e_a \leftrightarrow \Delta L_i$ enhance the effect. The feedback mechanisms are very important. According to Hansen et al. (1981) they even multiply the ΔL_i only due to the higher emissivity of the trace gases themselves by a factor of about 15. It should be borne in mind that with the 'radiative-convective' models it is not possible to calculate ΔL_i from ΔT_a. This is due to the feedback mechanism between these variables.

Table IV: Qualitative analysis of the effect of the increasing trace gas concentrations on some meteorological variables and surface fluxes.

	Glacier (G)	Location beyond thermal influence of glaciers (N)
2. Atmospheric level beyond the thermal influence of glaciers:		
temperature	ΔT_a	ΔT_a
vapour pressure	Δe_a	Δe_a
1. Atmosphere (1.5 m):		
temperature	$0 < \Delta T < \Delta T_a$	$\simeq \Delta T_a$
vapour pressure	$0 < \Delta e < \Delta e_a$	$\simeq \Delta e_a$
wind speed	$\Delta u > 0$	$\simeq 0$
0. Surface:		
temperature	$= 0$	$\simeq \Delta T_a$
vapour pressure	$= 0$	$\simeq \Delta e_a$
Fluxes:		
incoming long wave rad.	$\simeq \Delta L_i$	$= \Delta L_i$
outgoing long wave rad.	$= 0$	$\simeq -\Delta L_i$
sensible heat flux	> 0	$\simeq 0$
latent heat flux	> 0	$\simeq 0$
total energy flux	> 0	$= 0$

Calculated changes in the global radiation as a consequence of changing trace gas concentrations are insignificant compared to the changes in the long wave radiative fluxes (Hansen et al., 1981). Because the turbulent fluxes and the short wave radiative fluxes hardly change, the larger energy gain corresponding to ΔL_i must be balanced by an increase of the outgoing long wave radiation of the same amount. At the same time the long wave outgoing radiation is related to the surface temperature. Thus:

$$\Delta L_i \simeq \Delta(\sigma T_a^4) \qquad (3)$$

However, in the literature only ΔT_a is given in most cases, so that ΔL_i had to be recalculated for this study using Equation (3).

Let us now turn to the glacier site. In summer many locations on glaciers have a surface at the melting point. Such a location is considered here. Increasing trace gas concentrations cannot cause surface temperatures above the melting point, of course. However, temperature and vapour pressure in the adjacent air layer will increase, though by smaller amounts than ΔT_a and Δe_a. As on the average the katabatic wind speed increases with T_a, the wind may also be expected to become stronger. These changes lead to larger turbulent fluxes while the incoming long wave radiation increases by nearly ΔL_i . A small reduction of ΔL_i as calculated from Eq. (3), neglected in this study, is caused by the fact that temperature and vapour pressure just above the glacier change by smaller amounts than ΔT_a and Δe_a . Thus, both the turbulent fluxes and the incoming long wave radiation contribute to a larger total energy flux. So ablation increases, since parallel to the surface temperature the outgoing long wave radiation does not change.

4.2 Quantitative estimate of the effect on ablation

In this sub-section a best estimate will be made of the effect of the rising trace gas concentrations on ablation in an imaginary case. The field data from GLACSTAT will be used. According to Ramanathan (1985) the CO_2 effect is magnified by the rising concentrations of other trace gases. Therefore, experiments were done both for an increase of the CO_2 concentration only and for an increase of the concentrations of all relevant trace gases. For 1880 (representative for the pre-industrial level of concentrations) and for 2036, ΔT_a relative to 1986 was inferred from literature. Then, starting from the field data the resulting changes in the surface energy fluxes were estimated. In some more detail the following steps were taken (see Table V):

Table V: Input and output of the trace gas experiments with the 1986-GLACSTAT data.

		$[CO_2]$ in ppm	ΔT_a in °C	ΔL_i in Wm^{-2}	ablation in mm w.e.	relative to 1986
only CO_2	1880	275	-0.52	-2.8	531.1	-3.0%
	1986	339*	0.0	0.0	549.6	
	2036	449*	0.71	3.8	578.9	5.0%
all trace gases	1880		-0.79	-4.3	521.7	-5.1%
	1986		0.0	0.0	549.6	
	2036		1.54	8.4	613.8	11.3%

* values for 1980 and 2030 from Ramanathan (1985)

a. Estimate of the equilibrium ΔT_a's relative to 1980 for 1880 and 2030. These values are provided by Ramanathan et al. (1985)

b. The heat capacity of the oceans causes a time lag of the real temperatures relative to the equilibrium temperatures as they are calculated by Ramanathan et al. (1985). Although a reliable estimate of this time lag is not possible according to Ramanathan et al. (1985), Hansen (1981) gives a value of 6 years that is used in this study. Thus, the equilibrium ΔT_a-values for 1980 and 2030 correspond to real ΔT_a-values for 1986 and 2036. ΔT_a is the independent variable from which the changes in those energy fluxes, that are affected by the greenhouse warming, are derived.

c1. ΔL_i is simply obtained from Eq. (3). For each sample ΔL_i is added to the measured L_i.

c2. Estimate of the disturbed sensible heat flux (H').
If we would have a functional relationship $H(T_a)$, H' could simply be obtained by substituting $T_a + \Delta T_a$ for T_a. However, T_a was not measured. Therefore, the temperature from the highest level of the measurements (4.5 m) was taken instead of T_a. Figure 7 gives a schematic picture of the temperature profiles for T_a and $T_a + \Delta T_a$ at level 2. Profile measurements showed that the thermocline generally is below 4.5 m. In that case $T_{4.5}$ and $\Delta T_{4.5}$ should be close to T_a and ΔT_a, respectively.

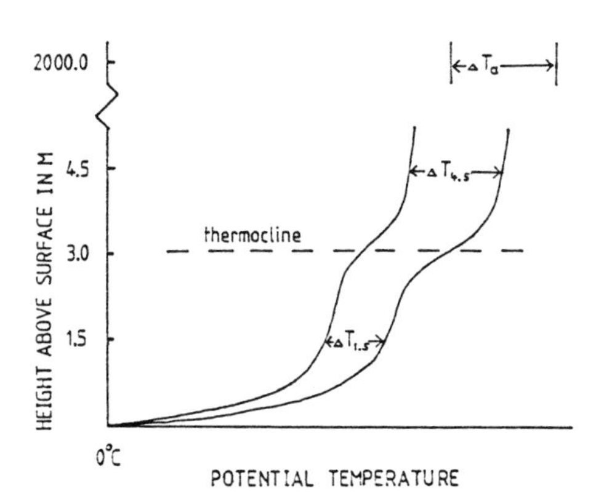

Figure 7. Schematic picture of temperature profiles for T_a and $T_a + \Delta T_a$ at a level beyond the thermal influence of the glacier. The surface is melting. ΔT decreases monotonically towards the surface and $\Delta T_{4.5}$ is almost equal to ΔT_a.

From hourly means of the data and the calculations we obtained:

$$H = 2.23 + 1.098\ T_{4.5} + 0.5164\ T_{4.5}^2 \qquad (4)$$

H' for each sample is then calculated by replacing $T_{4.5}$ by $T_{4.5} + \Delta T_a$ in this equation. In fact, by using this method to obtain H' it is assumed that a future climate disturbance (ΔT_a) will affect the sensible heat flux in the same way as variations in $T_{4.5}$ (representative of T_a) affected the sensible heat flux during the 10 days of these measurements.

c3. Estimate of the disturbed latent heat flux (E'). Again we are looking for a relation between E and T_a. As for the sensible heat flux, $T_{4.5}$ has to be used instead of T_a. There is, however, an extra complication. In the field data high temperatures are correlated with low relative humidities. This is not the kind of correlation expected for averages over longer periods (let us say one year). Relative humidities are expected to remain unchanged. Thus, a regression formula between E and $T_{4.5}$ based on the calculations and measurements of this study should not be used for the prediction of long term fluctuations in E. Instead, hourly means of $T_{4.5}$, and of temperature (T) and wind speed (u) at 1.5 m were used to obtain:

$$T = 0.19 + 0.861\, T_{4.5} \quad \text{(corr. coeff.} = 0.96) \tag{5}$$

$$u = 1.84 + 0.223\, T_{4.5} \quad \text{(corr. coeff.} = 0.52) \tag{6}$$

These relations may be expected to hold also for the averages over longer periods. Now, for each sample T' and u' were obtained by substituting $T_{4.5} + \Delta T_a$ in these equations. The disturbed vapour pressure (e') is then calculated from T' and fixed relative humidity, whereafter E' is obtained from e' and u'.

Individual values of L_i', H' and E' obtained this way certainly have no significance. Only their mean values over the whole period should be considered. This should be no disadvantage because we are dealing with climatic, thus long term average changes.

d. Determination of the total ablation for the 10-day period.

This way ablation reductions of -3.0% and -5.1% were obtained for the '1880-CO_2' and the '1880-all trace gases case', respectively, while for the corresponding '2036 cases' ablation increases by +5.0% and +11.3%, respectively. It is clear that the magnification of the CO_2-effect by the other trace gases is expected to increase. In the '2036-all trace gases case' the mean total energy flux increases by 24.5 Wm^{-2} relative to the '1986-case'. The contributions of the incoming long wave radiation, the sensible heat flux and the latent heat flux to this increase are 34%, 41% and 25%, respectively.

It is evident that the calculations contain a lot of uncertainties and assumptions. A few of them were mentioned before, but is seems worthwhile to add some remarks:

1. Though the Northern Hemisphere surface air temperature rose by about 0.5°C over the period of instrumental recording (Ellsässer et al., 1986), most authors treating this subject conclude that 'CO_2-warming did not yet rise above the noise level of natural climate variability' (Hansen et al., 1981). One might ask whether glaciers are more appropriate as a detector of the greenhouse warming. They are particularly sensitive to higher trace gas concentrations and the noise to signal ratio of snout variations is relatively low.
2. The calculations were made for the specific 10-days period and location of this study. Certainly, the Eqs. (4), (5) and (6) are sensitive to changes in time and location. Therefore, conclusions may not unthinkingly be used at other times and locations. Furthermore, according to 3-dimensional climate models (e.g. Manabe and Wetherald, 1980) the CO_2-warming will exhibit regional differences while global means were used here.
3. According to Hansen et al. (1981) climate models do not yet accurately simulate cloudiness. Therefore, it was assumed that cloudiness remains unchanged.

4. H' and E' are estimated by a kind of black box method. Hardly any physical insight into the relevant processes is used.
5. Retreat of the glacier in question and neighbouring glaciers results in a smaller glacierized to non-glacierized area ratio. The width of the glacier will also decrease. These geometric changes enhance the turbulent fluxes, even if the climate would remain unchanged (see Oerlemans, 1987). This positive feedback mechanism is not considered here.

5. CONCLUSIONS

In the case studied in this paper the radiative fluxes cause about 90% of the total ablation. This is due to the abnormally low albedo (0.16) and sunny, but relatively cold weather. Calculated and measured ablation differ by about 10%, probably due to an error in the measurements of the incoming total radiation.

While the mean turbulent fluxes are positive and relatively small at GLACSTAT (3% of the incoming total radiation), a large part of the energy gained by the incoming total radiation (-23%) is lost by turbulent fluxes at ROCKSTAT. The loss of energy by the outgoing long wave radiation is also larger at ROCKSTAT (-68% compared -57% at GLACSTAT). The extra energy available at GLACSTAT is mainly consumed by melting of the ice (-38%).

In Section 4.1 the consequences of the rising trace gas concentrations on ablation were explained in a qualitative way. It was shown that the mean incoming long wave radiation, the mean sensible heat flux and the mean latent heat flux will all increase. Thereafter, the relative contributions of these fluxes to the increased ablation for the imaginary '2036-all trace gases case' were estimated to be about 34, 41 and 25%, respectively. In that experiment ablation increased by 11%. Two kinds of problems force to interpret the results carefully. Firstly, there are a lot of uncertainties in the chain future release of trace gases → their atmospheric concentrations → climate → energy fluxes at the atmosphere - glacier interface. Secondly, the calculations were made for a specific 10-days period at a single location. The same kind of calculations should be done for longer periods and other places. The last part of the first problem and the second problem especially concern glaciologists and they might be topics for future studies.

ACKNOWLEDGEMENTS

This research has been sponsored by the Ministry of Housing, Physical Planning and Environment (The Netherlands), under contract 611003.01. We kindly thank prof. Kuhn and dr. Kaser from the University of Innsbruck for their help in organizing the field work and placing at our disposal some of their field work facilities including the use of a helicopter. Furthermore, we would like to thank Huib de Swart for critically reading and discussing Section 4.

REFERENCES

Businger, J.A., 1973. Turbulent transfer in the atmospheric surface layer. In workshop on micrometeorology, D.A. Haugen, Editor, published by the American Meteorological Society.

Carslaw, H.S. and J.C. Jäger, 1947. Conduction of heat in solids. Oxford: Clarendon Press.

Dirmhirn, J. und E. Trojer, 1955. Albedountersuchungen auf dem Hintereisferner. Arch. Met. Geoph. Biocl., Ser. B, 6, 400-416.

Ellsässer, H.W., 1986. Global climatic trends as revealed by the recorded data. Reviews of Geophysics, Vol. 24, 745-792.

Greuell, W. and Oerlemans, J. 1986. Sensitivity studies with a mass balance model including temperature profile calculations inside the glacier. Zeitschrift f. Gletscherkunde und Glazialgeologie, Band 22 (2), 101-124.

Hansen, J. et al., 1981. Climate impact of increasing atmospheric carbon dioxide. Science, Volume 213, 957-966.

Harding, R.J. et al., 1987. Energy and mass balance studies in the firn area of the Hintereisferner. This volume.

Hogg, I.G.G. et al., 1982. Summer heat and ice balances on Hodges Glacier, South Georgia, Falkland Islands Dependencies. Journal of Glaciology 28 (99): 221-238.

Hoinkes, H. and R. Steinacker, 1975. Zur Parametrisierung der Beziehung Klima-Gletscher. Rivista Italiana di Geofisica e Scienze Affini, vol. I, S. 97-104.

Holmgren, B., 1971. Climate and energy exchange on a sub-polar ice cap in summer (6 parts). Meteorologiska Institutionen Uppsala Universitetet, Meddelande nr. 107.

Kuhn, M., 1979a. On the computations of heat transfer coefficients from energy balance gradients on a glacier. Journal of Glaciology, vol. 22, no. 87, 263-272.

Kuhn, M. et al., 1979b. 25 Jahre Massenhaushaltsuntersuchungen am Hintereisferner: Institut für Meteorologie und Geophysik, Universität Innsbruck, 80 pp.

Manabe, S. and R.T. Wetherald, 1980. On the distrubution of climate change resulting from an increase in CO_2 content of the atmosphere. Journal of the Atmospheric Sciences, volume 37, 99-118.

Mannstein, H., 1985. The interpretation of albedo measurements on a snowcovered slope. Arch. Met. Geoph. Biocl., Ser.B, 36, 73-81.

Meier, M.F., 1984. Contribution of small glaciers to global sea level. Science 226, 1418-1421.

Oerlemans, J., 1987. On the response of valley glaciers to climatic change. Same volume.

Patzelt, G., 1970. Die Längenmessungen an den Gletschern der Österreichischen Ostalpen 1890-1969. Zeitschrift für Gletscherkunde und Glazialgeologie, Bd. VI, Heft 1-2, S. 151-159.

Ramanathan, V., 1981. The role of ocean-atmosphere interaction in the CO_2 climate problem. Journal of the Atmospheric Sciences, volume 38, 918-930.

Ramanathan, V. and R.J. Cicerone, 1985. Trace gas trends and their potential role in climate change. Journal of Geophysical Research, volume 90, 5547-5566.

Reynaud, L., 1983. Recent fluctuations of alpine glaciers and their meteorological causes: 1880-1980. In: Variations in the global water budget, A. Street-Perrott et al. (eds.), Reidel, 197-205.
Rott, H., 1979. Vergleichende Untersuchungen der Energiebilanz im Hochgebirge. Arch. Met. Geoph. Biokl., Ser. A. 28, 211-232.

ENERGY AND MASS BALANCE STUDIES IN THE FIRN AREA OF THE HINTEREISFERNER

R.J. Harding[1], N. Entrasser[2], H. Escher-Vetter[3], A. Jenkins[1], G. Kaser[2], M. Kuhn[2], E.M. Morris[1] and G. Tanzer[3]

[1] Institute of Hydrology, Wallingford, U.K.
[2] University of Innsbruck, Austria
[3] Institute of Radiohydrometry, Munich, F.R.G.

1. INTRODUCTION

In order to make quantitative predictions of the response of glaciers to climatic change it is necessary to have a mathematical model of the many interacting physical processes which occur in the glacial system. One of the most important of these is the transfer of energy between the atmosphere and the upper surface of the glacier. This controls the mass balance of the glacier and, together with dynamical processes, its length and volume. Physics-based "snowmelt" models of mass and energy transfers at a snow or ice-covered surface have been in existence for some time, but they are based on equations for turbulent transfer in the atmosphere which strictly only apply when the underlying sruface is an extensive flat plane and the geostrophic wind is constant in magnitude and direction. Neither of these conditions will generally hold in an alpine glacier valley. This paper discusses how far the physics-based models can be applied in these circumstances.

In July 1986 an international program of field studies was undertaken on the Hintereis Glacier, in the Oetztal Alps (Austria) to investigate the interactions between the glacier and the circulations in the lower atmosphere. The energy and mass transfers to the surface of the glacier were studied at two sites, in the firn area and on the ice tongue, and further meteorological measurements were made in the pro-glacial area. The experiment formed part of a long-term research programme on the Hintereisferner led by the University of Innsbruck. A comparison of the wind systems and energy balances at all the sites will be reported elsewhere; in this paper an analysis is made of the energy balance at a single site in the firn area. The field data are used to assess how accurately a physics-based snowmelt model (the Institute of Hydrology Distributed Model) may be used to predict snowmelt, evaporation/condensation and snow temperature.

In the model turbulent transfers are calculated using bulk transfer coefficients which depend on scaling lengths z0, zH and zE. The length z0 is normally identified with the aerodynamic roughness length of the

J. Oerlemans (ed.), Glacier Fluctuations and Climatic Change, 325–341.

snow surface. zH and zE are scaling lengths for the turbulent transfer of sensible and latent heat respectively. A comparison of predicted with measured data allows the optimum value of these parameters to be determined. It is shown that at the Hintereisferner firn site good simulations can be obtained using the IHDM physics-based snow melt model. However, the parameters z0, zH and zE are not equal (as is often assumed) and z0 cannot be regarded simply as the roughness length of the surface. The optimum values of zH and zE are found to be considerably larger than the roughness lengths determined from wind profiles measured over extensive, flat, snow-covered areas by other workers. Direct measurements of evaporation/condensation and surface temperature allow and unusually detailed check to be made on the calculation of turbulent transfers. It appears, although the field data are not conclusive, that the transfers of sensible and latent heat may not be controlled by the same bulk turbulent transfer coefficients at this site.

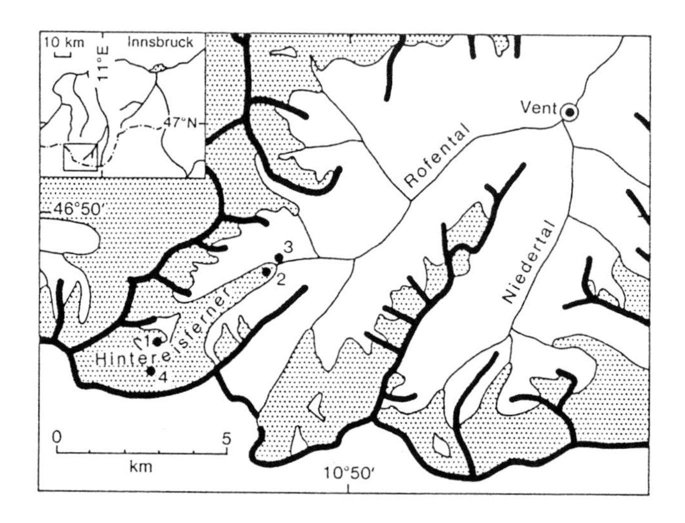

Figure 1. The Hintereisferner showing (1) the Hintereis meteorological station (3026 m) and energy balance study sites; (2) on the glacier tongue (2500 m); (3) in the pro-glacial area (2442 m) and (4) in the firn area (2960 m, site 29).

2. FIELD MEASUREMENTS

The firn site (also known as site 29) was located at the upper end of the Hintereisferner at an altitude of 2960 m (Fig. 1). The surface here is gently sloping, approximately 5°, to the northwest. At the time of the experiment (between 11th and 22nd July 1986) the transient snow line on the glacier was between altitudes 2650 m and 2750 m and the snow depth at site 29 was approximately 1 m. During the early part of the experimental period the weather conditions were dominated by high radiation levels and very low wind speeds. Between the 18th and mid-day on the 20th cloudy conditions prevailed with a snowfall of approximately 8cm. On the night 17/18th the wind increased to over 4 m/s for a number of hours but then dropped to almost zero. On the 20th the cloud-free conditions returned and very high radiation levels were again observed on the 21st and 22nd. During the day the air temperature ranged between 0°C and 7°C dropping at night to as low as -8°C. Rapid melting of the snowpack took place during late morning and afternoon with refreezing at night.

Groups from three different organizations were involved in the data collection at site 29. The Institute of Hydrology team, made measurements of radiation components, air temperature, humidity and wind speed at a singe level, at 2 m above the snow surface, and the melt rate, density, grain size, temperature and water content of the snow cover. The group from the Institute of Radiohydrometry, Munich, measured air temperature and wind speed at three heights together with incoming and outgoing total radiation. The data were recorded as 2 minute values on a magnetic cartridge recorder (Microdata M 1600 L) and then read into a personal computer where further processing took place. In addition, a black and white photograph was taken of the site twice a day, thus recording - although rather crudely - the condition of the snow beneath the reflected short-wave radiometer. This technique has been used for several years on the Vernagtferner to document the condition of the glacier surface by daily photographs (Escher-Vetter, 1985). Members of the Geographical and Meteorology and Geophysics Institutes of the University of Innsbruck made direct measurements of evaporation and condensation rates, air temperature and the surface temperature of the snow. A complete list of variables measured, equipment deployed and probable errors is given in Table I (see pages 339-340)

Air temperature was measured using platinum resistance probes (Pt100). The Munich group probes were shielded in Baumbach huts and were unventilated. The instrument heights were 2.7 m, 3.7 m and 4.6 m above the snow surface. The probe on the Institute of Hydrology AWS was shielded by a Didcot Instruments weather station screen modified to allow aspiration of the temperature sensors. The degree of aspiration using the standard fan has been estimated at 2 m/s, which has proved quite adequate in most applications. However, in the extreme conditions found on the glacier, with very high radiation, both incident and reflected, and low wind speeds, the radiation errors can be appreciable. Figure 2 shows a comparison of the air temperature recorded by the IH sensor at a height of 1.13 m, by the unaspirated Munich sensor at 2.6 m and temperatures at 1.3 m read manually by the Innsbruck group using a very well aspirated sensor. It is evident that with the standard fan the air temperatures at 1.3 m recorded by the IH AWS were overestimated by 2°C during periods of high radiation and low wind speed. In order to imporve the aspiration a larger fan, which gives a flow rate of 6 m/s, was installed for short periods. The higher aspiration reduced the radiation error in the air temperature but it was still between 0.5°C and 1.0°C at very low wind speeds. Unfortunately it was not possible to employ this larger fan continuously during the 11 days of the field

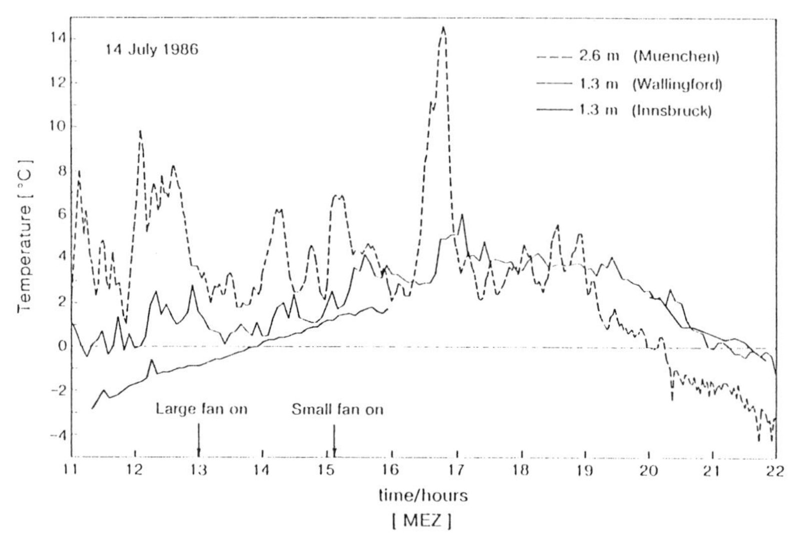

Figure 2. The effect of aspiration on measurements of air temperature. The Innsbruck sensor is well aspirated, the Munich sensor is unaspirated and the Wallingford sensor is aspirated by a large or small fan as shown in the figure.

experiment because of its large power consumption. The shielding of the temperature sensors from reflected radiation was not adequate and this probably accounts for the very high radiation errors observed over snow. In this paper turbulent transfer of convected heat is calculated using the IH data with a simple correction equation for the true air temperature, TA', at 1.3 m:

$$TA' = TA - 2.0°C \tag{1}$$

for wind speeds $W < 1.0$ m/s and incoming solar radiation $R > 400$ W/m^2. (At other times the measurement in temperature is assumed to be correct.)

The humidity of the air was measured using an aspirated wet bulb (Pt100) thermometer and a gold-mirror dew point hygrometer. Figure 3 shows the difference between the specific humidity calculated from the dew point temperature and that calculated from the wet bulb temperature as a function of time. The dew point temperatures have been corrected by the equation:

$$DPT' = DPT + 1.0°C \tag{2}$$

to adjust for a zero error in the hygrometer and the corrected dry bulb temperature, TA', has been used in the calculation of specific humidity from the measured wet bulb temperatures. Large differences between specific humidities occur when the air temperature is below 0°C and the wet bulb thermometer does not function properly. There are also problems at low wind speeds.

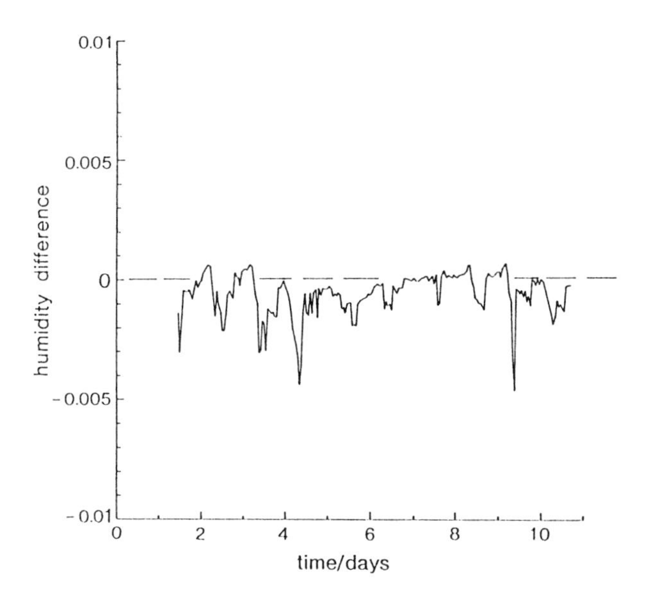

Figure 3. The difference between specific humidities calculated from measurements of dew point temperature and wet bulb temperatures as a function of time.

Snow temperatures were monitored using three small, 1 mm diameter thermistors. They were installed at 5, 50 and 100 cm below the surface. As the snow melted the 5 cm probe became exposed and had to be reinstalled at 5 cm below the new surface on occasions. Since the thermistors were not held on a rigid framework there was some uncertainty in the absolute position with respect to the underlying ice. Their depth below the surface at any given time had to be calculated from measurements of snowmelt which introduced further uncertainty. There was a considerable radiation error in the temperatures recorded by the upper thermistor even when this was beneath a full 5 cm of snow. This error, although large, has been fairly easy to correct because it generally only occurred when the pack was melting and hence known to be a 0°C. Surface temperature was measured using an infra-red thermometer at hourly intervals when people were on site.

The accuracy of the wind speed observations is extremely important for calculation of turbulent transfers. The anemometer used on the IH AWS is accurate to ± 0.1 m/s at speeds above 1 m/s. Below this the wind speed is underestimated and there is a stall speed of approximately 0.3 m/s. The wind speed at site 29 was less than 1 m/s for approximately half the experimental period and hence the stalling error must be taken into account. In this paper turbulent transfer have been calculated using the IH data and a simple correction equation

$$W' = 0.6 + 0.4 \text{ m/s} \tag{3}$$

for wind speeds $W < 1$ m/s. The Munich group used three Casella sensitive anemometers which have a stall speed of 0.15 - 0.2 m/s. The error in velocity is considerably lower than 0.1 m/s especially at low wind speeds and corrections are not necessary. The data from these anemometers, which were installed at heights of 2.5 m, 3.5 m and 4.4 m above the snow surface at the beginning of the experiment, have been used to investigate the form of the wind profiles at site 29.

A Funk type net radiometer was used by the IH team; the Munich group used Schenk radiometers for incoming and outgoing radiation. Funk radiometers are usually considered more accurate than Schenk radiometers since they have thinner plastic domes and a larger collecting surface. The accuracy of the Funk type net radiometers is generally accepted as ca. 5%. Measurements of diffuse and reflected short-wave radiation were also made by the IH group using Kipp solarimeters; the detailed description of these data will be the subject of a separate paper.

The evaporation measurements were made by the Innsbruck group using small lysimeters (surface area 400 cm^2 , depth 25 cm) made of plexiglas. These were weighed manually every hour (Kaser, 1982) to a precision of ca. 1 g. The precision of an individual hourly measurement of evaporation or condensation is therefore ca. 0.0125 mm/hr. The accuracy of the measurements is very difficult to estimate; errors will be introduced by edge effects at the double plexiglas wall and by the sample becoming unrepresentative of the surrounding snow through different ageing processes within the lysimeter. However, these errors can be reduced by using two or more lysimeters, with the samples being removed at different times. Previous work has shown that the mean value determined using several lysimeters represents the processes on the surrounding snow fairly well (Kaser, 1983; Fohn, 1978). Clearly, the relative accuracy rapidly decreases as the absolute value of evaporation or condensation decreases.

Density profiles within the snowpack were measured regularly using both gravimetric and neutron probe techniques (Harding, 1986). The gravimetric densities were accurate to ca. 0.05 g/cc. Taking measurements using a neutron probe lowered into a fixed metal access tube is considerably less labour intensive than digging a snow pit and taking cores. At depth the method is also considerably more precise since the instrumental error is equivalent to ca. 0.01 g/cc. However, near the surface there are two major sources of error. The sphere of influence of the neutrons emitted by the source is about 50 cm so that within this distance from the surface density will be underestimated. In addition, the access tube suffered from considerable radiational heating, causing snowmelt around the tube and formation of an air gap. This is a further reason for underestimation of the density. The solution to the air gap problem seems to be to remove the access tubes after each reading and reinstall them, in the existing holes, prior to the next reading. This technique worked successfully, despite the obvious risk that the hole might be enlarged when the tube was reinserted.

It was hoped that accurate measurements of density profiles using the neutron probe would allow snowmelt rates to be determined. The inaccuracies in neutron probe measurements near the surface meant that in the end the melt was best estimated from measurements of snow depth and gravimetric measurements of density. However, the daily neutron probe observations are well able to demonstrate the consistency of snow densities over the study period; this was confirmed by the less frequent snow pit data. Snow depths were monitored at least twice a day using the access tubes as a network of snow stakes. The surface of the snow was very uneven so the random error in the depth at a given stake could be fairly large (ca. 0.5 cm). The variation between stakes gave an error in the calculated melt rate of the order of 0.2 mm snow/hr.

The snow water content was measured using a brine technique (Morris, 1981) and using a dielectric probe. The dielectric probe used in this field experiment has been developed at IH for soil water content measurements and is not very suitable in its present form for work on snow. However, it has been established by Ambach and others (e.g. Ambach and Denoth, 1975) that dielectric measurements are a promising and potentially extremely useful method of measuring the liquid water content of snow. Both methods gave low water contents (3-5%) with only small changes through the experimental period.

3. THE IHDM ENERGY BUDGET MODEL

The Institute of Hydrology Distributed Model, (IHDM), is a physically-based mathematical model for hydrological processes within a catchment. It contains a sub-model for the distribution of meteorological data and a suite of three models for calculating snowmelt which were originally developed for the Système Hydrologique Européen (SHE), (Morris, 1982). These sub-models have been used to predict snow depth, evaporation and surface temperature at station 29 from the measured meteorological data. The meteorological data have first been corrected for the gradient and aspect of the slope at station 29. Then the energy budget approach has been used to calculate mass and energy transfers at the snow surface and changes in temperature of the snow. The most important features of the IHDM energy budget snowmelt sub-model are:

1. that the snow density is assumed to remain constant over time,
2. that the surface temperature of the snow is estimated using an analytical solution of the heat conduction equations, and
3. that the bulk coefficients of turbulent transfer are calculated from "effective" aerodynamic roughness and scaling lengths.

In neutral conditions the turbulent transfer coefficient for momentum is given by the logarithmic expression

$$Dn(t) = W(t)\ (k/\ln(Z(t)/z0(t)))^2 \qquad (4)$$

z0 is the aerodynamic roughness length, k van Karmans constant, W (t) the wind speed and Z the height of the anemometer above the snow surface. Similar expressions relate the coefficients for transfer of sensible heat vapour with scaling lengths zH and zE. Stability corrections may be made to equation (4) using the Monin-Obhukov functions or, more easily, the bulk Richardson number, Ri. Moore (1983) for example suggests the expression

$$Ds(t) = Dn(t)\ (1-5\ Ri(t))^2 \qquad 0.0 < Ri < 0.2 \qquad (5)$$

for the transfer coefficients for momentum, sensible heat and water vapour in stable conditions assuming that z0, zH and zE are equal. It is normal in energy budget models to use equations of the same form as (4) to calculate average transfers over a given time period given average values of the meteorological data. For example, the IHDM is run on an hourly time step using the hourly average wind speed, W, and the effective turbulent transfer coefficient, Dn, is calculated from the expression

$$Dn = W\ (k/\ln(Z/z0))^2 \qquad (6)$$

where Z is also averaged over an hour and z0 is the effective roughness for the period. However, unless the weather conditions are extraordinarily stable this approach must introduce errors. In variable conditions equation (6) will not be a good approximation to equation (4) integrated over one hour. The value of z0 will then be a measure not only of the physical nature of the snow surface but also of the changing structure of the atmospheric boundary layer. The same argument also applied to zH and zE. This problem is particularly important during periods of light, variable katabatic winds.

The IHDM allows the use of different values for z0, zH and zE if desired. The efficiency of the model in predicting snow melt is determined by calculating the [parameter

$$Fz = (z' - z)^2\ /\ (c - \langle z \rangle)^2 \qquad (7)$$

where z measured values of the snowpack depth, $\langle z \rangle$ is their average and z' are the corresponding predicted depths.

4. RESULTS

The profile measurements made by the Munich group showed that the maximum wind speed occurred at 2 m to 3 m above the ground, that is at the height of the two lower anemometers. This observation is generally in good agreement with other measurements (Weber, 1987). The maximum velocities are larger than those measured on the nearby Vernagtferner during the LUZIFER experiment as a result, amongst other things, of the longer passage of the air masses to site 29 on the Hintereisferner. The direction of the gradient wind was approximately from the west during the whole measuring period but nevertheless it could not, on the whole, fortify the velocities observed in the glacier wind layer. Thus wind speeds in this lowest layer were very low during part of the experiment. Because of this, and because there were no anemometers very close to the snow surface, it was not possible to investigate the velocity profile of the glacier wind layer. However, it is clear that the assumption of a logarithmic wind profile is not valid. The temperature profile was found to be approximately of the log-log type with the corrected air temperature from the IH AWS fitting rather well on the curve. Conditions were extremely stable, with temperature gradients of more than +1°C/m. Hourly average temperature ranged from colder than -8°C at 2.7 m in the late night hours of 14th July to nearly +10°C at 4.6 in the late afternoon of 17th July.

As was expected the profile measurements at site 29 did not suggest a clear value of aerodynamic roughness length z0 which could be used as an estimate of zH and zE. The appropriate "effective" values of zH and zE could only be determined by optimisation of the IHDM energy budget model. Since the energy transferred by latent heat flux is minimal it is possible to determine zH separately from zE by optimising first using measured snowmelt to determine zH and then using measured mass transfer to determine zE. z0 values were optimised for the first 6 days of the data, thus the second period can be regarded as a verification period, although there is some uncertainty over the changes between day 7 and 9 because of uncertainties in the inputs.

Figure 4 shows the variation in Fz, the efficiency of the IHDM in predicting snowmelt, as a function of effective scaling length zH for station 29 calculated using measurements of snow depth z, made over the period 13th to 18th July. Measurements made after the bad weather on the 19th and 20th are not included since there is no independent measurement of the amount of snowfall in that period. Curve (a) was obtained using the correct, hourly average snow density over the whole pack of 0.55 g/cc. The value of zE was set equal to zH but the results are not in fact sensitive to the value of zE. The minimum at zH = 40 mm is well defined although the sensitivity of the melt prediction to the scaling length is low, as would be expected since the snowmelt over the experimental period was dominated by radiation which contributed 60% of the total energy input. The minimum value of Fz = 0.046 is equivalent to a correlation coefficient for linear regression between predicted and measured values for r^2 = 0.954. Thus the energy budget model can produce a good simulation of snow melt although as will be discussed in the next section, the optimum value of zH is very high. Values of zH between 37 and 45 mm will produce acceptable simulations with Fz less than 0.05 and r^2 greater than 0.95.

Curve (b) shows the variation in Fz if the average density of the snow is assumed to be 0.5 g/cc. This change is within the precision of a single gravimetric measurement of snow density. While there is no evidence that the snow density changed during the period because of the limitations of measuring techniques there is a possibility of a systematic error of ± 0.05 g/cc. The optimum value of zH is now 27 mm and the minimum value of Fz is reduced to 0.041 (r^2 = 0.96). A further demonstration of the sensitivity of the model to errors in the input data is given by curve (c). This curve shows the variation in Fz for a density of 0.55 g/cc if the measured net radiation values are increased by 5%. Further reductions in zH (to 17 mm) and Fz (to 0.035) are produced by this variation which is within the precision of the net radiometer. The relative importance of accurate estimation of zH and correct radiation input data can be quantified by saying that the change in minimum Fz produced by changing the input data, that is ΔFz (min) = 0.011, is equivalent to the change in Fz produced by an uncertainty in zH of approximately $\pm$ 10 mm.

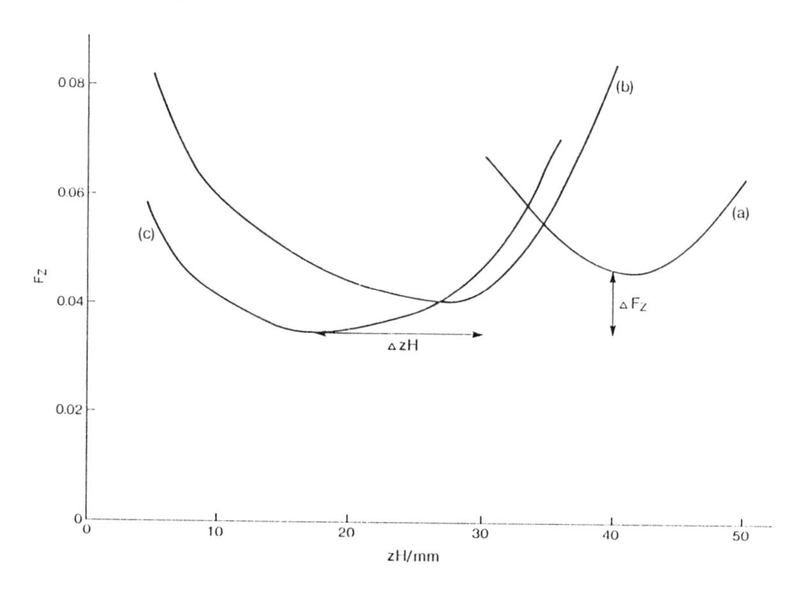

Figure 4. The efficiency of the energy budget model in predicting snowmelt, Fz, as a function of the effective scaling length zH (a) snow density 0.55 g/cc; (b) snow density o.55 g/cc and (c) snow density 0.55 g/cc with 5% increase in net radiation. No corrections for the stability of the boundary layer have been made.

The effect of stability corrections is interesting. Although it cannot really be expected that equation (5) should apply to averaged hourly data a similar equation

$$\begin{aligned} Ds &= Dn\ ((1 - 5\ Ri)^2 + K) && Ri < 0.2 \\ &= Dn && Ri > 0.2 \end{aligned} \qquad (8)$$

was tested since this approach has been followed by previous authors (eg. Dunne, Price and Colbeck, 1976). The constant K was allowed to take non-zero values to allow for the possibility of remnant turbulent transfer even in conditions of strong stability. A minimum value of Fz occurred at zH = 32 mm and K = 0.6 but the value of 0.075 was much higher than the minimum of 0.046 for the corresponding input data with no stability correction. Matters were not improved by making K a function of wind speed. Better results were obtained using a very simple stability correction at low wind speeds only.

$$Ds = Dn\ K \qquad W < 1\ m/s \qquad (9)$$

The optimum value of K was found to be 0 with the optimum value of zH unchanged from its original value (Curve (a) Fig. 4). of 40 mm. The minimum Fz fell to 0.38 which indicates that the simulation can be improved by the simple assumption that turbulent transfer is suppressed when wind speeds are below 1 m/s. However, it should be remembered that the measurements of wind speed below 1 m/s have been corrected for stalling of the anemometer. Given the possible errors in low wind speeds it is not possible to draw definite conclusions about the usefulness of equation (9) for stability corrections.

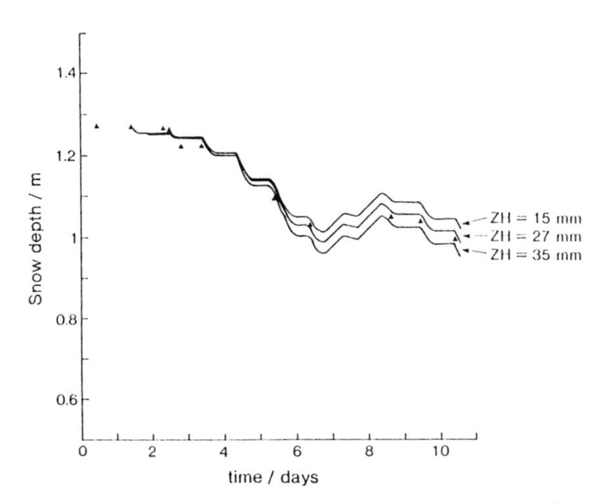

Figure 5. Measured and predicted snowpack depths for zH = 15 mm, 27 mm and 35 mm, and snow density 0.5 g/cc.

Figure 5 shows the measured and predicted snowpack depths for zH = 15 mm, 27 mm and 35 mm and an average density of 0.5 g/cc. A steady snowfall of 1.8 mm water equivalent per hour has been assumed for 28 hours with sub-zero temperatures between the 18th and 20th July. This amounts to a total snowfall of 50.4 mm water equivalent and is sufficient to bring the predicted snow depth close to the measured value of the 20th. This assumption was necessary because detailed measurements of snow input were not made, other than the snow depths and densities before and after (which include melt as well as input). By the time measurements of snow depth and density resumed on 20th July most of the new snow had melted and so by this time this new fall did not make a major contribution to the snow pack and the assumption of the average density of 0.55 g/cc was still correct. Two further measured values which have not been used in the optimisation of zH lie close to the predicted curve. Note that most of the total turbulent transfer occurs on one day (day 5, 17th July).

Although the value of zH = 40 mm gives a good simulation for snowmelt when the density is 0.55 g/cc the simulation of mass transfer using zE = 40 mm is not good at all. In order to improve the simulation a lower transfer coefficient must be used for mass transfer. Figure 6 shows the measured and predicted evaporation and condensation for zH = 40 mm, zE = 5 mm and snow density 0.55 g/cc. Error bars have not been shown on the diagram for the sake of clarity, but as mentioned in section 2, these are expected to be of the order of ± 0.0125 mm/hr for the measured mass transfers. The efficiency of the energy budget model in predicting mass transfer at the snow surface, Fe, can be defined in the same was as Fz. The value for the simulation shown in Fig. 6 is high (1.58) although the simulation looks reasonably good to the eye. This is the best that can be achieved by optimising zE. It should be noted, however, that the measured values of mass transfer are a biased selection from the population of values over the experimental period.

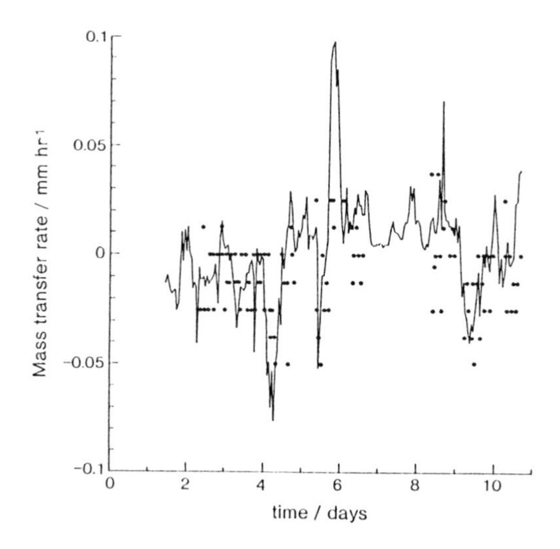

Figure 6. Measured and predicted evaporation and condensation rates for zH = 40 mm, zE = 5 mm and snow density of 0.55 g/cc.

For obvious reasons there are more day-time measurements of evaporation than night-time measurements of condensation. This means that Fe will be higher than would be the case if the full variability of mass transfer had been recorded. It appears that there are periods when the model gives good estimates of mass transfer rate, e.g. from 16th to 18th July but there are other days (days 6, 8 and 10 - 18th, 20th and 22nd July) when evaporation is measured in the afternoon but condenstation is predicted.

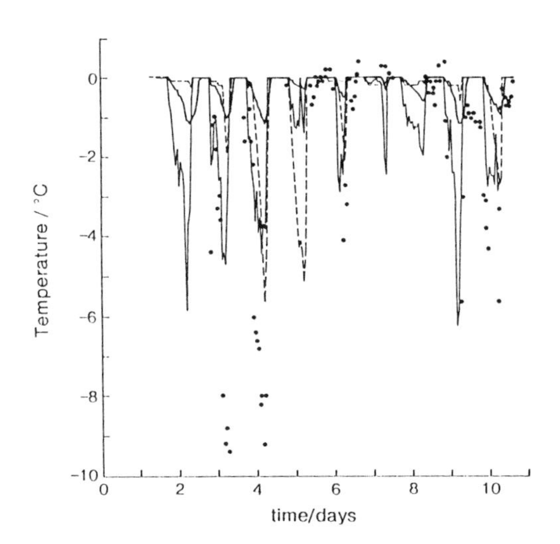

Figure 7. Measured and predicted snow temperatures. (a) surface temperature measured by infra-red thermometer (.); (b) temperature in the top 5 cm of the pack measured by the thermistor (- - -); (c) predicted surface temperature (————) and (d) predicted average temperature (———).
Predicted values were obtained using zH = 40 mm, zH = 5 mm and density 0.55 g/cc.

5. DISCUSSION

One possible reason for this is that the model is failing to predict the surface temperature of the snow correctly. This temperature is used to calculate the specific humidity gradient between air and snow which controls the mass transfer rate. Figure 7 shows measured and predicted snow temperatures for zH = 40 mm, zE = 5 mm and density 0.55 g/cc. The thermistor data gives temperature at a varying depth within the surface (0-5 cm) layer of the snow. There may not always be good thermal contact between the thermistor and the snow and during the day there are radiation errors so these data should be treated with considerable caution. The spot measurements of snow surface temperature made with an infra-red thermometer have a precision of ± 0.5°C but the total error is likely to be greater. Values above 0°C are clearly not correct. The order of magnitude of these temperatures (i.e. down to -10°C and ever lower) agrees with calculations of surface temperature made using the outgoing long-wave radiation data collected by the Munich group. However, this cannot be regarded as strong confirmation since there are always difficulties in comparing surface temperatures determined by these two methods. The predicted average and surface temperatures are also shown. The correspondence between measured and predicted surface temperature is not very good during periods of low temperature, especially on the nights of 15th and 16th July (days 3 and 4). However, the periods of bad prediction of evaporation occur when the predicted surface temperature is 0°C but the measured values are about -0.5°C. This means that predicted evaporation rates will be too large and predicted condensation rates too small. This does not help to explain why condensation is predicted for periods when evaporation is measured. It appears that the source of error in the mass transfer calculations is not the method of estimating surface temperature used in the model.

The same problem of evaporation occurring when condensation is predicted was found by de Quervain (1951) and Kaser (1982, 1983) when comparing mass transfers measured using lysimeters with measured vapour pressure gradients. Unpublished data collected by Ohmura (pers. comm.) near the mean equilibrium line of the Rhône Glacier may lead to an explanation. There, synchronous variations in air temperature and dewpoint of the order of 1°C occurred with a time scale of seconds when there was a glacier wind. It has been suggested (Kaser, 1983) that for similar sites and conditions a change in mass flow direction is always connected with a changbe in stability. Condensation is connected with a high stability, evaporation with a lower stability. Therefore, given the same magnitude of vapour pressure gradient, there will be less condensation than evaporation. The average vapour pressure gradient over a relatively long period, such as the one hour time step used in the IHDM, may be zero, or even may suggest condensation, while the average mass transport over the period is evaporation.

6. CONCLUSIONS

The range of values of z0 for ice and snow surfaces determined from wind profile experiments on extensive, flat sites with steady geostrophic winds is 10^{-3} mm to 1 mm. We have shown that the effective value of zH for site 29 in the firn area of the Hintereis Glacier, determined by optimisation of an energy budget model, is considerably larger, that is zH = 40 ± 20 mm. The considerable uncertainty in zH arises not from the model but because the input meteorological data and the initial conditions of the snow cannot be measured with complete accuracy. The precision of our instruments is fairly standard for micrometeorological studies (for example ± 5% for net radiation) so we would suggest that a similar level of uncertainty applies to values of zH determined by other authors in energy budget studies. The effective value of zE is even less well determined but appears to be approximately 5 mm.

We suggest that the effective value of zH for site 29 is large for three reasons

1. the site is sloping and surrounded by mountains so the wind profile is not solely controlled by the roughness of the snow,
2. the winds are variable on a time scale less than the hourly time step of the model so that averaging of input data introduces bias and
3. for very light winds the transfer of sensible heat is more effective than the transfer of momentum.

Table I. Instruments deployed at Station 29 (for ref. see page 327)

Institute of Hydrology

Variable	Instrument	Precision	(Other) Errors
incoming solar	Kipp	± 10 W/m^2	radiometers
reflected solar	Kipp	± 10 W/m^2	must be
diffuse solar	Kipp	± 10 W/m^2	horizontal
net radiation	Funk	± 5%	
humidity	aspirated wet bulb	± 0.1 g/kg	only operates above 0°C radiation error (of up to 2°C)
	dew point hygrometer	± 0.1 g/kg	unknown zero error of 1°C
air temperature	aspirated Pt100	± 0.1°C	radiation error error of up to 2°C
snow temperature	thermistor	± 0.1°C	large error from radiation near surface
Wind speed	cup anemometer	± 0.1 m/s	high stalling speed, 0.3 m/s
snow depth	stakes	± 2 mm	random error depends on surface roughness

Table I continued

Variable	Instrument	Precision	(Other) Errors
snow density	snow tube	± 0.05 g/cc	error from compaction at low densities
	neutron probe	± 0.05 g/c	large error near surface
liquid water content	freezing point depression	± 5%	error from compaction at low densities
	dielectric probe	± 5%	error from air gap around probe
	University of Innsbruck		
evaporation	lysimeter	± 0.01 mm/hr	edge effects
surface temp.	infra-red thermometer	± 0.5°C	
air temperature	aspirated thermometer	± 0.1°C	radiation error small
	Institute of Radiohydrometry		
wind speed	light weight anemometers	< ± 0.1 m/s	
incoming radiation	Schenk radiometer	± 15 W/m	
outgoing radiation	Schenk radiometer	± 15 W/m	
air temperature	unaspirated Pt100 thermocouple	± 0.1°C	radiation error ⩽ 12°C
albedo	time lapse photography		

REFERENCES

Ambach, W. and Denoth, A. 1975. On the dielectric constant of wet snow. Proc. Grindelwald symposium. IAHS Publ. 114, 136-142.

Dunne, T., Price, A.G. and Colbeck, S.C. 1976. The generation of runoff from sub-arctic snowpacks. Wat. Resources Res. 12 (4), 677-685.

Escher-Vetter, H. 1985. Energy balance calculations from five years' meteorological records at Vernagtferner, Oetztal Alps. Zeitschrift für Gletscherkunde und Glazialgeologie 21, 397-402.

Fohn, P. 1978. Schneeverdunstung im alpinen Gelände: Die Verdunstung in der Schweiz. Beiträge zur Geologie der Schweiz-Hydrologie 25, 35-42.

Harding, R.J. 1986. Exchanges of energy and mass associated with a melting snowpack. Proc. Budapest Symposium. IAHS Publ. 155, 3-15.

Kaser, G. 1982. Measurements of evaporation from snow. Arch. Met. Geoph. Biokl., Ser. B 30, 333-340.

Kaser, G. 1983. Verdunstung von Schnee und Eis. Dissertain Universität Innsbruck, 153 pp.

Moore, R.D. 1983. On the use of bulk aerodynamic formulae over melting snow. Nordic Hydrol., 193-206.

Morris, E.M. 1981. Field measurement of the liquid water content of snow. J. Glaciol. 27 (95), 175-178.

Morris, E.M. 1982. Sensitivity of the European Hydrological System snow models. Proc. Exeter symposium, IAHS Publ. 138, 221-231.

de Quervain, M. 1951. Zur Verdunstung der Schneedecke. Arch. Met. Geoph. Biokl., Ser. B 3, 27-64.

Weber, M. 1987. Vertikalsondierungen der eisnahen Luftschicht am Vernagtferner (Luzifer, 1983). In: "Arbeiten aus dem Abteilungen des Meteorologischen Instituts der Universität München". Norbet Beier, Münchener Universitätsschriften, Fakultät für Physik, no. 56.

A SIMPLE METHOD FOR DETERMINING THE RESPONSE TIME OF GLACIERS

T. Jóhannesson[1], C.F. Raymond & E.D. Waddington
Geophysics Program, AK-50
University of Washington
Seattle, WA 98195, U.S.A.

ABSTRACT

The response time of temperate glaciers is estimated by the following continuity argument: The difference in the steady state ice volume of a glacier, before and after a mass balance perturbation must be accumulated (or ablated) before the glacier can reach a new steady state. This leads to a time scale which is termed the "volume time scale" of the glacier. It is argued that the response time of the glacier can be expected to be equal to the volume time scale. The volume time scale can be derived by a simple argument and is under fairly general conditions expressed by the formula $T_V = H/(-b_t)$ where H is a thickness scale of the glacier and b_t is a scale of the ablation along the terminus of the glacier. This estimate of the response time of glaciers is of the order of decades for small maritime glaciers, which is in reasonable agreement with experience. This is much shorter than the theoretical long response time of the order of several hundred or a thousand years which has been derived in the past from kinematic wave theory.

1. INTRODUCTION

The response time of a glacier is a measure of the time it takes a glacier to complete its adjustment to a climatic change. The response time is one of the most important of the physical variables characterizing the dynamics of a glacier. It is of fundamental importance when interpreting past glacier variations in terms of climatic history. It is also an invaluable tool for glaciologists trying to explain or predict the response of glaciers to known or postulated climatic changes.

[1] present address:
National Energy Authority, Grensásvegi 9
108 Reykjavík, Iceland

J. Oerlemans (ed.), Glacier Fluctuations and Climatic Change, 343–352.

Theoretical analyses (Nye, 1963a; Paterson, 1981) have indicated that glaciers have a very long memory. Even relatively small valley glaciers (of the order of 10 km long) are expected to have response times of the order of several hundred or even a thousand years. This result is, however, not evident in the records of past glacier and climatic fluctuations. This apparent discrepancy has been discussed by various authors (Lliboutry, 1971; Paterson, 1981; Hutter, 1983). No satisfactory explanation of the discrepancy has been suggested nor has an alternative theoretical response time been proposed.

2. THE VOLUME TIME SCALE

When a non-surge type glacier in a steady state experiences a sudden and permanent change in its mass balance, ice is added to or removed from the glacier until it reaches a new steady state of equilibrium with the new mass balance distribution. It is well-known that the volume difference of the two steady state glacier profiles, i.e. the volume change caused by the mass balance perturbation, is the most important factor controlling the response time of the glacier. This arises because the response time can be no less than the time it takes to accumulate or ablate the necessary volume of ice to bring the glacier to its new steady state. This time may be termed the "volume time scale" of the glacier. The fact that the response time must be greater than or equal to the volume time scale was used by Paterson (1981) to explain the long response time predicted by Nye's (1963a) kinematic wave theory.

The volume time scale can be defined without reference to kinematic wave theory or in fact any time-dependent analysis of glacier dynamics. It is derived from the difference in volume of the datum steady state profile and the perturbed steady state profile of the glacier and represents a lower limit for any response time which can be correctly derived from a time-dependent analysis. In the case of Nye's well known analytical examples, which have provided valuable insight into the response of glaciers to climatic changes (Nye, 1963a; Paterson, 1981; Hutter, 1983), it turns out that the response time predicted by the time-dependent analysis is in fact exactly equal to the volume time scale. The response times of glacier models calculated from glaciological data from South Cascade Glacier, U.S.A. and Storglaciären, Sweden (Nye, 1963b; Nye, 1965) using kinematic wave theory, are also fairly close to the volume time scales of these models. That is, the response times derived from time-dependent solutions of the equations for the models of the South Cascade Glacier and Storglaciären are accurately predicted by the volume time scales derived from the corresponding steady state versions of the same equations. This result arises because redistribution of ice by kinematic waves and diffusion occurs on a shorter time scale than the volume time scale (orders of magnitude shorter in the case of Nye's analytical examples). The volume time scale, therefore, becomes the dominating factor determining the response time of the glacier. This suggests that the volume time scale can not only be used to explain the response time obtained from time-dependent analyses, but that it can also be regarded independently as an estimate of the response time. Since steady state analysis is much simpler than time-dependent analysis, and since it is likely to be based on fewer physical and mathematical simplifying assumptions, one might even hope to obtain a more general estimate of the response time in this way than is provided by a time-dependent analysis.

3. DEFINITION

The volume time scale, T_v of a steady state glacier with respect to a mass balance perturbation, δb, may be defined as

$$T_v = \delta V/\delta B \tag{1}$$

where δV is the difference in the steady state volume of the glacier before and after the mass balance perturbation, and δB is the integral of the mass balance perturbation over the area of the glacier.

Real glaciers rarely reach a steady state in their response to ever-changing climate. A time scale that measures the duration of the transition from one steady state to another steady state after a step change in climate, may therefore seem to be of limited value for the analysis of the response of glaciers to realistic climatic fluctuations. If the climatic fluctuations in question are small (so that the equations describing glacier motion can be linearized), the response to an arbitrary time-dependent climatic history can be viewed as a combination of the response to many infinitesimal changes happening at different times, expressed as a convolution of the actual climatic history with the response to a step climatic change. The response to a step climatic change will measure the length of time over which a particular climatic event affects the glacier. Thus an estimate of the response time of glaciers, which is derived from analysis of a step change in climate, will be of general value for the analysis of the response of glaciers to time-dependent climatic fluctuations.

4. DERIVATION FOR A TWO-DIMENSIONAL GLACIER

Consider a temperate non-surge type datum glacier of length L_d, and maximum thickness H_d (subsequently the subscript "d" will refer to the datum steady state). For simplicity we will assume that the flow is two-dimensional and thus ignore any effects from the finite width of the glacier. Assume that the glacier has reached a datum steady state thickness, $h_d(x)$, with respect to a datum mass balance, $b_d(x)$. The volume per unit width is then

$$V_d = L_d \langle h_d \rangle = f_d L_d H_d \tag{2}$$

where $\langle . \rangle$ represents an average over the length of the glacier and f_d is a geometrical factor which is equal to the ratio of the average ice thickness to the maximum ice thickness.

After a mass balance perturbation, $\delta b(x)$, the glacier will eventually reach a new steady state, $h_p(x)$, with respect to the new mass balance, $b_p(x) = b_d(x) + \delta b(x)$ (subsequently the subscript "p" will refer to the perturbed steady state). The length will be L_p and the maximum thickness will be H_p. The volume per unit width of the profile, $h_{p(x)}$, is given by an equation similar to equation (2) and the volume change may thus be expressed as

$$\delta V = V_p - V_d = f_p L_p H_p - f_d L_d H_d \tag{3}$$

This equation may be written

$$\delta V = (f_d + \delta f)(L_d + \delta L)(H_d + \delta H) - L_d H_d$$

where $\delta f = f_p - f_d$, $\delta L = L_p - L_d$ and $\delta H = H_p - H_d$. If the mass balance perturbation is relatively small, the geometrical factors, f_d and f_p, will be close to each other, since we do not expect the shape of the glacier to be significantly altered by a small mass balance perturbation. Then the perturbation δf in the above equation may be ignored. Ignoring higher order terms in the small perturbations δL and δH yields

$$\delta V \approx f_d \, (L_d \delta H + H_d \delta L) \tag{4}$$

The advance or retreat, δL, is easily seen to be given by (again correct to linear terms)

$$\delta L = L_d \, \langle \delta b \rangle / (-b_d(L_d)) \tag{5}$$

This expresses the fact that after the glacier reaches its new steady state, the extra ice that is accumulated by the mass balance perturbation upstream from the datum terminus must be ablated below the datum terminus.

The perturbation δH is more difficult to obtain. From theory one knows that the most important variable controlling the maximum thickness of a glacer on a relatively flat bed is its length. In the extreme case of perfect plasticity (Nye, 1951) the thickness is proportional to the square root of the distance from the terminus and independent of the mass balance. Hyper-elliptical glacier profiles (Nye, 1959) exhibit a near square root relation between maximum ice thickness, H, and glacier extension, L, with the power in the relationship between H and L being 7/12 (for the value m = 2.5 of a sliding coefficient m) instead of 1/2 for a square root relation. Similar profiles derived from a flux relationship based on Glen's flow law (Paterson, 1972, 1980) show a square root relationship between H and L. In the two latter cases the maximum ice thickness is very insensitive to mass balance perturbations that do not change the length of the glacier. We will therefore write

$$\delta H / H_d = 1/2 \; \delta L / L_d \tag{6}$$

as an approximation that should include the most important contribution to the perturbation, δH. This equation is a linearization of a square root relationship between H_d and L_d. The validity of equation (6) is limited to temperate glaciers flowing on relatively flat beds. The thickness of steep glaciers increases more slowly with the length of the glacier than indicated by equation (6). We shall discuss the applicability of our results to steep glaciers below.

The adjustment of cold ice caps and ice sheets to a climatic change necessarily involves an adjustment of the temperature of the ice to the climatic change. The resulting changes in the viscosity of the ice invalidate the present derivation and we shall therefore restrict our discussion to temperate glaciers and ice caps.

The geometrical factor, f_d , of equation (4) is clearly between 0.5 (triangular shape) and 1.0 (rectangular shape). For perfect plasticity profiles it is equal to 2/3 and for the hyper-elliptical profiles and the profiles based on Glen's flow law mentioned earlier, it is also very close to 2/3. We will therefore use $f_d = 2/3$ as a reasonable approximation. Using this value of f_d and combining equations (4), (5) and (6) we find

$$\delta V = H_d \delta L = H_d L_d \langle \delta b \rangle / (-b_d(L_d)) \qquad (7)$$

The term on the left of equation (7) has a simple physical interpretation. As a consequence of the mass balance perturbation the glacier advances by δL. Assume that the thickness of the glacier at a certain point below the point of maximum thickness is mainly a function of the distance from the terminus (this is exactly true in the case of perfect plasticity and fairly close to being true for the hyper-elliptical profiles and the profiles based on Glen's flow law). The volume that must be added to the glacier to bring it to the new steady state corresponds roughly to the slice, of length δL and thickness H_d, that must be added to the datum glacier profile at the point of maximum thickness to shift the portion of the datum profile, which is below the point of maximum thickness, to the new advanced position of the terminus.

This argument is illustrated in Fig. 1 for a glacier on a flat bed. The figure shows three steady state profiles of the type discussed by Paterson (1980). The profiles are based on a (non-dimensional) flux relationship derived from Glen's flow law for a two-dimensional symmetrical ice cap on a flat bed. The fully drawn profile corresponds to a (non-dimensional) datum mass balance of +1 for $0 < |x| \leqslant 1/2$ and -1 for $|x| > 1/2$. The dashed profile corresponds to the perturbed mass balance after a uniform mass balance perturbation $\delta b = 0.1$ (for a detailed description of the equations that lead to profiles of this kind see Paterson (1980)). The volume perturbation per unit width, δV, is equal to the area between the perturbed profile and the datum profile. The dotted profile is the datum profile shifted to the right by δL, so that the two termini coincide. The figure shows that the perturbed profile and the shifted datum profile nearly coincide for the entire length of the datum profile. It follows that the shaded area in the figure is nearly equal to δV, or in other words that $\delta V \approx H_d \delta L$.

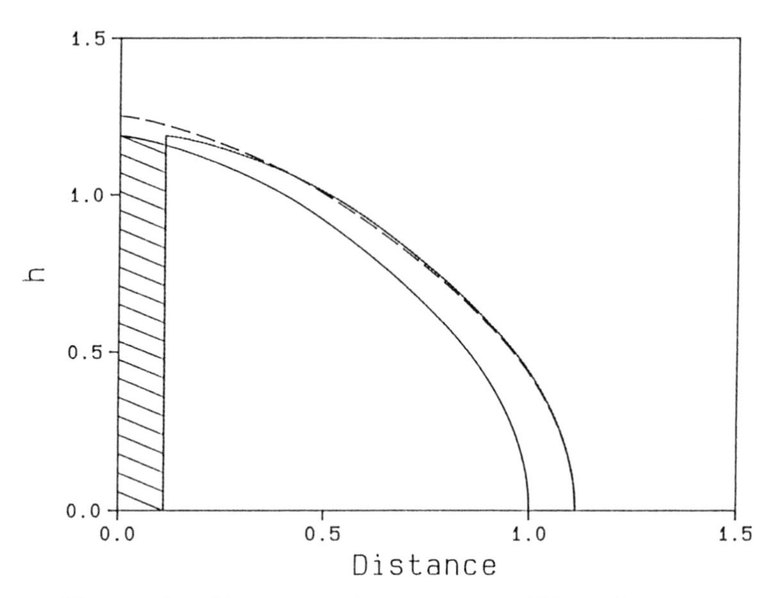

Figure 1. Three steady state profiles for a two-dimensional symmetrical ice cap. The fully drawn curve is the datum profile (see text), the dashed curve is the perturbed profile and the dotted curve is the datum profile shifted to the right. The shaded area to the left is the volume perturbation $\delta V \approx H d \delta L$.

For glaciers on a sloping bed a similar "shifting" argument may be applied to the portion of the steady state profile that is below the point where the ice thickness reaches its maximum. This makes equation (7) plausible, even for steep glaciers, for which equations (6) and the estimate, $f_d \approx 2/3$, cannot be expected to hold. The equation on the far right of (7) is a simple consequence of continuity and holds independent of the steepness of the glacier.

The mass balance perturbation, δB, of equation (1) is given by $\delta B = L_d \langle \delta b \rangle$ in the case of two-dimensional geometry. Equations (1) and (7) may be combined to give the following estimate of the volume time scale

$$T_v = H_d/(-b_d(L_d)) \tag{8}$$

As espected $\langle \delta b \rangle$ cancels out with the result that the volume time scale is independent of the size of the small mass balance perturbation. As an estimate of the response time this simple equation has structure similar to a number of equations for the time scales of different kinds of reservoirs in many branches of physics. It we interpret the length of the glacier as an indication of its "storage", equation (8) states that the response time of the glacier is the ratio of the "storage coefficient" of the glacier reservoir (note that the maximum thickness measures the volume added to the glacier per unit increase in its length) to the "flux coefficient" or transmissivity of the reservoir (note that the mass balance at the terminus measures the increase in the rate of volume removal from the glacier per unit increase in its length).

Among well known equations from other fields similar to equation (8) is the equation for the time scale of an electrical circuit with a capacitance, C, and a conductance, 1/R, $T = C/(1/R)$, the time scale of a linear magazine (of the type used in hydrological models) with a storage coefficient, S, and a flux coefficient, k, $T = S/k$; the time scale of the water level changes in a well drilled through a cylindrically symmetric hydrological aquifer with radius, R, a storage coefficient, S, and transmissivity, Kh, $T = SR^2/(2.40^2Kh)$; and the time scale of the one-dimensional heat equation, $T = \rho CL^2/(\pi^2 k)$, where ρ, C and k are the density, the heat capacity and the thermal conductivity of the material, respectively, and L is the length scale of the problem.

It is worth noting that the analyses that lead to the time scales in the preceding paragraph are all based on conservation equations (conservation of electrical charge, water mass or thermal energy). In addition these reponse times can all be interpreted as the time needed for a flux induced by the deviation from a steady state to import the necessary "material" to build up a new steady state. It is not surprising that a similar equation should arise in the analysis of glacier movement, which is based on the conservation equation for mass.

The derivation of equation (8) was based on a simplified model of the geometry of a glacier and the question arises whether it can be expected to hold for other geometries. The simple form of the equation and its similarity to known equations for time scales in other fields suggests that the validity of equation (8) is not restricted to this simple case. This question is addressed in the next section.

5. OTHER GEOMETRIES

Let us first consider a temperate, cylindrically symmetrical ice cap on a flat bed. The argument of the preceding section can be extended to this case if one assumes that the maximum thickness, H, is again proportional to the square root of the extent, R, and relatively independent of small perturbations in the mass balance distribution. The geometrical factor, f, becomes $f = 8/15$ if the shape of the profile is not far from a square root shape as is observed in reality (Paterson, 1981). This leads to the equation

$$T_v = (2/3)\ H_d/(-b(R_d)) \qquad (9)$$

From this it follows that glaciers flowing in diverging valleys will have somewhat shorter volume time scales than given by equation (8), perhaps closer to T_v given by equation (9).

A more general heuristic argument may be presented as follows. Consider a temperate, steady state glacier or an ice cap with the area, A_d, and the volume, V_d. A mass balance perturbation, δb , with an average, $\langle\delta b\rangle$, over the area of the glacier, will lead to a change, δV, in the volume of the glacier and to a change, δA, in its area. Assume that the volume of the glacier is not critically dependent on (small) changes in the mass balance distribution, so that δV, is mainly a consequence of the change, δA, in the extension of the glacier. Then the volume perturbation, δV, may be written as

$$\delta V \approx (dV/dA)\ \delta A \qquad (10)$$

where the derivative (dV/dA) indicates how the volume of the glacier increases in relation to its area. By continuity $\delta B = \langle\delta b\rangle A_d = -\langle b_t\rangle_d\ \delta A$, where $\langle b_t\rangle_d$ is the average of the (negative) mass balance over the area δA along the terminus of the glacier. Equations (1) and (10) now yield the following estimate of the volume time scale

$$T_v = \delta V/\delta B \approx (dV/dA)/(-\langle b_t\rangle_d)$$

$$= (dV/dA)\ (A_d/V_d)\ \langle h\rangle_d/(-\langle b_t\rangle_d) \qquad (11)$$

where $\langle h\rangle_d = V_d/A_d$ is the average thickness of the glacier. Equation (11) is similar in form to equations (8) and (9) but the factor (dV/dA) (A_d/V_d) needs to be estimated. This factor describes how a relative change, $\delta A/A_d$, in the area of the glacier, leads to a corresponding relative change, $\delta V/V_d$, in its volume. For a two-dimensional, perfectly plastic glacier this factor is equal to 3/2 because of the square root relationship between the average thickness of the glacier and its extension. For a cylindrically symmetrical, perfectly plastic glacier this term is equal to 5/4 and in general one would expect the term to be between 1 and 3/2 because the average thickness of glaciers is assumed to increase gradually with their extension. Paterson (1972; 1981) discusses the volume of ice sheets and ice caps on the basis of data from six ice sheets and ice caps whose volume varies by four orders of magnitude. He finds that the volume increases proportional to the area to the power 1.23 to a fair degree of accuracy, so that (dV/dA) (A_d/V_d) is in this case very close to the theoretical value of 1.25 for a cylindrically symmetrical, perfectly plastic ice sheet.

The times scales of equations (8), (9) and (11) should be regarded as very crude measures of the time- and space-dependent response of glaciers to mass balance perturbations. They provide an estimate of the duration of the response to a climate change, but do not describe the response in any detail. They are thus only useful as order of magnitude estimates and their precise numerical value, say within a factor of 2, is not relevant. Considering the nature of the volume time scale as an order of magnitude estimate, neither the factor 2/3 in equation (9) nor the factor (dV/dA) (A_d/V_d) in equation (11), which may be expected to be between 1 and 3/2, are significantly different from 1. By the same argument the maximum thickness of equations (8) and (9) and the average thickness of equation (11) should both be regarded as convenient scales of the thickness of the glacier. We may therefore write

$$T_V = H/(-b_t) \tag{12}$$

where H is a thickness scale of the glacier and $(-b_t)$ is a scale of the ablation along its terminus, as a fairly general estimate of the volume time scale and of the response time of glaciers.

6. DISCUSSION

As stressed earlier the response time of equation (12) should not be regarded an exact statement incorporating all of the dynamics of a glacier. Rather it should be used, as are similar scales in other fields of physics, as a tool providing simple physical insight into the complicated problem of determining the time- and space-dependent response of glaciers in detail. It is also useful as a natural time scale for non-dimensionalization of the time co-ordinate in the equations that describe the flow of glaciers and ice caps.

The response time of equation (12) is based on a mass balance perturbation which is constant in time. In reality the positive mass balance feedback, which is caused by the fact that the mass balance of glaciers increases with increasing elevation, will make this assumption somewhat unrealistic (Bödvarsson, 1955). A positive mass balance perturbation will lead to an increase in the surface elevation of the glacier. The increased elevation will add to the mass balance perturbation and the final mass balance perturbation can be much larger than the initial one. The advance of the glacier and the associated change in its volume will also be greater than in the absence of mass balance feedback. This leads to an increase in the volume time scale of equation (1), $T_V = \delta V/\delta B$. This effect is to some extent counteracted by an increase in the average mass balance perturbation over the time of adjustment, since the increased volume perturbation will be built up by an increased mass balance perturbation. Although mass balance - elevation feedback can alter the response time significantly, preliminary numerical experiments indicate that it does not change the order of magnitude of the response time. An analysis of mass balance - elevation feedback requires time-dependent numerical modelling and will not be discussed further within the simple heuristic framework of this paper.

The response time of equation (12) is of the order of decades for small maritime glaciers for which $H \approx 200$ m and $b_t \approx -(5-10)$ m/a are reasonable estimates. This agrees with expectation (Lliboutry, 1971) and provides an alternative to the long response time of the order of several hundreds or even a thousand years derived from kinematic wave theory, which has been a matter of debate among glaciologists for some time.

How then does the long response time of Nye (1963a) relate to this result? As discussed at the beginning of this paper, the response time of the analytical glacier models, from which the long response time of glaciers was derived, equalled the volume time scale for those glacier models. The volume time scale derived for the models thus correctly predicts their long response time. The long response time must therefore be a consequence of these particular models of glacier dynamics. The difference as compared with the estimates proposed in this paper arises because the volume changes predicted for the analytical models are much larger than the changes derived here. The reason for this difference appears to come from the description of the dynamics near the terminus of the glacier in terms of the kinematic wave velocity and the diffusion coefficient used by Nye. This will not, however, be discussed further in this paper, but is addressed separately in another paper in preparation.

REFERENCES

Bödvarsson, G. 1955. On the flow of ice sheets and glaciers. Jökull 5, 1-8.

Hutter, K. 1983. Theoretical glaciology. Dordrecht etc. D. Reidel Publishing Co.

Lliboutry, L.A. 1971. The glacier theory. In: Advances in Hydrodynamics, Ven Te Chow (Ed.), New York, Academic Press 7, 81-167.

Nye, J.F. 1951. The flow of glaciers and ice sheets as a problem in plasticity. Proceedings of the Royal Society A 207 (1091), 554-572.

Nye, J.F. 1959. The motion of ice sheets and glaciers. Journal of Glaciology 3 (26), 493-507.

Nye, J.F. 1963a. On the theory of the advance and retreat of glaciers. Geophysical Journal of the Royal Astronomical Society 7 (4), 431-456.

Nye, J.F. 1963b. The response of a glacier to changes in the rate of nourishment and wastage. Proceedings of the Royal Society A 275 (1360), 87-112.

Nye, J.F. 1965. The frequency response of glaciers. Journal of Glaciology 5 (41), 567-587.

Paterson, W.S.B. 1972. Laurentide Ice Sheet: estimated volumes during late Wisconsin. Review of Geophysics and Space Physics 10 (4), 885-917.

Paterson, W.S.B. 1980. Ice sheets and ice shelves. In: Dynamics of snow and ice masses, S.C. Colbeck (Ed.), New York, Academic Press, 1-78.

Paterson, W.S.B. 1981. The physics of glacier, second edition. Oxford etc, Pergamon Press.

ON THE RESPONSE OF VALLEY GLACIERS TO CLIMATIC CHANGE

J. Oerlemans
Institute of Meteorology and Oceanography
University of Utrecht, UTRECHT, The Netherlands
and
Alfred-Wegener-Institut für Polar- und Meeresforschung
BREMERHAVEN, F.R.G.

ABSTRACT

In many cases the response of a glacier to changing climatic conditions is complicated due to the large number of feedback loops that play a role. Examples are: ice thickness - mass balance feedback, nonlinearities arising from complicated geometry, dependence of ablation on glacier geometry, coupling between debris cover, ice flow and ablation etc.

In this paper an attempt is made to quantify such processes by carrying out numerical experiments with an ice-flow model. Some conclusions and suggestions are:

(i) The longitudinal bed profile is very important. Apart from the well-known fact that glaciers are more sensitive when the bed slope is small, a reversed slope (slight overdeepening) creates branching of the equilibrium states, i.e., for the same climatic conditions two glaciers of different geometry can both be in a stable steady state.

(ii) Due to the height-mass balance feedback, glaciers on a smaller slope react slower to climatic change.

(iii) The mass balance gradient as observed on long valley glaciers is to a substantial part determined by systematic changes (along-valley) in glacier width and surface albedo. The balance gradient is thus coupled to the dynamics, and this should be studied further.

1. INTRODUCTION

In historic times, glacier variations have drawn the attention of many people inhabiting mountaneous regions. This applies in particular to those events that have brought damage to farmlands and buildings, either by direct advance of a glacier snout (e.g. Ostrem et al., 1977) or by blocking of rivers with subsequent flooding when the resulting lakes break through (e.g. Hoinkes, 1969). Such events have been documented by reports of all kind. For a discussion of glacier hazards, with many examples and additional references, see Tufnell (1984).

J. Oerlemans (ed.), Glacier Fluctuations and Climatic Change, 353–371.

Painters have also contributed significantly to our knowledge of glacier fluctuations. From the 18th and 19th centuries, a wealth of drawings, etches and paintings exist, making it possible, in combination with other investigations, to reconstruct front variations of large valley glaciers like for instance the Grindelwald Gletschers, the Rhône Gletscher, the Vernagtferner, and the Glacier d' Argentière. In more recent times, the last 100 years say, more systematic measurements have been carried out on a more global scale (Kasser, 1967, 1973; Müller, 1977; Patzelt, 1970; Reynaud, 1983). Some long records of front positions are shown in Fig. 1. In these curves the earlier parts are more uncertain, but extreme positions are probably fairly reliable.

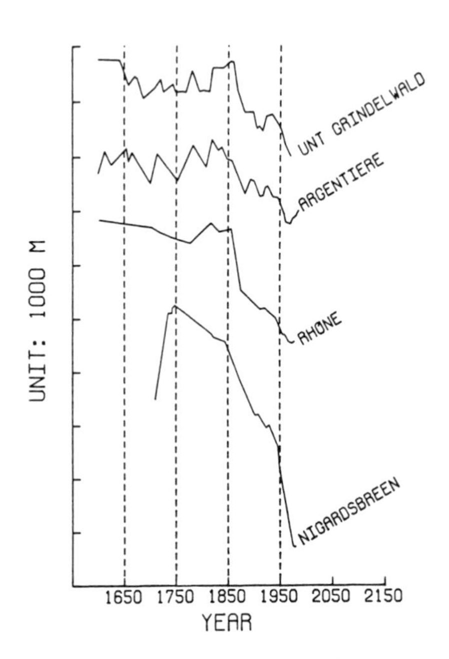

Figure 1. Variations in the ice-front positions of four valley glaciers. Data from Messerli et al. (1975), Vivian (1975), Aubert (1980), Ostrem et al. (1977).

The state of a glacier is of course not only determined by the position of the front, but also by its volume. Records of changes in volume are hardly available, however. Most systematic observations have been made in the last decades. The same applies to direct measurements of the mass balance. It is only since the use of hydroelectric power plants in glacierized areas that many glaciers are monitored on a regular basis. Series of mass balance measurements are at best about 30 years long, but most are shorter (Letréguilly, 1984). Nevertheless, it is the mass balance that relates glacier variations to climatic change. So a proper understanding of how the mass balance depends on climatic parameters like temperature, radiation and precipitation is crucial to the study of front variations.

It is generally assumed that glaciers have a relatively long time scale, typically 100 years. Although this is a realistic figure for the global mechanics of a large valley glacier, the front can react much quicker. A strong increase in melting rate on a glacier tongue, for instance, will lead to almost instantenous retreat of the snout. The record of the Brenva glacier (descending eastwards from the Mont Blanc) provides an interesting illustration. In 1926 a large rock slide covered part of the glacier and led to substantially lower melting rates. The ice front showed a quick reaction: in contrast to the fronts of other glaciers in the nearby regions, it advanced rapidly (Orombelli and Porter, 1982). Of course, the response time depends on the geometry; smaller glaciers react quicker.

The purpose of this paper is to draw attention to a number of feedback processes that make valley glaciers very sensitive to climatic change, and in particular to small changes in the global radiation balance. Changing glacier geometry will turn out to be an important factor when melting rates are considered. Such rates can be quite large, typically in the 5 to 10 m/yr range for large valley glaciers descending into regions where mean summer temperature is about 10 °C or so. Daytime surplus of the radiation balance of the ice surface and turbulent heat flux form the most important sources of energy for the melting process (e.g. Paterson, 1981). The relative importance depends on the particular geometric and climatological conditions. At lower elevation, where melting rates are higher, the contribution from the turbulent heat flux is comparable to that from radiation. In higher regions radiation becomes relatively more important.

To set the stage for some model experiments and the more detailed discussions on specific feedback loops in subsequent sections, we consider what might happen if, for some reason, the planetary radiation balance is perturbed. From models of the earth's climate we expect that a + 6 W/m^2 perturbation will lead to an increase of the mean surface temperature of about 2 K. However, the thermal intertia of the world oceans will damp the response; it may take a decade or longer before the new equilibrium temperature is actually approached. However, this 'oceanic damping' will be less significant when the energy balance of a specific site depends less on horizontal advection of heat. Surface temperature in the continental interiors should thus be expected to react more quickly to a change of the global radiation budget, and this probably also applies to some extent to deep valleys in mountanous regions.

If we now turn to valley glaciers, the first thing to note is that the increase in air temperature is the secondary effect (but not necessarily less important on the long run). The mass balance will react directly and <u>immediately</u> to the change in the radiation balance. Since, during the ablation season, the temperature of the glacier surface is at the melting point, the extra radiation will be used entirely for additional melting. As pointed out recently by the author (Oerlemans, 1986), there is another important aspect associated with the fact that the glacier temperature cannot increase. Any increase in the radiation balance will lead to a larger temperature difference between glacier and direct surroundings, causing a larger advective flux of sensible heat towards the glacier tongue ('oasis effect'). This mechanism is particularly effective when the glacier tongue is narrow: the increase in melting rate may be doubled.

So it appears that a glacier can make use of surplus radiative energy in a valley in a very efficient way. Important is the fact that the reaction of the mass balance is almost immediate. To this then comes the effect of the gradually increasing global air temperature, retarded by the thermal inertia of the climate system.

There are other factors, of a more general nature (i.e. not directly related to changes in the radiation balance), that make valley glaciers extremely sensitive to climatic change. For instance, the width of a glacier generally decreases when the ice thickness decreases. However, it is reasonable to assume that the contribution of the advective heat flux to melting is roughly inversely proportional to the glacier width. Thus, once retreat has been initiated and the ice thickness becomes smaller, additional melting will be caused by the decreasing glacier width.

In the following sections, an attempt is made to assess the potential importance of the feedback loops mentioned above. In many cases this cannot be done in a rigorous way, because observations on the mass and energy budgets of glaciers do not give a complete picture of the energy fluxes in a valley-glacier system. Order-of-magnitude estimates can be made, however. To link changes in the mass balance to front variations, a dynamic glacier model is needed. Such a model is briefly described in section 2, and it will be used as a tool throughout the rest of this paper. In section 3 the effect of geometry and topography of the glacier bed on the response to climatic change will be discussed. Section 4 deals with the feedbacks involving ablation at the glacier tongue. Then, in section 5, an attempt is made to tie the results together and to perform a simple 'carbon-dioxide experiment'. Here, perturbations of the global radiation balance and the mean air temperature are imposed as time-dependent functions to the glacier model.

2. A SIMPLE DYNAMIC GLACIER MODEL

To simulate the transient behaviour of glaciers for any valley profile and mass balance, a numerical model is required. We use a model that treats the vertically-integrated ice flow as a direct response to the driving stress. Such models have been used in glacier studies by various workers (e.g. Budd and Jenssen, 1975; Kruss, 1984; Oerlemans, 1986). The model used here is a flow-line model, in which the cross section is trapezoidal and may depend on distance x along the flow line, and on ice thickness H. The geometry is shown in Fig. 2. Here only a brief description is presented, for a general discussion on numerical modelling of ice flow the reader is referred to Oerlemans and Van der Veen (1984).

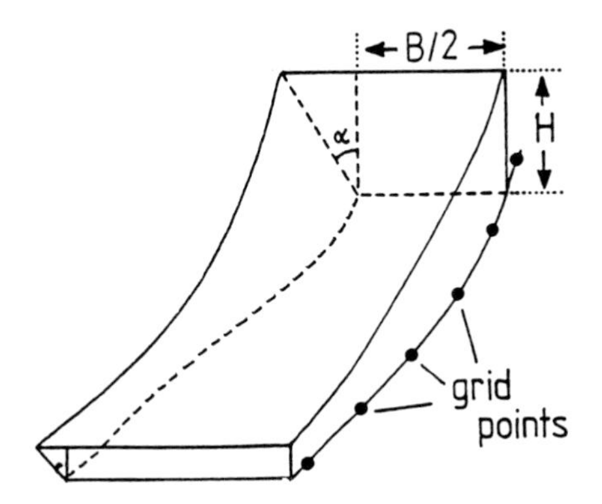

Figure 2. Geometry of the model glacier. The x-axis is aligned along the central flowline. At each grid point, width of the valley floor (B) and mean slope of the side walls has to be prescribed. The latter determines how glacier width is related to ice thickness.

The appropriate continuity equation reads:

$$\frac{\partial S}{\partial t} = -\frac{\partial (US)}{\partial x} + MB_s \tag{1}$$

S is the cross-sectional area, U the vertical mean ice velocity parallel to the bedrock, M mass balance, B_s glacier width at the surface [B_s = B+2Htg(α) = B+2μH]. Expressing S in H and B yields, after insertion in (1):

$$\frac{\partial H}{\partial t} = \frac{-1}{B+2\mu H}\left[(B+2\mu H)\frac{\partial (UH)}{\partial x} + UH\frac{\partial}{\partial x}(B+\mu H)\right] + M \tag{2}$$

This equation can be used to calculate the transient behaviour of a glacier for any geometry, provided that the ice velocity is locally related to the driving stress τ, defined as:

$$\tau = -\rho g H \frac{\partial h}{\partial x} \tag{3}$$

Here h is surface elevation. The total velocity U is made up of a sliding part U_s and a deformational part U_d, related to the driving stress as follows:

$$U_d = F_1 H \tau^3 \qquad U_s = \frac{F_2 \tau^3}{N} \tag{4}$$

N is the normal load, F_1 and F_2 are flow parameters. Since the effect of basal water pressure is not taken into account in this study, the normal load is simply set equal to the overburden ice weight. The values of the flow parameters actually used are: $F_1 = 0.95 \times 10^{-22}\ m^6\ s^{-1}\ N^{-3}$, $F_1 = 0.9 \times 10^{-14}\ m^5\ s^{-1}\ N^{-2}$.

Substituting the expressions for the velocity components in the continuity equation leads a nonlinear diffusion equation for H. A forward time-differencing scheme is used for the integration, together with a staggered grid to evaluate the ice flux divergence. For a sufficiently small time step, depending on ice thickness and surface slope, this scheme is absolutely stable without additional smoothing. All experiments discussed in this paper were carried out on a grid of 50 points along the flow line, spaced at 300 m. No attempt was made to treat the glacier snout in a sophisticated way (interpolation between grid points), because this does not effect in any significant way the response of the model glacier to climatic change. The glacier length thus appears as a discrete variable.

3. GEOMETRIC EFFECTS

In this section we investigate how valley width and longitudinal profile influence the response of a glacier to changing environmental conditions. It has of course been recognised for a long time that the glacier front position will be particularly sensitive to a change in the equilibrium-line altitude when:

(i) the accumulation area is large and the glacier tongue narrow;

(ii) the longitudinal slope of the bed is small.

For a recent discussion, see for instance Furbish and Andrews (1984).

Before discussing some experiments with the numerical model, we first identify from an extremely simple analysis how the bed slope effects the sensitivity of a glacier. We consider a glacier of constant width resting on a bed with constant slope γ (see Fig. 3). The mass balance is assumed to increase linearly with height relative to the equilibrium-line altitude E, i.e. $M=\alpha(h-E)$ Here α is a positive constant. If the length of the glacier is denoted by L, equilibrium requires that:

$$\int_L M dx = \alpha \int_L (H+b_o-\gamma x-E)dx = 0 \qquad (5)$$

Integrating and solving for L yields:

$$L = \frac{2(H^*+b_o-E)}{\gamma} \qquad (6)$$

Here H^* is the mean ice thickness. We now assume that the base stress is more or less constant, implying that $H(dh/dx) = \Lambda$, where Λ is about 11 m for a base stress of 1 bar. It follows that $\gamma H^* = \Lambda$, so the solution for the equilibrium length of the glacier becomes:

$$L = \frac{2(\Lambda/\gamma+b_o-E)}{\gamma} \qquad (7)$$

From this expression a few things can be noted. First of all, the height-mass balance feedback, reflected in the term Λ/γ , becomes more important when the bed slope is smaller. For a base stress of 1 bar and a slope of 0.05, for instance, the mean ice thickness would be 220 m, which, for many glaciers, would not be negligible as compared to b_o -E. Secondly, the sensitivity of glacier length to changes in the equilibrium-line altitude is inversely proportional to the bed slope (i.e. $\partial L/\partial E=-2/\gamma$). So this simple calculation indeed illustrates that glaciers resting on a bed with a small slope should preferable be considered when studying climatic change.

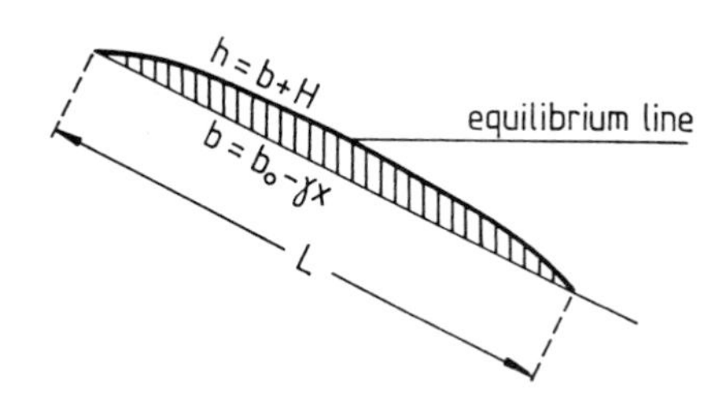

Figure 3. Geometry for the simple calculation of how glacier sensitivity depends on the slope of the bed. The x-axis is pointing in the direction of flow.

It is possible to do a similar analysis for a more complex geometry or a more detailed parameterization of the mass balance (upper limit for accumulation, for instance). However, this falls outside the scope of the present paper. Instead we now turn to some experiments carried out with the numerical glacier model.

In the following the mass balance is again parameterized in terms of elevation relative to the height of the equilibrium line, but now with an upper limit:

$$M = \min[1.5,\ 0.01(h-E)] \text{ m ice depth/yr} \qquad (8)$$

Here h and E are in m. With E between 2500 and 3500 m, this could represent midlatitude conditions in a climate that is not too dry.

The results of two integrations with different bed profile are shown in Fig. 4. The glacier width is a prescribed function of x, see plan view in the figure, and does not yet depend on ice thickness ! It is a valley glacier with a narrow tongue, and a wider accumulation basin. The integrations extend over 1000 yr of simulated time, and the equilibrium-line altitude is prescribed as follows:

$$\begin{aligned} t &< 300 \text{ yr}: & E &= 2700 \text{ m} \\ 300 < t &< 550 \text{ yr}: & E &= 3000 \text{ m} \\ 550 < t &< 800 \text{ yr}: & E &= 3300 \text{ m} \\ t &> 800 \text{ yr}: & E &= 2700 \text{ m} \end{aligned}$$

The changes in snowline elevation are stepwise. In the first experiment the bed has a steep upper slope (up to 3500 m) with a flat plateau at its foot, and then a slope making a constant angle with the horizontal. The resulting equilibrium glacier profile (the one shown is for E=2700 m) is simple. Ice thickness in the lower part is small and decreases very smoothly towards the snout. In such a situation one expects a simple and quick respons to changes in the snowline elevation as is indeed shown by the plot of glacier length versus time. A 300 m increase of E leads to a retreat of about 1500 m within 25 yr. The 600 m lowering of the equilibrium line at the end of the simulation brings the glacier quickly back to its original shape.

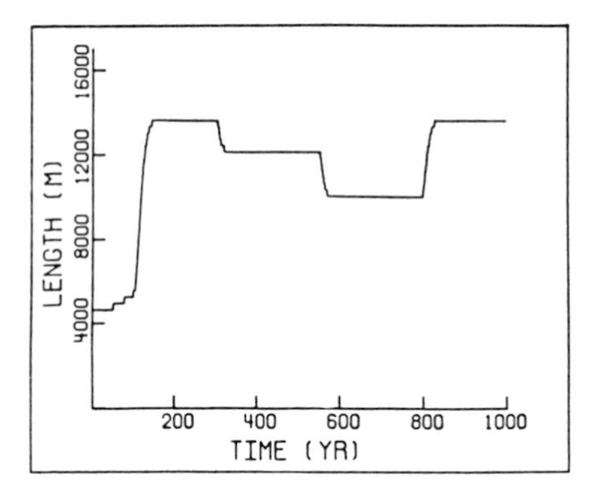

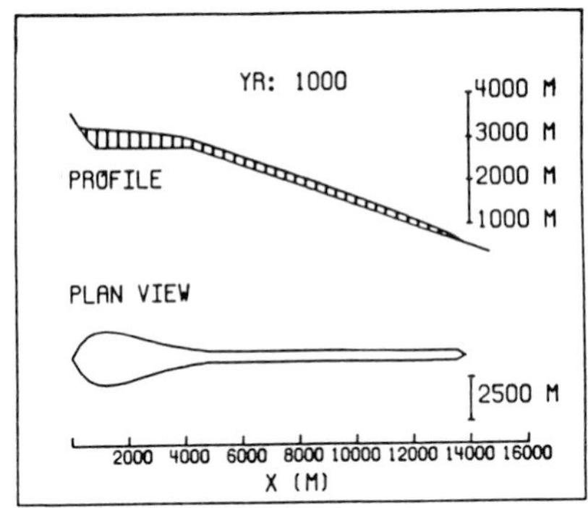

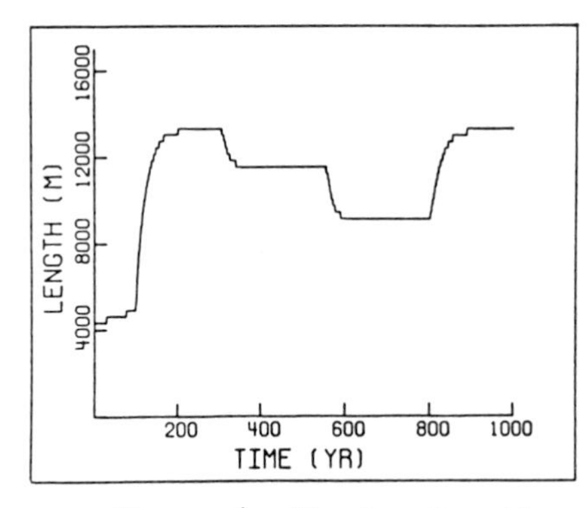

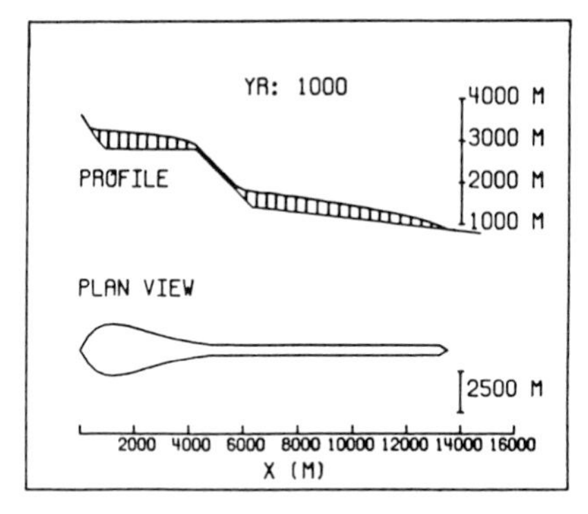

Figure 4. Glacier length as a function of time (left) and steady-state geometry (right) for a sensitivity experiment with different bed topography. A stepwise change in snowline elevation is imposed three times (see text). Note differences in response time due to differences in bed profile.

In the second experiment (lower part of the figure), the bed profile is somewhat more complicated. After 5 km there is an 'ice fall', changing abruptly into a rather flat valley bottom. This allows the glacier to grow thicker, leading to a situation in which the sensitivity to changes in E has increased (see analysis given above), and in which the response time is notably longer. Firstly, it takes about 50 years more to reach a steady state, but the reaction to the rising equilibrium line is also slower (compared to the first experiment, it roughly doubles). These effects are also a consequence of the height-mass balance feedback, in which the growing or shrinking glacier effects its own mean surface elevation such that it causes a significant change in the mass balance.

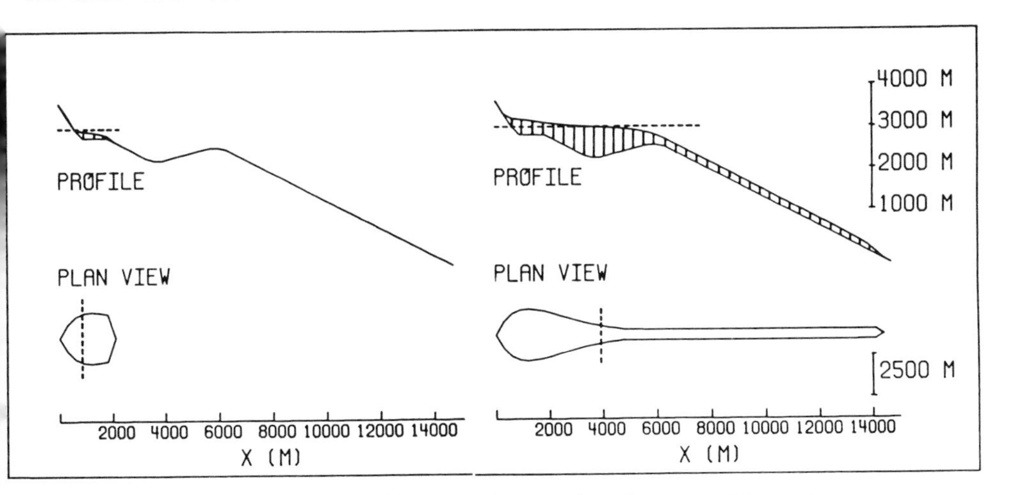

Figure 5. An illustration of the nonlinearity introduced by the height-mass balance feedback. These glaciers are both in equilibrium, under the same climatic conditions (E=3000 m). The equilibrium line is indicated by the dashed line.

In case of a bed where the slope changes sign, the height-mass balance feedback introduces essential nonlinear behaviour. An example is shown in Fig. 5. Both states shown represent stable equilibria for an equilibrium-line altitude of 3000 m. So although the glacier geometries are very different, they both have a zero net balance. The small glacier was obtained with a no-ice initial state , keeping E at 3000 m all the time. The large glacier, however, can only be simulated by starting with a lower equilibrium line and raising it after 100 yr or so to 3000 m. This experiment clearly shows that one should be careful with the interpretation of former glacier-front positions in terrain with complex topography (this point has recently been stressed also by Burbank and Fort, 1985). In such cases, using a relatively simple glacier model as the one presented here will certainly be helpful.

4. ABLATION

4.1 Introduction.

In recent years, ablation measurements have been carried out on many glaciers and a number of methods have been developed to parameterize ablation in terms of meteorological elements. For a general discussion see for instance Kuhn (1980).

The simplest method to estimate ablation from weather data is to 'integrate positive temperatures', i.e.:

$$A = \Sigma \int \max(0,T)dt \qquad (9)$$

Here A is total ablation, T air temperature of a nearby station and Σ the 'degree-day factor'. Methods based on this approach have been used for a long time (for a recent discussion, see for instance Braithwaite, 1984). For a particular location the method is succesful in explaining variations in ablation rate, but it is not very universal: the degree-day factor varies widely. There have also been attempts to derive statistical relations between ablation (and net balance) and temperature, radiation and precipitation (e.g. Martin, 1977; Letreguilly, 1984; Pollard, 1981). Although some general results have shown up [for instance: summer temperature and early summer precipitation (snowfall) are the most important factors concerning the net balance of glaciers in the Alps], the constants of proportionality again vary widely from place to place.

A more accurate calculation of ablation requires a consideration of the energy budget at the glacier surface. This has been done for instance by Ambach (1965), De la Casinière (1974) and Hogg et al. (1982). The basic assumption is that, in the ablation season, the ice is at the melting point, and that any positive energy balance will thus lead to melting immediately. A more detailed calculation carried out recently by Greuell and Oerlemans (1986), in which vertical heat fluxes in- and outside ice and/or snowpack are taken into account explicitly, has shown that this assumption is correct. Substantial differences only occur in regions where the ablation season is short.

So there is a complete hierarchy of models relating ablation to meteorological conditions. However, it is not so obvious that relations thus found are meaningful instruments in studying how glaciers react to climatic change. One should be aware of the fact that other factors, in particular those associated with glacier geometry, can become important once substantial changes occur in the shape of a glacier. One of these involves the horizontal exchange of energy between the atmospheric boundary layers above the glacier and above the ice-free surrounding grounds.

4.2 Glacier width and ablation rate.

Although the glacier-wind circulation and the thermal contrast between air just above a glacier and just above surrounding rock are well-known phenomena, very little attention has been paid to the implications for the energy budget of the glacier surface. To the knowledge of the author, systematic investigations of the contribution of the advective heat flux to melt energy at glacier tongues have not been carried out [A pilot study undertaken by Wendler (1974) suggests that it could be very important]. In estimating the contribution from the turbulent heat flux to the glacier surface, air temperature is normally prescribed (taken from observations). However, when the glacier geometry changes substantially, the advection of heat from the surroundings will change, and so will air temperature over the glacier (apart from any global climatic perturbation).

In a recent paper (Oerlemans, 1986), the author presented a simple model to calculate ablation rates, in which the advective heat flux was taken into account. Two important results emerged, namely, (1) the ablation rate depends in a significant way on the width of the glacier, and (2) the increase in ablation due to a change in the net radiation balance is further enhanced by the advective heat flux from ice-free grounds towards glacier. In this subsection we elaborate further on (1).

The importance of advective heat fluxes is directly reflected by the fact that on many glaciers ablation increases when going from the centre to the edge. It should thus be possible to use a transverse ablation gradient to examine the role of the advective heat flux, and to find a relation between glacier width and mean ablation. This could further support the theoretical result referred to above, which was obtained with a model in wich parameter values have to be chosen in a rather ambiguous way.

Suppose that the transverse ablation profile can be written as:

$$A = A_1 + A_2\, e^{-(W+y)/L} \tag{10}$$

The y-axis is perpendicular to the central flowline, and the region of interest is from y=-W (edge of glacier) to y=0 (middle of glacier). So the ablation rate at the edge equals $A_1 + A_2$. The length scale determines to what extent the ablation on the glacier is affected by the surrounding ice-free grounds. Integrating over the half-width of the glacier yields for the mean ablation A^* :

$$A^* = A_1 + A_2 \frac{L}{W} \left\lfloor 1-e^{-W/L} \right\rfloor \quad , \tag{11}$$

showing that the ablation goes to A_1 when the glacier width goes to infinity.

Although it is evident that the intensity of the thermal convection on the valley determines to a large extent the value of L, it is not so clear what this value actually should be. Data on cross-glacier ablation profiles are too scarce to derive a value for L; something of the order of 1000 m seems to be reasonable, however.

One of the few valley glaciers on which the ablation pattern has been measured in some detail is the Hintereisferner (Oetztaler Alpen, Austria). On its tongue, ablation increases substantially towards its sides. To arrive at an order-of-magnitude estimate, we apply ablation measurements from stakes at a section of the glacier where the surface elevation is about 2670 m for the period 1971-1973 (three ablation seasons). This elevation was choosen because here the stakes had favourable positions for the present purpose for a number of years (a few very close to the glacier margin, a few in the middle). Data were taken from Kuhn et al. (1979). In the period referred to, the mean annual ablation was 2.65 m water eq. in the middle, and 3.25 m water eq. at the sides. The difference is of the same order as that reported by Wendler (1974). At this location on Hintereisferner, the half width is about 340 m. Matching these data with the cross-glacier ablation profile proposed above yields (with L = 1000 m):

$$A^* = 1.18 + 2.07 \frac{L}{W} \lfloor 1-e^{-W/L} \rfloor \text{ m water eq.} \tag{12}$$

From this expression we find for example ablation values (m water eq.) of 3.21, 2.78 and 2.07 for half-width's of 50, 500 and 2000 m, respectively.

Whether the relation between ablation on glacier width effects the climatic sensitivity in a significant way depends on the geometry of the valley, of course. Here we consider just one numerical experiment in a situation where $\mu = \frac{1}{2}\, \partial B_s/\partial H$ is large. A simple linear bed profile is used (see Fig. 6) and μ is set to 16. The basic width B of the valley is 100 m. The mass balance is now written as the sum of ablation and accumulation according to:

$$M = Acc + Abl$$

$$Abl = \min \lfloor 0,\ A_1 + 2A_1 \frac{L}{W} (1-e^{-W/L}) \rfloor \tag{13}$$

$$A_1 = \min \lfloor 0,\ (h-h_o)a \rfloor$$

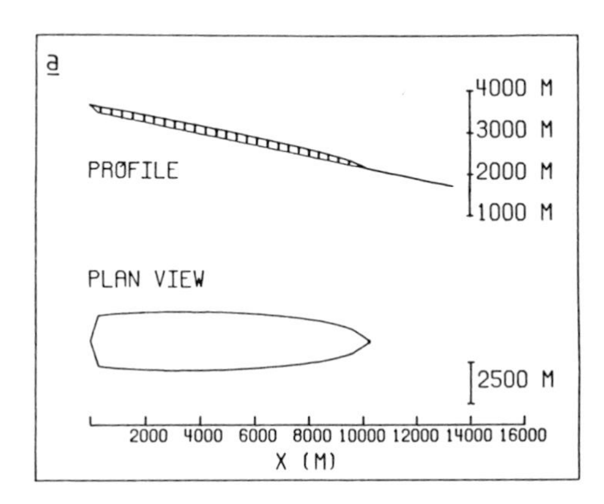

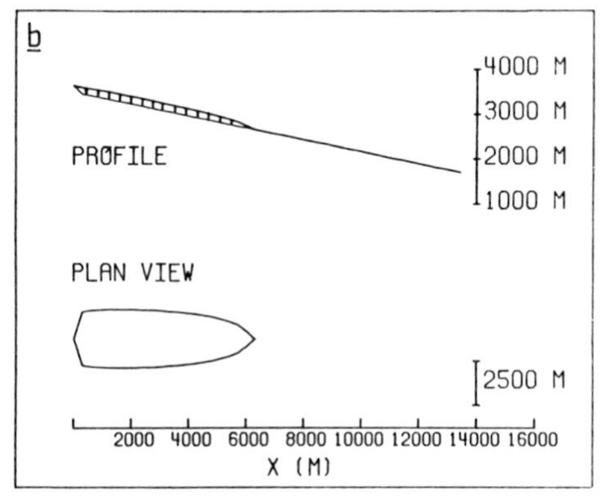

Figure 6. An experiment in which glacier width depends on ice thickness. In <u>a</u> a steady state is shown, in <u>b</u> the new steady state after a sudden 300 m increase of the 'melt line' (see text).

The accumulation is assumed to be constant (here, Acc = 1.5 m/yr), and the 'reference ablation' ($B \rightarrow \infty$) decreases linearly with surface elevation (here a = 0.007 m ice depth/m; h_0 = 3200 m; this implies an equilibrium-line altitude of 2915 m)

The glacier model was integrated in time until a steady state was reached. This state is shown in Fig. 6a. Then the value of h_0 was increased instantaneously by 300 m, and this of course leads to substantial retreat of the glacier (exp. 1). It takes about 100 yr before a new equilibrium has been established, but 75 % of the retreat occurs within 50 yr. The final state after the 'climatic warming' is shown in Fig. 6b.

For comparison, the experiment was repeated with a mass balance formulation independent of glacier width (exp. 2). This was done by replacing the ablation in equation (13) by: Abl=min [0, A_1], and through adjustment of A_1 in such a way that the initial equilibrium state is the same as in the former run. Retreat for some selected times after the stepwise change in h_0 are given in Table I. As expected, it appears that in the case of ablation depending on glacier width the sensitivity is larger. However, the difference decreases with time progressing, and the equilibrium states are rather similar. Anyway, this example shows that the ablation - glacier width coupling may be important, but it is difficult to make a general statement since so much depends on the actual geometry of the bed.

Table I. Retreat of the model glacier after a sudden climatic warming. The difference between exp. (1) and exp. (2) is described in the text.

time (yr)	exp. 1 (m)	exp. 2 (m)	difference (m)
10	290	180	90 (31%)
20	1110	780	330 (30%)
40	2690	2270	420 (16%)
80	3770	3480	390 (10%)
150	4210	3960	250 (6%)

4.3 Variations in albedo.

Systematic variations in surface albedo may well contribute to the longitudinal ablation gradient. The amount of exposed morainic material generally increases towards the snout, causing large differences in effective (area-averaged) short-wave albedo. For a large valley glacier a typical value for the albedo (during the ablation season), in the vicinity of the equilibrium line, may be 0.5. At the glacier tongue, however, this figure may go down to 0.25 (see Fig. 7 to illustrate the point). The resulting difference in ablation can easily be a few m ice depth. It depends on the surface slope of the glacier how large the actual contribution to the balance gradient is. For a 500 m altitude difference between equilibrium line and glacier front, the 'albedo effect' thus typically contributes 0.4 m ice depth per 100 m, i.e. about 40 % of a characteristic balance gradient!

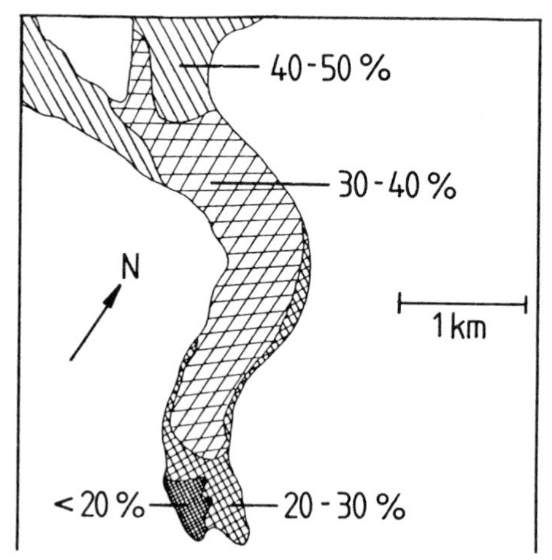

Figure 7. Variations in albedo over the lower ablation zone of a valley glacier (Nigardsbreen, 15 August 1972). From Tvede (1974).

The complicating factor now is that the amount of exposed morainic material (and atmospheric dust) interferes with the transient behaviour of the glacier. On a retreating glacier with an almost stagnant tongue (low ice velocities), the amount of debris on the surface increases in time, and so probably does the ablation rate, until the debris layer become so thick that the isolating effect starts to dominate. For an advancing glacier, on the other hand, morainic material is deposited more efficiently at the glacier snout (and at the sides) and ablation may decrease. It is outside the scope of this paper to discuss this mechamism in detail, but the order-of-magnitude estimate given above points to a certain importance.

4.4 Summary.

In virtually all studies in the literature, the mass balance of a glacier is represented in terms of surface elevation, and the balance gradient has generally been accepted as a parameter characterizing climatic conditions. For many purposes this may be acceptable, but when studying the effects of climatic change one has to be careful.

The objection raised above is that part of the (longitudinal) gradients as observed in the field stem from:

(i) differences in glacier width along the flow line (through the effect of advective heat transfer);

(ii) systematic increase of surface albedo with altitude on the glacier tongue (varying amount of exposed morainic material).

Depending on the particular geometry, these factors will increase the sensitivity of glacier length to climatic change slightly or strongly. The uncertainties involved call for further study of the energy budget of the entire valley-glacier system.

5. GREENHOUSE WARMING AND GLACIER RETREAT

As discussed in the Introduction, ablation on a glacier surface reacts to a change in the radiation balance as well as to a change in air temperature. There are a number of effects to be taken into account when the concentration of carbon dioxide (and other radiatively active gases) changes. For the present purpose, it seems reasonable to distinguish between:

(1) A change in the radiation balance at the earth's surface, all other things (distribution of moisture, atmospheric temperature profile, etc.) being equal. This effect is immediate.

(2) The feedback on the radiation balance. The increasing humidity associated with rising tropospheric temperatures is probably the most important effect (Ramanathan, 1981; Hansen et al., 1981) and leads to substantial magnification of the initial perturbation of the radiation balance. Over a land surface, the major part of the increasing downward longwave flux is cancelled by an increasing upward flux due to higher surface temperature. Over a melting ice surface this is not the case. However, since humidity is related to ocean surface temperature, a lag will occur.

(3) Changing air temperature associated with the perturbed radiation balance of the entire atmosphere, also with a lag.

For a 1 K increase in surface air temperature, the increase of the net longwave balance at the surface is estimated to be about 1 W/m^2 (Luther & Cess, 1985; Table B.1), whereas the downward component increases by about 6 W/m^2. The difference is large and suggests that melting ice bodies could be good indicators of changes in atmospheric longwave emissivity! Indications on former levels of carbon dioxide have recently been reviewed by Gammon et al. (1985), and the conclusion has been reached that atmospheric amounts of both methane and carbon dioxide were already increasing in the first half of the nineteenth century. The best evidence for this comes from carbon isotopes in tree rings (e.g. Stuiver et al., 1984), and substantial changes in land use are assumed to be the cause. However, the burning of fossil fuels soon took over.

It is not the intention to review here scenarios for the greenhouse warming as presented in the literature. The one shown in Fig. 8 is a kind of average picture accepted by many climatologists. It is fairly simple to apply this to a model of a schematic glacier, to arrive at an order-of-magnitude estimate of glacier retreat. The changes in air temperature and radiation balance are first translated into changes of h_o (the 'melting altitude'). Assuming accumulation to be constant, this is equivalent to shifting the equilibrium line. Kuhn (1980) analyzed glacier mass balance in slightly different climatic conditions in basically the same geographical region. His work suggests:

$$h'_o = 6.5\,Q' + 125\,T' \qquad (15)$$

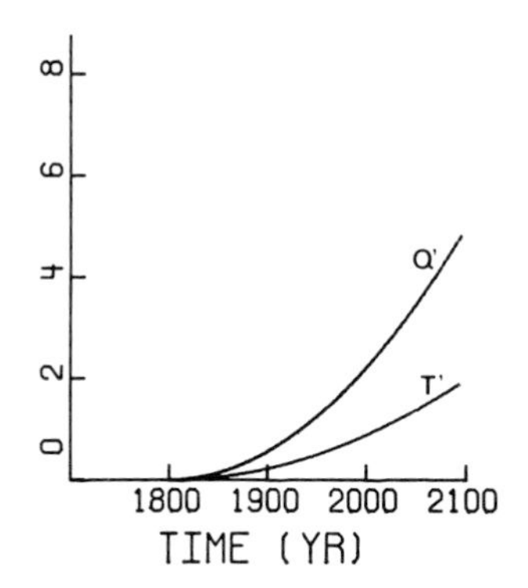

Figure 8. Forcing imposed on the glacier model, representing in a schematic way the effect of increasing atmospheric carbon dioxide and other greenhouse gases. Q' is the perturbation of the radiation balance (W m^{-2}), T' of air temperature (K).

The analysis did not contain the feedbacks discussed in the previous section, however.

Figure 9 shows the result of an integration for a 'typical' large valley glacier. The model was first run to a steady state and then a growing perturbation according to Fig. 8 and eq. (15) was imposed. As expected, the initial retreat is slow and small. Still, 'Greenhouse-warming retreat' predicted for the year 2000 is significant. It seems that for large valley glaciers a 1000 m retreat is typically what one expects to be the result of increasing carbon dioxide up til now. Or, to put it another way, a significant fraction of the observed world-wide retreat of valley glaciers could be due to the carbon dioxide warming.

6. FINAL REMARKS

Some authors, notably Meier (1984), have suggested that part of the observed sea-level rise over the last century can be attributed to melting of mountain glaciers. The sensitivity tests carried out here add that this melting probably is the result of the increasing concentration of carbon dioxide (and other radiatively active trace gases) in the atmosphere. So there seems to be increased evidence that the rise of world mean sea level is related to the carbon dioxide warming. This is an important point, because in many projections of future sea level the rate at which the sea rises presently (that is, over the last 100 years) is taken as basic trend, to which carbon dioxide effects are added.

It is quite obvious that many uncertainties still exist concerning the response of glaciers to a slightly changing climate. I hope to have identified some of them. Although many energy-balance and mass-balance studies have been carried out, it seems worthwhile to direct some effort to two specific points, namely:

(i) The role of the advective heat flux, related to a changing glacier geometry. This certainly involves the study of local wind systems.

(ii) The dependence of effective surface albedo on the dynamic history of the glacier ('debris dynamics').

Since snout variations are so well documented for a number of glaciers, and since there seem to be many reasons why glaciers should be very sensitive to a changing radiation balance, it seems worthwhile to employ them as basic climate indicators. However, a better understanding of the feedback loops discussed in this paper should first be achieved.

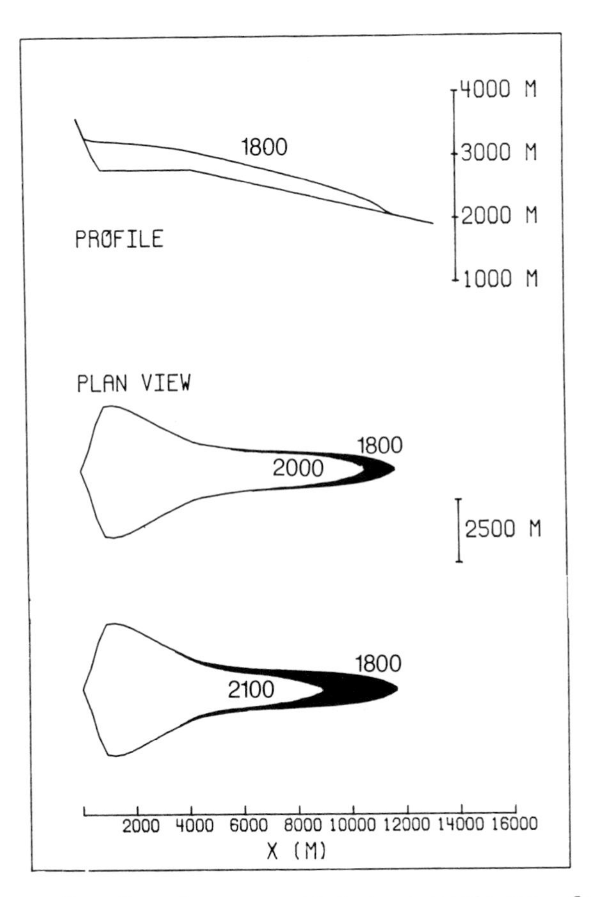

Figure 9. Greenhouse warming experiment for a schematic glacier. The black area shows the difference between the 1800 equilibrium profile and years as indicated.

REFERENCES

Ambach, W., 1965. 'Untersuchungen des Energiehaushaltes und des freien Wassergehaltes beim Abbau der winterliche Schneedecke', Archiv Meteor., Geophys. Bioklim. B14, 148-160.

Aubert, D., 1980. 'Les stades de retrait des glaciers du Haut Valais', Bull. Murithienne 97, 101-169.

Budd, W.F., and Jenssen, D., 1975. 'Numerical modelling of glacier systems', IAHS Publ 104, 257-291.

Braithwaite, R.J., 1984. 'Calculation of degree-days for glacier-climate research', Z. Gletscherk. Glazialgeol. 20, 1-8.

Burbank, D.W. and Fort, M.B., 1985. 'Bedrock control on glacial limits: examples from the Ladakh and Zanskar Ranges, north-western Himalaya, India', J. Glaciol. 31, 143-149.

De La Casinière, A.C., 1974. 'Heat exchange over a melting surface', J. Glaciol. 13, 55-72.

Furbish, D.J. and Andrews, J.T., 1984. 'The use of hypsometry to indicate long-term stability and response of valley glaciers to changes in mass transfer', J. Glaciol. 30, 199-211.

Gammon, R.H., Sundquist, E.T. and Fraser, P.J., 1985. 'History of carbon dioxide in the atmosphere', in: Atmospheric Carbon Dioxide and the Global Carbon Cycle (Ed.: J.R. Trabalka), U.S. D. of Energy, 26-62.

Greuell, W., and Oerlemans, J., 1986. 'Sensitivity studies with a mass balance model including temperature profile calculations inside the glacier', Zeitschr. Gletscherk. Glazialgeol. 22, 101-124.

Hansen, J.E., Lee, P., Rind, D. and Russell, G., 1981. 'Climate impact of increasing atmospheric CO_2', Science 213, 957-966.

Hoinkes, H.C., 1969. 'Surges of the Vernagtferner in the Oetztal Alps since 1599', Canadian J. Earth Sci. 6, 853-861.

Hogg, I.G.G., J G Paren, J.G. and Timmes, R.J., 1982. 'Summer heat and ice balances on Hodges glacier', South Georgia, Falkland Islands Dependencies. J. Glaciol. 28, 221-238.

Kasser, P., 1967. Fluctuations of Glaciers, Vol. 1., UNESCO, Int. Association of Scientific Hydrology (Paris).

Kasser, P., 1973. Fluctuations of glaciers, Vol. 2., UNESCO, Int. Association of Scientific Hydrology (Paris).

Müller, F., 1977. Fluctuations of glaciers, Vol. 3., UNESCO, Int. Association of Scientific Hydrology (Paris).

Kruss, P.D., 1984. 'Terminus response of Lewis glacier, Mount Kenya, Kenya, to sinusoidal net balance forcing', J. Glaciol. 30, 212-217.

Kuhn, M., 1980. 'Die Reaktion der Schneegrenze auf Klimaschwankungen', Zeitschr. Gletscherkunde Glazialgeol. 16, 241-254.

Kuhn M., G Kaser, G. Markl, H.P. Wagner and H. Schneider, 1979. 25 Jahre Massenhaushaltuntersuchungen am Hintereisferner, Universität Innsbruck, 79 pp.

Letréguilly, A., 1984. Bilans de masse des glaciers alpins: methodes de mesure et repartition spatio-temporelle. Publ. 439 Laboratoire de Glaciologie (Grenoble).

Luther, F.M and Cess R.D., 1985. Review of the recent Carbon dioxide-climate controversy. In: Projecting the Climatic Effects of Increasing Carbon Dioxide (eds.: M.C. MacCracken, F.M. Luther).

U.S. Dept. of Commerce, 34-335.

Luther, F.M. and Ellington, R.G., 1985. 'Carbon dioxide and the radiation budget', in: Projecting the Climatic Effects of Increasing Carbon Dioxide (Eds.: M.C. MacCracken, F.M. Luther), U.S. Dept. of Commerce, 25-55.

Martin, S., 1977. 'Analyse et reconstitution de la série de bilans annuels du glacier de Sarennes. Fluctuations du niveau de 3 glaciers du Massif du Mont-Blanc, Bossons, Argentière, Mer de Glace', Zeitschr. Gletscherk. Glazialgeol. 13, 125-163.

Meier, M.F., 1984. 'Contribution of small glaciers to global sea level', Science 226, 1418-1420.

Messerli, B., Zumbühl, H.J., Ammann, K., Keinholz, K., Oescher, H., Pfister, C. and Zurbruchen, M., 1975. 'Die Schwankungen des unteren Grindelwaldgletschers seit dem Mittelalter', Zeitschr. Gletscherk. Glazialgeol. 11, 3-110.

Oerlemans, J. and Van der Veen, C.J., 1984. Ice Sheets and Climate, Reidel (Dordrecht).

Oerlemans, J., 1986. 'Glaciers as indicators of a carbon dioxide warming', Nature 320, 607-609.

Oerlemans, J., 1986. An attempt to simulate historic front variations of Nigardsbreen, Norway. Theor. Appl. Climatol. 37, 126-135.

Orombelli, G. and Porter, S.C., 1982. 'Late holocene fluctuations of Brenva glacier', Geografia Fisica e Dinamica Quaternaria 5, 14-37.

Ostrem, G., Liestol, O. and Wold, B., 1977. 'Glaciological investigations at Nigardsbreen, Norway. Norsk Geogr. Tidsskr. 30, 187-209.

Paterson, W.S.B., 1981. 'The Physics of Glaciers', sec. ed., Pergamon Press, New York.

Patzelt, G., 1970. 'Die Längemessungen an den Gletscher der Österreichischen Ostalpen 1890 bis 1969.' Zeitschr. Gletscherkunde Glazialgeol. 6, 151-159.

Pollard, D., 1980. 'A simple parameterization for ice sheet abaltion rate', Tellus 32, 384-388.

Ramanathan, V., 1981. 'The role of ocean-atmosphere interactions in the CO_2 climate problem. J. Atmos. Sci. 38, 918-930.

Reynaud, L., 1983. 'Recent fluctuations of alpine glaciers and their meteorological causes: 1880-1980. In: Variations in the Global Water Budget (eds.: A Street-Perrott et al.), 197-205. Reidel (Dordrecht).

Stuiver, M., Burk, R.L. and Quay, P.D., 1984. '$^{13}C/^{12}C$ Ratios and the transfer of biospheric carbon to the atmosphere', J. Geophys. Res. 89, 1713-1748.

Tufnell, L., 1984. 'Glacier Hazards' Longman (London).

Tvede, A.M., 1974. 'Glasiologiske Undersökeler i Norge 1972', Norges Vassdrags-Og Elektrisitetsvesen Rapp. 1-74, Oslo.

Vivian, R., 1975. 'Les Glaciers des Alpes Occidentales', Allier, Grenoble.

Wendler, G., 1974. 'A note on the advection of warm air towards a glacier. A contribution to the international hydrological decade. Zeitschr. Gletscherk. Glazialgeol. 10, 199-205.

NUMERICAL MODELLING OF GLACIER D'ARGENTIERE AND ITS HISTORIC FRONT VARIATIONS

Ph. Huybrechts, P. de Nooze and H. Decleir
Geographical Institute
Free University of Brussels
Pleinlaan 2
B-1050 Brussels, Belgium

ABSTRACT

A numerical glacier model has been developed for Glacier d'Argentière (France) in order to study its relation with climate and investigate possible causes for the observed variations in the terminus record since the beginning of the Little Ice Age.

At first results are presented from a basic sensitivity investigation, with plots of steady state glacier length versus perturbations in mass balance and glacier reaction with respect to sinusoidal net balance oscillations. An attempt is then made to simulate the historic front variations. The mass balance history of the glacier is constructed assuming a linear relationship with (i) summer temperature anomalies and (ii) mean annual temperature anomalies for Basel dating back to the beginning of the 16th century.

Although model run (ii) turns out to yield better agreement with the observations, both simulations have in common that the observed glacier retreat comes too late. Improved simulations can only be obtained assuming an additional negative mass balance perturbation of around 0.1 m/year over roughly the last 150 years.

These results indicate that the assumption of a linear relationship between summer temperature and the glacier's mass balance may not be valid anymore when extrapolated to past environments. This might be evidence of additional micrometeorological and glacier surface conditions prevailing in valleys at maximum glacier extent, that are not absorbed well in the climatic records.

1. INTRODUCTION

Glacier d'Argentière, a valley glacier in the French Mont Blanc area (45.9°N, 7.0°E), is one of the few Alpine glaciers that have a fairly reliable historic record of front variations dating back to about 1600, i.e. since the beginning of the Little Ice Age (Vivian, 1975). This

J. Oerlemans (ed.), Glacier Fluctuations and Climatic Change, 373–389.

record is characterized by several rapid advances, most noticeably between 1625-1640, 1700-1720 and 1765-1780, culminating in the Neoglacial maximum in 1825. Since 1860 the glacier has been generally receding, alternatively slow and discontinuously (1883-1945: 1.1 m/year) and fast and uniformly (1866-1883: 37.1 m/year; 1945-1967: 21.4 m/year) over about 1600 m, reaching a glacial minimum in 1968. Since that time and up to the mid-eighties, Glacier d'Argentière is reported to have recovered some 300 m (Reynaud, 1986).

Glacier d'Argentière is also of special interest because it is quite well documented and surveyed, both in terms of its mass balance and ice depth, with velocity measured at about 5 locations along its length, in particular since 1975 by the Laboratoire de Glaciologie et de Geophysique de l'Environnement, Saint Martin d'Hères, France (Hantz, 1981; Reynaud, 1986). This relative wealth of data and the glacier's fairly simple geometry - it is a 'classical' Alpine non-surging valley glacier with a wide accumulation basin and a long narrow tongue - make Glacier d'Argentière a prime candidate for a modelling study.

Glaciers respond to climatic change in a complex way. Fluctuations in climatic parameters cause changes in snow accumulation and ice melt on the glacier surface. Over a sufficiently long period of time then, the terminus reponse will reflect the integrated changes in climatic input via the overall mass balance. Each glacier behaves as a particular case, with a dynamic response determined in essence by its own geometry. Important here are glacier length, bed topography and area-elevation distribution. Since the historic record goes back further than the instrumental record, an investigation of the relationship between terminus position and climatic input should aid in the reconstruction of past climate.

This suggests the use of a dynamical glacier model relating changes in ice thickness and hence, terminus position, to changes in the mass-balance. There have been some model studies in this direction, based on the numerical models developed by Budd and Jenssen (1975). In an early paper an attempt was made to estimate climatic change from the retreat of small mountain glaciers in Irian Jaya, Indonesia by Allison and Kruss (1977). Smith and Budd (1981) made an estimate of the difference in conditions between the present and the peak of the Little Ice Age by comparing known terminus histories and the reaction of the modelled glaciers (Storglaciärens, Hintereisferner, Vernagtferner, Aletschgletscher) to a simple sinusoidal function. Similar studies have been conducted by Kruss and Smith (1982) and Kruss (1983, 1984) on Vernagtferner and Lewis Glacier (Kenya).

In the present paper, as in Oerlemans (1987) in a study on Nigardsbreen (Norway), climatic series in part based on proxy data will be used to drive the glacier's mass balance history. Studying Alpine front position series in this way may then lead to a better understanding of the causes of glacier variations in Europe. At first results are discussed from a basic sensitivity investigation, in which glacier length versus perturbations in the mass balance are studied and in which the model is submitted to sinusoidal net balance forcing of various periods and amplitudes. An attempt is then made to simulate the historic front variations of Glacier d'Argentière. We performed experiments in which climatic data for Basel dating back to the first half of the 16th century are used, as compiled by Pfister (1984) on the basis of weather descriptions, environmental data and available instrumental records.

It will turn out, however, that the attempt is not very successful. In spite of the fact that a model of the present type is expected to reproduce the low frequency response only, in particular the glacier retreat since about 1860 is not modelled well by any of the performed runs. In a final section, possible causes of the discrepancy between modelled and observed front variations are discussed.

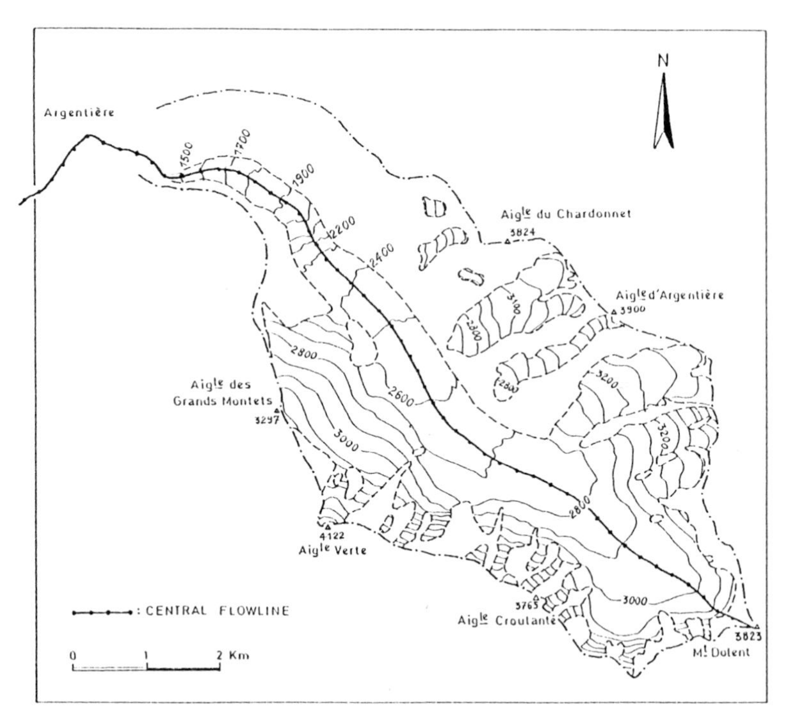

Figure 1. Map of Glacier d'Argentière showing present (1980) extent and the central flowline with gridpoint spacing of 250 m.

2. THE NUMERICAL MODEL

The dynamical glacier model employed in this study describes ice flow along a central flowline. Regarding the fairly simple geometry, this flowline is easily constructed and follows approximately the centre of the glacier, see Fig. 1. Every point x along this line, spaced 250 m apart, is assumed to represent mean lateral conditions. The most important parameter concerning three-dimensional geometry, namely the varying width distribution, is then accounted for in the continuity equation. Since ice density is assumed to be constant, we have:

$$\frac{\partial H}{\partial t} = -\frac{1}{b}\frac{\partial(HUb)}{\partial x} + M = -\frac{\partial(HU)}{\partial x} - \frac{HU}{b}\frac{\partial b}{\partial x} + M \qquad (1)$$

where H is local ice thickness, U is the mean velocity parallel to the bedrock, b the glacier width and M the annual mass balance, expressed in m ice depth per year.

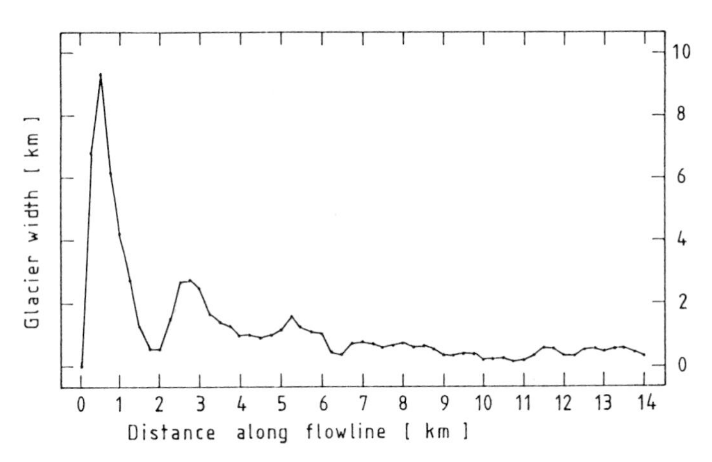

Figure 2. Present glacier width distribution (b_{ref}) as used in the ice flow model.

The gridded width values b_{ref}, displayed in Fig. 2, have been constructed respecting the present area-elevation distribution. This approach may at first glance present some peculiarities. For instance, it explains the remarkably large width gradient and apparent flow convergence found between x = 1000 m and x = 2000 m, since the surface slope along the centre line (between the 3000 and 2900 m contour lines) is here not representative for the entire flowband. Obviously, conditions at grid points high up in the accumulation area are less well defined.

The width distribution should also depend on ice thickness. Especially in the lower tongue area, this may become important, since a thinning glacier generally implies a decreasing width and consequently, a smaller ablation area. We assumed a simple parabolic cross section leading to:

$$b = b_{ref}\left[\frac{H}{H_{ref}}\right]^{\frac{1}{2}} \qquad (2)$$

where H_{ref} is taken from the present (1980) modelled glacier and b_{ref} read from a map (IGN topographical map on scale 1/25,000, 1980). Values of b_{ref} downstream the 1980 front position were generated with a constant ice thickness H_{ref} of 100 m.

Other input data, namely bed topography is known at 5 location along the glacier length (Hantz, 1981). Since surface topography is read from a topographical map, remaining bed values, or, alternatively ice thickness, can in first approximation be estimated from plastic ice flow theory, relating ice thickness to the local surface gradient, with the factor of inverse proportionality given by a specific choice of the yield stress τ_0 (e.g. Paterson, 1981). Calculated driving stresses (with a surface gradient taken over 1500 m in order to smooth local surface irregularities) at these 5 locations were then inter/extrapolated over the whole glacier to obtain the unknown ice thicknesses. Except for the glacier part below the ice fall (Sẽracs de Lognan at x = 8250 m), these driving stresses (as defined in eq. (4)) appear to be in the range 150-200 kPa. As calculated 'shape factors' (assuming a parabolic cross section) for these same 5 locations all are between 0.5 and 0.6 (the difference with 1 gives the part of the driving stress opposed by side drag), this corresponds with basal stresses of the order of 80-120 kPa, i.e. within the expected value range (Paterson, 1981). In what follows, however, distinction between side and basal drag will not be made, in effect putting forward a constant 'shape factor', whose value is then most easily incorporated in the flow parameter.

To arrive at an equation for the mean horizontal velocity, the flow law is written as (e.g. Oerlemans and Van der Veen, 1984):

$$U = A \tau_b^3 H \tag{3}$$

with A the flow parameter and where the driving stress is given by the basal shear stress τ_b:

$$\tau_b = - \rho g H \frac{\partial(H+h)}{\partial x} \tag{4}$$

Here ρ is ice density (taken as 870 kg m^{-3}), g acceleration of gravity and h bedrock elevation. Hence, the velocity is a locally defined quantity, depending on ice thickness and surface slope. Strictly speaking, equation (3) describes the velocity contribution resulting from internal deformation only. However, since there is with respect to (vertical) mean velocity not so much difference between deformation (concentrated near the base) and basal sliding, any basal sliding may be assumed to be reflected in the value of the flow parameter. As in the Smith and Budd (1981) model, preliminary experiments have been conducted in which the geometry of the glacier cross-section was taken into account with shape and velocity factors (relating the centreline velocity to the cross sectional 'mean'). It was found, however, that contrasts in these geometric parameters within realistic bounds were by no means crucial to the general behaviour of the model. For simplicity then, these effects are also assumed to be represented by the value of the flow parameter, that now essentially serves a tuning purpose. Good results are obtained while setting A = 0.8 10^{-16} $year^{-1}$ Pa^{-3}.

Substitution of (3) and (4) in the continuity equation (1) then leads to an expression that can be regarded as a diffusion equation for ice thickness H. The resulting equation is most easily solved with a straightforward explicit finite difference scheme. To ensure computational stability a staggered grid in space is used, in effect calculating volume fluxes in between grid points with a mean diffusivity. Details of this scheme can be found in Oerlemans and Van der Veen (1984). With a spatial resolution of 250 m, this scheme allows time steps up to 0.1 years.

Finally, a simple procedure is adopted to follow the glacier terminus position in between grid points, because the maximum length contrast in the period considered amounts to roughly 1600 m, i.e. about 6-7 grid points only. This problem has been adress by Kruss (1984) in a more sophisticated way by calculating the volume of ice past the last grid point in use, and then computing the length necessary to contain this volume within a specified longitudinal snout shape. The procedure employed here is given in by the observation that the model glacier will jump to the next grid point whenever the absolute value of the ratio of the flux divergence and mass balance in that point equals unity. An approximately 'linear' transition during glacier evolution from one grid point to another is then obtained by raising this ratio to some power m:

$$L = (N_g - 1) \cdot \Delta x + \frac{\left[1/b\ \partial(HUb)/\partial x\right]_{Ng+1}}{M_{Ng+1}}^{m} \cdot \Delta x \qquad (5)$$

where L is total glacier length, N_g the index number of the last grid point in use and Δx the grid point distance. We used m = 0.2.

3. BASIC SENSITIVITY EXPERIMENTS

The basic mass balance parameterization versus surface elevation, to be perturbed in the experiments, is taken as a linear fit to the observed mean mass balance for the period 1976-1983 (Reynaud et al., 1986), in m ice depth/year:

$$M = 0.0071\ (H + h - 2910) \qquad \text{if } (H + h) > 1750 \text{ m} \qquad (6a)$$

$$M = 0.02\ (H + h - 2910) + 15 \qquad \text{if } (H + h) \leqslant 1750 \text{ m} \qquad (6b)$$

$$M = 3.60 \qquad \text{if } M \geqslant 3.60 \qquad (6c)$$

During this period the mean equilibrium line altitude (ELA) was about 2910 m a.s.l. Here it is assumed that the mean balance gradient of 0.71 m ice depth/100 m, as observed in the 1800-2800 m altitude interval, also applies higher up in the accumulation zone. The maximum value of 3.60 m/year is the observed 1954-1971 mean for Vallée Blanche (Mont Blanc area) at an altitude of aroung 3600 m. In the lower tongue area below around 1750 m, Hantz (1981) mentions a single year balance measurement in 1957, with a significant larger balance gradient (about 2 m/100 m). We feel this is a real feature, possibly related to radiative side wall effects as the glacier enters a more narrow valley, and, hence, should no be left out.

One can essentially distinguish two ways to express variations in the net balance curve following changes in the climatic state. Either one assumes a balance shift or an elevation shift, respectively shifting the net balance curve parallel to the balance or elevation axis. In the first approach the balance gradient at a specified location is considered to be constant, so that the imbalance is independent of altitude. It appears that this is the case for Alpine glaciers (Kuhn, 1984), who postulates that most likely increased ablation gradients in the long ablation period of negative years are balanced by increased gradients of solid precipitation and of albedo due to summer snow falls in positive years. Also, with respect to the limited mass balances measurements available, the linear balance model of Lliboutry (1974) seems to work well for Glacier d'Argentière (Hantz, 1981; Reynaud et al., 1986).

Since it is not known how far the present glacier is out of equilibrium, the flow law parameter A has been chosen such that the basic mass balance distribution produces a steady state length close to the 1980 position, that will be considered as a reference state. This glacier with total length = 10.49 km is shown in Fig. 3. Except for grid points in the upper accumulation area, where the central flowline represents less well mean lateral conditions anyway, the model is very well capable of reproducing the right thickness and surface elevation. For the present glacier, mean modelled velocities then generally range between 40 and 80 m/year, with a maximum of 150 m/year in the strong convergence zone at x = 2000 m and 140 m/year at the ice fall at x = 8250 m. Taking into account that these velocities should be interpreted as lateral and vertical mean values, they appear to be slightly higher (although by a factor 1.5 at most) than the observations (Hantz, 1981).

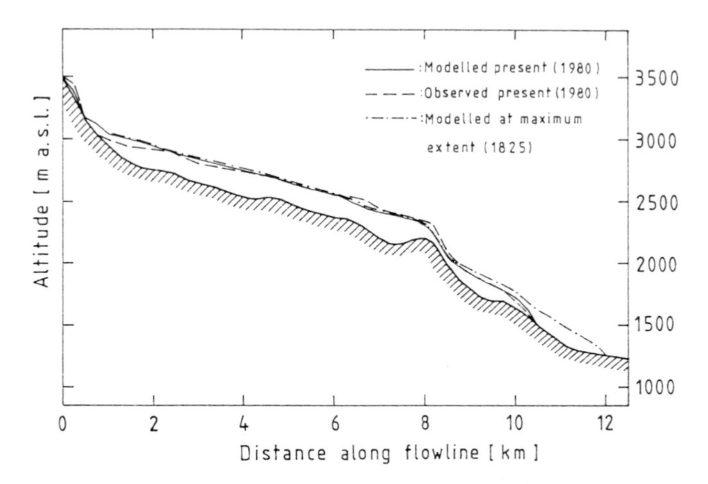

Figure 3. Longitudinal glacier profiles corresponding with the present glacier (10.49 km) and a glacier at maximum extent (11.98 km). Both profiles differ in the steady state with a balance shift of 0.37 m/year or an ELA shift of 52 m. The bed profile along the flowline is shown hatched

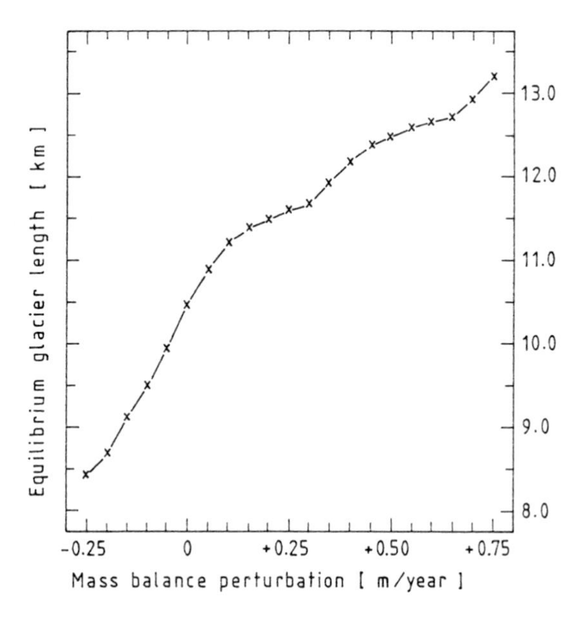

Figure 4.Steady state glacier length versus perturbations in the mass-balance with respect to the 1980 reference state. The total abscissa range of 1 m/year corresponds with a total ELA shift of 140 m.

Basic model sensitivity versus perturbations in the mass balance is reflected in Fig. 4. A noteworthy feature here is the differential behaviour with respect to positive or negative shifts in mass balance. The decreasing sensitivity encountered for positive balance perturbations appears to be, apart from an increased ablation gradient below 1750 m, essentially a geometric effect: the longer the glacier gets, the less important becomes the relative weight of the wide accumulation surface area in the total glacier area distribution. The asymmetric width distribution (large firn zone area and narrow glacier tongue) also explains the large sensitivity of the glacier terminus position with respect to the climatic state. A Little Ice Age glacier of maximum length 11.98 km and the 1968 minimum state of 10.30 km differ in the steady state by a mean change in net balance of 0.40 m/year only, or accordingly, a change in ELA or around 55 m. Of course, the concept of steady state is a theoretical one, and climatic inferences of this ELA-shifts are highly uncertain. Following theory developed by Kuhn (1980), a change in ELA of 100 m would correspond to about a 0.8 K temperature difference, a value derived for the eastern Alps. Typical thickness changes between these two glaciers are then about 10 to 20 m above the ice fall, and increasing to almost 150 m further below, as shown in Fig. 3.

Another basic parameter of Glacier d'Argentière concerns its response time, i.e. the order of time it takes to adjust to a change in its mass balance. This is investigated in Fig. 5, showing the response of the glacier following a stepwise change in mass balance. Defining this time as the time it takes for the glacier terminus to reach its final position within a fraction e^{-1}, the response time appears to be in the range 27-33 years for a positive perturbation and, the response being somewhat slower, 40-45 years for a negative mass balance change.

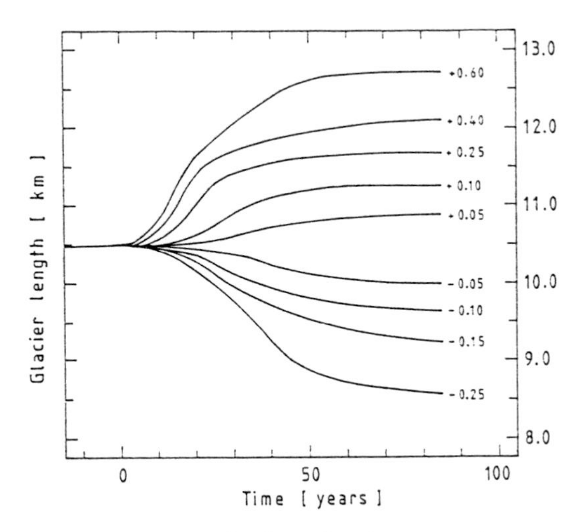

Figure 5. Reaction of the glacier terminus position to a stepwise change in mass-balance of given magnitude at time = 0.

Sensitivity of Glacier d'Argentière with respect to sinusoidal forcing around the 1980 reference state, is summarized in Figs. 6 and 7. Here again, the time lage between applied net balance forcing and terminus response (measured from minimum/maximum net balance extreme to the corresponding terminus extreme, Fig. 6) depends on the sign of the perturbation, but appears to be virtually independent of applied amplitude (with a maximum difference of 10% between amplitudes B = 0.25 m/year and B = 0.50 m/year). The interesting parameter here is the time lag at a 1000 year period, as it has been assessed for other glacier (Kruss, 1984). In the present case it is about 50 years (mean of max. and min. values), compared to 110 years for Hintereisferner (Austria) and 30 years for Lewis Glacier (Kenya). As in similar experiments discussed by Kruss (1984), the terminus response amplitude (Fig. 7) reaches a maximum for the longer periods, and depends linearly on applied amplitude. However, a doubling of the applied net balance amplitude does not result in a two-fold increase in terminus response, but rather by a mean factor of 1.75. So, the experiments discussed above clearly demonstrate basic sensitivity of Glacier d'Argentière with respect to changing climatic conditions. They also provide little evidence of basic model failures that could lead to fully wrong results.

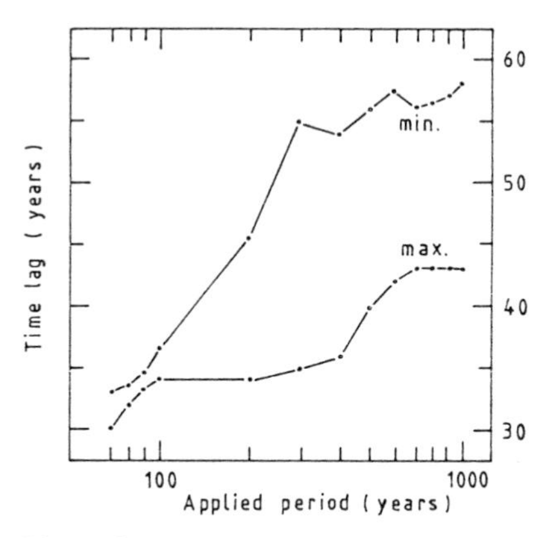

Figure 6. Time lage between sinusoidal net balance forcing and terminus position, respectively following a net balance maximum (max.) and minumum (min.). The net balance amplitude is 0.25 m/year here.

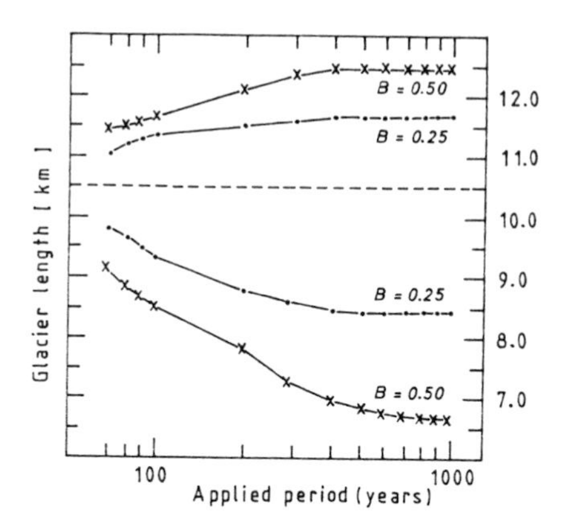

Figure 7. Amplitude of the Glacier d'Argentière's terminus response. Shown are the maximum and minimum glacier positions against applied period of net balance sinusoidal oscillations of amplitude B = 0.25 and 0.50 m/year respectively.

4. SIMULATION OF HISTORIC FRONT VARIATIONS

In order to get more insight in possible causes that might explain the variation in the terminus record and, in particular, the general retreat of Glacier d'Argentière since the middle of the last century, and attempt is then made to simulate the historic record. As mentioned in the introduction, a comparison between modelled and observed terminus histories is then expected to tell more about the type of climatic variables decisively affecting the glacier's mass balance. Naturally, mass balance measurements are not available over the time span covering the Little Ice Age. The same applies to instrumental climatic records for the first part of this period. Hence, a proper forcing function, in part based on proxy data, must be carefully chosen and relied upon.

In this study climatic data dating back to 1530 for the north face of the Alps are used (Pfister, 1984, personal communication). These data (seasonally and yearly precipitation and temperature) are given as 10-year running means and expressed as anomalies with respect to the 1901-1960 mean for Basel. It may be assumed that long-term climatic trends in the Mont Blanc Area are well described by this series, as the climate in the two locations is governed predominantly by the same large scale oceanic circulation types (a point also taken up in section 5).

Due to the fairly long characteristic response time scale of the glacier (of the order of 50 years) and the long time the glacier takes to reach equilibrium (approximately 300 years), calculations started at 1000 A.D. with zero ice thickness. The forcing function until 1530 is taken from the central England series (Lamb, 1977) and is made to match the Pfister series. It essentially shows a general cooling trend since about 1200. The last value of the Basel series (1970-1979 mean) is then kept constant until the year 2000, when calculations end.

Assuming that the linear balance model also applies to the past, the mass balance is perturbed in the experiments according to:

$$\Delta M = C_1 \ (\Delta T + C_2) \qquad (7)$$

where ΔM is equal for all altitudes, ΔT is temperature anomaly (°C), and C_1 and C_2 are constants to be optimized. C_1 essentially controls the terminus response amplitude and the constant C_2 mainly influences the 'mean' glacier length. They are specified below.

It is generally accepted that mass balances of Alpine glaciers are to a large extent controlled by meteorological conditions in summer, in particular summer temperature (e.g. Reynaud, 1983). On the basis of a 27-year record, Martin (1977) found that 58% of the total variance in the mean specific net balance record for Glacier de Sarennes could be explained by summer temperatures (July and August) at the nearby Lyon station, with minor contributions in the total variance from winter precipitation (October to May) and early summer conditions (June precipitation).

Hence, in a first experiment summer temperature anomalies were used to force the glacier model. Results with $C_1 = -0.6$ m/°C and $C_2 = -0.3$ °C are shown in Fig. 8. Striking features in the forcing function (upper panel) include a marked cooling around 1600, a very cold interval around 1816 (year 'without summer') and to a lesser extent in 1886 and 1912, and a steep temperature rise between 1912 and 1947. The observed front record, based on continuous measurements over roughly the last 100 years and on dated moraines and sporadic written sources before that, has been described in the introduction (middle panel). The lower panel shows that the modelled response does not resemble the observed length variations very well, even when taking into account that only the long term climatic trend is expected to be reflected in the glacier response, as the glacier effectively acts as a low-pass filter. The model then shows a marked glacier increase towards 1600, approximately coinciding with the beginning of the Little Ice Age. Between 1600 and 1800, the glacier is, unlike the observed record, almost continuously receding, reaching a terminus position around 1810 rather close to the present state. Note that the forcing does not display remarkably cold conditions during this period either. Thereafter, the modelled glacier response exhibits an important growing phase, obviously related to the 1816 cold summer spell. The observed glacier retreat since 1850 then comes almost a century later.

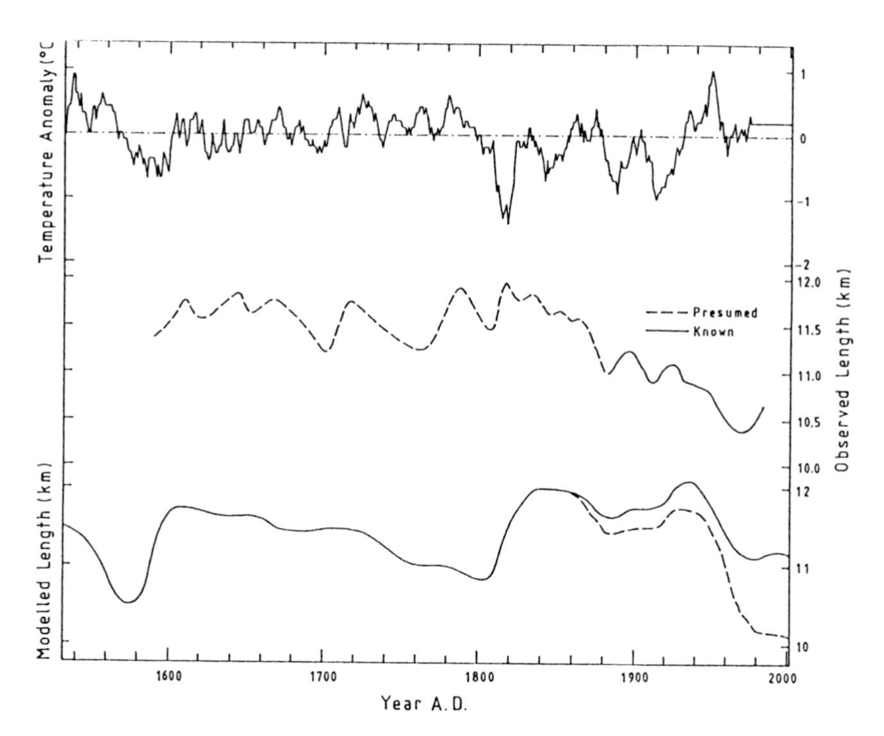

Figure 8. The experiment in which summer temperature anomalies (upper panel) were used to construct the mass balance history. The middle curve shows the observed length variations. The lower curve shows computed glacier length with full line: C_1 = -0.6m/°C, C_2 = -0.3°C and dashed line: C_2 = -0.1°C from 1850 onwards.

As displayed in Fig. 9, somewhat better agreement between modelled and observed histories can be obtained when relating mass balance perturbations to more general atmospheric conditions, in casu mean annual temperature. Nevertheless, both simulations appear to have in common that the Neoglacial maximum and concomitant retreat are out of phase by 50-100 years.

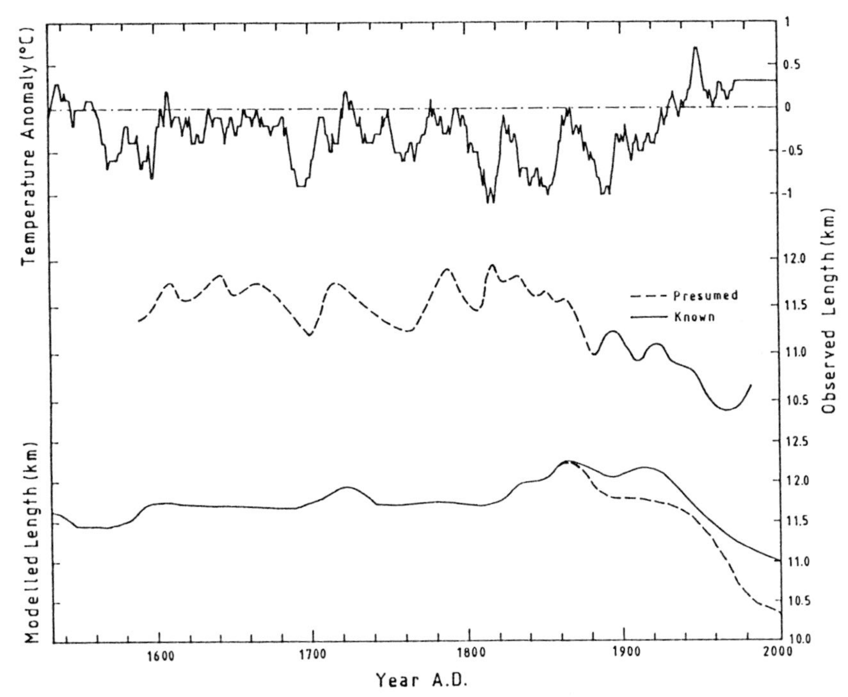

Figure 9. As in Fig. 5, but now with mean annual temperature anomalies as forcing function. C_1 = -0.45 m/°C, C_2 = -0.4°C (full line) and C_2 = -0.2°C from 1850 onwards (dashed line).

This point has been taken up in an experiment in which a (moderate) additional mass balance perturbation of around -0.1 m/year was imposed from 1850 onwards (dashed line, see Figs. 8 and 9), which turned out to significantly improve the simulation, in particular in the model run forced with mean annual temperature anomalies. In a recent paper, Oerlemans (1986) demonstrated that melting rates at the lower glacier parts may be extremely sensitive to changes in the radiation budget (through increased advective heat transport from surrounding rock areas), providing a possible link between glacier retreat and increased carbon dioxide levels in the atmosphere. This is certainly not in contradiction with the improved results obtained here, when crudely mimicing increased ablation since the 19th century. Similarly, assuming increased balance gradients during the same period also turned out to yield better agreement between modelled and observed glacier retreat (not shown here). Many more model runs were conducted, involving forcing functions based on other precipitation and temperature data. In a similar way as in Martin (1977), also composed series were derived from a multiple linear regression analysis of the Basel-data with available specific net mass balance records (Glacier de Sarennes, Aletschgletscher). However, as these model runs do no add much to the general picture, they are not discussed here.

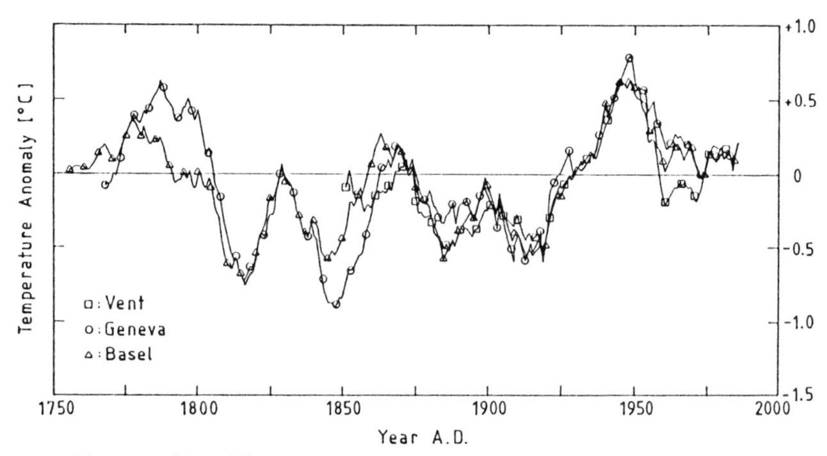

Figure 10. 15-year running mean summer temperature anomalies with respect to the 1901-1950 mean compared with the Basel series. The summer temperature is defined here as the mean over the three summer months (June, July and August) with contributions from May and September witha weighing coefficient of 0.5. Plot kindly provided by Wouter Greuell.

5. DISCUSSION

As is clear from the results in Figs. 8 and 9, none of the simulated glacier response curves compares really well with the recorded historic positions of the terminus. Numerical experiments, in which model parameters and data on the glacier were varied within realistic ranges (ice thickness, flow parameter, inclusion of variable shape and velocity factors) brought to light that the discrepancy between modelled and observed terminus histories cannot be due to errors in these parameters. Rather changes in these lead to re-adjustment of velocity and/or ice depth along the flowline, while the terminus response remains largely unaffected, in essence reflecting the integrated changes in overall mass balance only. For the same reason, also the inclusion of longitudinal stresses, providing a potential additional means of propagating disturbances along the glacier, is not expected to influence the glacier front response significantly. Moreover, one may anticipate local effects of extension and compression to cancel each other out, when changes in total glacier length are considered. This is corroborated by experiments conducted by W. Greuell on Hintereisferner (personal communication), showing the response time to be quasi unchanged by longitudinal stress effects.

This means that the unsuccessful simulation must be due to an inadequate description of the mass balance history. In Fig. 10 it is shown that there can be no doubt about the suitability of the Basel-series to describe climatic trends in the Mont Blanc area. In particular since 1875, there is a striking resemblance between summer temperature anomalies of Basel (north of the Alps), Geneva (west of the Alps) and Vent (Oetztal, central Alps). So, as the response curve shape turns out not to depend critically on the parameters C_1 and C_2, the obvious conclusion should then be that an oversimplified relation between climate and mass balance is put forward and that apparently, mass budget imbalances of Glacier d'Argentière are not linearly correlated with summer temperature anomalies on the time span considered.

6. CONCLUSION

In this study an attempt was made to investigate possible causes of glacier variations by simulating the historic record of Glacier d'Argentière with a dynamical model. Forcing the mass balance history linearly (spatially as well as in time) with summer and mean annual temperature anomalies for Basel, brought to light, that, in particular, the observed glacier retreat since about 1850 is not fully understood.

Apart from basic shortcomings in model formulation and glacier data, for which there is otherwise no direct evidence, it appears that the critical point in this study is connected with the assumption that the type of variation evident from a relatively short series of mass balance measurements can be extrapolated to past environments. Actually, the results of the simulations seem to suggest the contrary. Although that on the basis of available records, summer temperature seems to be highly correlated with mass balance variations that are in addition independent of altitude, this relationship does apparently not explain the dominant feature of the historic record, in casu the almost continuous retreat of the Argentière since the middle of the 19th century.

This result and the improved model simulations that could be obtained while assuming an additional negative mass balance perturbation during roughly the last 150 years, seems to point to additional features affecting the glacier's mass balance that are not captured well in the ambient climatic records. A thorough investigation of this point is far beyond the scope of the present study, but it may very well be that melting rates, and hence, balance gradients respond critically and non-linearly to changes in the radiation budget, valley geometry or glacier surface conditions much along the lines of what has been demonstrated in Oerlemans (1986).

ACKNOWLEDGEMENTS

We thank Louis Reynaud (Laboratoire de Glaciologie et Geophysique de l'Environnement) for communicating in part unpublished field data. Christian Pfister is acknowledged for providing us with numerical data on the Basel-series. Critical remarks and helpful comments by Wouter Greuell and Hans Oerlemans were very much appreciated. Philippe Huybrechts is supported by the Belgian National Fund for Scientific Research (N.F.W.O.).

REFERENCES

Allison, I. and Kruss, Ph. 1977. Estimation of recent climatic change in Irian Jaya by numerical modeling of its tropical glaciers. Arctic and Alpine Res. 9, 49-60.

Budd, W.F. and Jenssen, D. 1975. Numerical modelling of glacier systems. IAHS Publ. 104, 257-291.

Hantz, D. 1981. Dynamique et hydrologie du Glacier d'Argentière. These de docteur-ingenieur, Laboratoire de Glaciologie et Geophysique de l'Environnement, Grenoble, 181 pp.

Kruss, Ph. 1983. Climatic change in East Africa: a numerical simulation from the 100 years of terminus record at Lewis glacier, Mount Kenya. Zeitschr. für Gletscherk. und Glazialgeol. 19, 43-60.

Kruss, Ph. 1984. Terminus response of Lewis glacier, Mount Kenya, Kenya, to sinusoidal net balance forcing. J. Glaciology 30, 212-217.

Kruss, Ph. and Smith, I. 1982. Numerical modelling of the Vernagtferner and its fluctuations, Zeitschr. für Gletscherk. und Glazialgeol. 18, 93-106.

Kuhn, M. 1980. Die Reaktion der Schneegrenze auf Klimaschwankungen. Zeitschr. für Gletscherk. und Glazialgeol. 16, 241-254.

Kuhn, M. 1984. Mass budget imbalances as criterion for a climatic classification of glaciers. Geografiska Ann. 66A, 229-238.

Lamb, H.H. 1977. Climate: Present, Past and Future, vol. 2: Climatic history and the future. Methuen & Co., London.

Lliboutry, L. 1974. Multivariate statistical analysis of glacier annual balances. J. Glaciology 60, 371-392.

Martin, S. 1977. Analyse et réconstitution de la serie des bilans annuels du glacier de Sarennes, sa relation avec les fluctuations du niveau des trois glaciers du massif du Mont Blanc (Bossons, Argentière, Mer de Glace). Zeitschr. für Gletscherk. und Glazialgeol. 13, 127-153.

Oerlemans, J. 1986. Glaciers as indicators of a carbon dioxide warming. Nature 320, 607-609.

Oerlemans, J. 1987. An attempt to simulate historic front variations of Nigardsbreen, Norway. Theoret. Appl. Climatology, in press.

Oerlemans, J. and Veen, C.J. van der 1984. Ice Sheets and Climate. D. Reidel Publishin Co., Dordrecht), 217 pp.

Paterson, W.S.B. 1981. The Physics of Glaciers, 2nd ed. Pergamon Press, Oxford, 380 pp.

Pfister, Ch. 1984. Zehnjahrige Mittel von Temperatur und Niederschlag in den Jahreszeiten. In: W. Kirchhofer (ed.), "Klimaatlas der Schweiz". Verlag des Bundesamtes der Landestopographie, Bern.
Reynaud, L. 1983. Recent fluctuations of Alpine glaciers and their meteorological causes: 1880-1980. In: A. Street Perrot et al. (eds.), " Variations in the global water budget". D. Reidel Publishing Co., Dordrecht, 197-205.
Reynaud, L. 1986. Glacier d'Argentière 1985: Rapport sur les travaux executes pour Electricité d'Emosson s.a. (Annecy). Université Scientifique Technologique et Medical de Grenoble, 15 pp.
Reynaud, L., Vallon, M. and Letréguilly, A. 1986. Mass balance measurements: problems and two new methods to determine variations. J. Glaciol. 32, 1-27.
Smith, I. and Budd, W.F. 1981. The derivation of past climatic changes from observed changes of glaciers. IAHS Publ. 131, 31-52.
Vivian, R. 1975. Les glaciers des Alpes occidentales. Allier, Grenoble, 513 pp.

HISTORIC FRONT VARIATIONS OF THE RHONE GLACIER: SIMULATION WITH AN ICE FLOW MODEL

A. Stroeven, R. van de Wal and J. Oerlemans*
Institute of Physical Geography
University of Utrecht
The Netherlands

ABSTRACT

The Rhône Glacier (Switzerland) is one of the few valley glaciers for which the record of front variations goes back to the beginning of the 17th century. After the neoglacial maximum in the year 1602 there have been notable advances around 1818 and 1856. Since that time the glacier has shown steady retreat with some minor interruptions only.

In this contribution we make an attempt to simulate the historical front variations with a numerical glacier model. It is based on the continuity equation for ice mass, applied to the central flowline while taking into account the varying geometry. Both sliding and deformation are related directly to the local driving stress. The gridpoint spacing is 250 m.

After a general survey of the basic sensitivity of the Rhône glacier to changes in mass balance and geometry, some climatic series (temperature, precipitation, tree-ring width) were imposed as forcing to the model. A reasonable match of calculated and observed front positions could be obtained, except for the retreat of the last hundred years. To simulate this retreat, an additional 85 m increase of equilibrium-line altitude has to be imposed to the model.

1. INTRODUCTION

It is generally assumed that glaciers are good indicators of climatic change. A fairly good statistical relation can be established between series of summer temperature, winter precipitation and front position of a group of glaciers (e.g. Lamarche and Fritts, 1971; Aellen, 1981). However, such methods are less succesful when applied to a single glacier, because the specific geometry can lead to substantial

* Institute of Meteorology and Oceanography, University of Utrecht, The Netherlands

J. Oerlemans (ed.), Glacier Fluctuations and Climatic Change, 391–405.

Figure 1. The Rhône Glacier viewed from Gletsch (1800 m) in the middle of the 19th century (upper) and recently (lower picture).

deviations from the average response. This problem can partly be overcome by using a dynamic glacier model to find the relation between changes in mass balance and migration of the glacier snout. A few studies have been carried out now in this way (e.g. Lewis Glacier: Kruss, 1984; Nigardsbreen: Oerlemans, 1986; Glacier d'Argentière: Huybrechts et al., this volume).

In this contribution we present results from a study in which the Rhône Glacier was studied with a numerical ice flow model. The Rhône glacier has one of the best historic records. Front positions are known from the beginning of the 17th century (e.g. Aubert, 1980). The glacier has a wide accumulation basin and a narrow tongue, and according to the classification of Furbish and Andrews (1984) it should react in a sensitive way to climatic change. Several investigations have further been carried out since Mercanton (1916) started his famous surface velocity measurements in 1874 (see Jost, 1936; Müller at al., 1980; Wächter, 1983).

The Rhône Glacier is situated in the upper north-eastern part of Wallis (Switzerland). Figure 1 gives an impression of the glacier and illustrates the recent retreat. Exposure is to the south and the total glacierized area measures 17.7 km^2. During the last four centuries the length of the glacier varied between 10 and 12 km (see Fig. 2). The little ice age maximum was reached in 1602. From then to the beginning of the 19th century the glacier has been retreating slowly. Between 1800 and 1856 a new and fast advance occurred; after that time the glacier retreated continuously.

The glacier model employed here is a simple mass continuity model, which does take into account varying geometry (width of the bed, slope of the valley walls) along a central flowline. It is similar to the model used by Oerlemans(1986) in a study of Nigardsbreen, and therefore described here only briefly. Forcing is formulated in terms of variations in equilibrium-line altitude E, with fixed mass-balance gradients. E is related to climatic series like summer temperature, winter precipitation and tree-ring width. It turns out that simulation of the front variations is a difficult matter. A reasonable match is obtained for the period prior to 1850. However, the retreat after that time cannot be reproduced without additional forcing.

2. BRIEF DESCRIPTION OF THE MODEL

The model is one-dimensional (flowline along x-axis) and time dependent. It is assumed that the ice velocity is determined by the local driving stress (τ) only. The cross profile is of a trapezoidal shape, i.e. has two degrees of freedom (see Fig. 3). The relevant parameters are valley width at the base $W(x)$, bedrock height $b(x)$ and steepness of valley walls $\gamma(x)$. Values of $\gamma(x)$ for the two sides are averaged, so the model glacier is symmetric with respect to the x-axis. Values of W', valley width at the surface, and γ were taken from topographic maps in such a way that the area/elevation ratio is distorted as less as possible. This means for instance that, in assigning a glacier width to a particular gridpoint, not the width perpendicular to the flowline was taken, but rather the length of the contour line passing through the gridpoint.

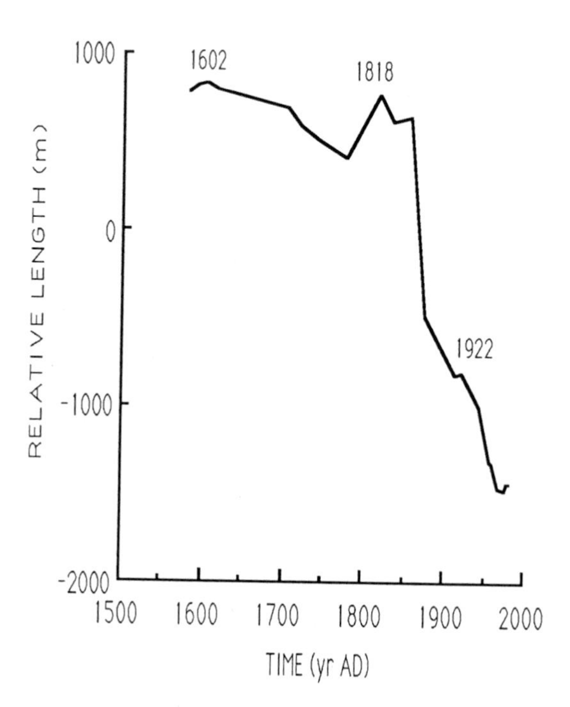

Figure 2. Front positions of the Rhône Glacier since 1602 (when the Neoglacial maximum was reached). From Aubert (1980).

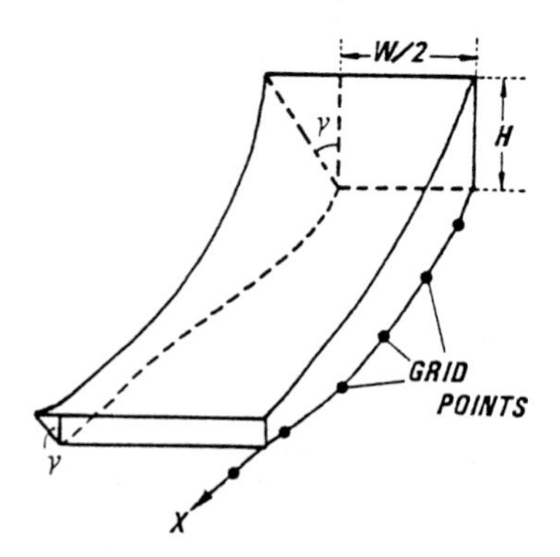

Figure 3. Geometry for the glacier model.

The model equations are:

continuity: $$\frac{\partial S}{\partial t} = - \frac{\partial (VS)}{\partial x} + MW' \qquad (1)$$

geometry: $$S = H\ (H \tan\gamma + w) \qquad (2)$$
$$W' = W + 2H \tan\gamma$$

ice velocity: $$V = f\ H\tau^3 + f^* \tau^3/N \qquad (3)$$

driving stress: $$\tau = - \rho g H \frac{\partial h}{\partial x} \qquad (4)$$

mass balance: $$M = \min \lfloor a\ (h-E);\ M^* \rfloor \qquad (5)$$

external forcing: $$E = C_1 (F(E) - C_2) \qquad (6)$$

In Eq. (1), S is the cross sectional area, t time, V the vertical mean ice velocity,and M the mass balance. Inserting the geometric relations in Eq. (2), a rate equation for ice thickness H is easily obtained. The vertical mean ice velocity V consists of two parts: one associated with internal deformation (first term) and one with sliding (second term). They both increase with the third power of the driving stress τ, while sliding is also inversely proportional to normal load N (overburden ice pressure minus basal water pressure; however, variations in water pressure are not considered in this study). For a general review on the description of vertically-integrated ice flow see e.g. Budd and Jenssen (1975), Paterson (1981), Oerlemans and Van der Veen (1984). In the expression for the driving stress, Eq. (4), ρ is ice density, g gravitational acceleration, and h surface elevation (i.e. b + H).

The flow parameters f and f* are set to $6 \times 10^{-25}\ m^6\ N^{-3}\ S^{-1}$ and $9.5 \times 10^{-17}\ m^5\ N^{-2}\ S^{-1}$, respectively. These values are just outside the range for flow parameters given in the literature (Paterson, 1981; Budd et al., 1979; Hooke, 1981). However, by using these values the predicted ice thickness and the field measurements match very well. No attempt was made to introduce a shape factor to correct for the drag of the valley walls, as has for instance been done by Bindschadler (1980) and Kruss (1984).

The mass balance M is linear in height relative to the equilibrium line altitude E, subject to an upper limit M*. The mass balance gradient is denoted by a. For the period 1885-1910, Mercanton (1916) obtained an ablation gradient of 0.008 yr^{-1} . More recently, Müller et al. (1980) found a value of only 0.003 yr^{-1} for the period August-September 1977. We decided to use a value of 0.006 yr^{-1}, together with M*=2.5 m/yr.

Finally the forcing is imposed to the model by moving the equilibrium line up and down. Equation (6) shows that two parameters are involved (F(t) is the forcing function). C_2 determines the mean value of E, while the range of variation is essentially determined by C_1 The general procedure we applied is to adjust the parameters in such a way that minimum and maximum glacier extent are well reproduced by the model. As will become clear later, this certainly does not constrain the model strongly: it is still difficult to obtain a good simulation of the historic front variations!

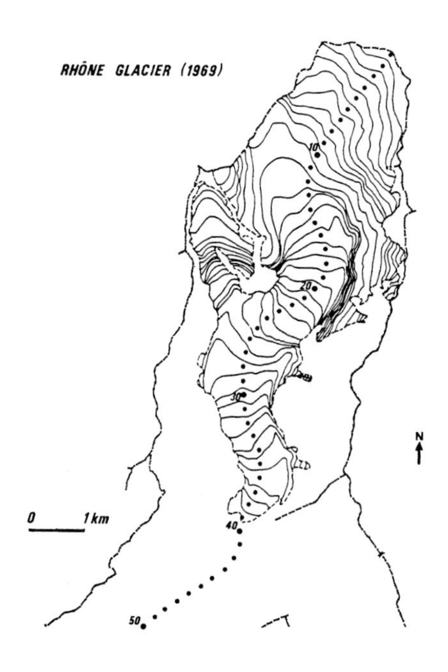

Figure 4. Map of the Rhône Glacier, with the applied grid point contour intervals. The grid points are shown as black dots along the central flow-line. Note: no height intervals are indicated.

The geometric input was obtained from a topographic map (1:25000, issued 1973, glacier stand 1969). Figure 4 shows the flowline, with gridpoints spaced at 250 m. The boundaries of the height intervals were determined by following isohypses midway to two adjacent gridpoints. This was done to retain the hypsometry as accurate as possible. Some data on the geometry are given in the Appendix. Present-day ice thickness, necessary to obtain the bed profile, is not known everywhere. The values used are partly based on Jost (1936) and on Wächter (1983). This concerns the lower part of the glacier. For the upper part an estimate of the bed slope was made by assuming constant basal shear stress, i.e. a constant value of surface slope and ice thickness averaged over ten times a typical ice thickness. The resulting bed profile is obvious from Fig. 5a (next section).

The angle of the side walls was also estimated from the topographic map. Values for both side were averaged; $\tan\gamma$ turned out to be in the range 0 - 1.9.

3. SOME BASIC CALCULATIONS

In a first experiment stationary states were calculated for many values of the equilibrium line altitude E (steps of 25 m in the range of interest). It turned out that for E = 2875 m the calculated glacier

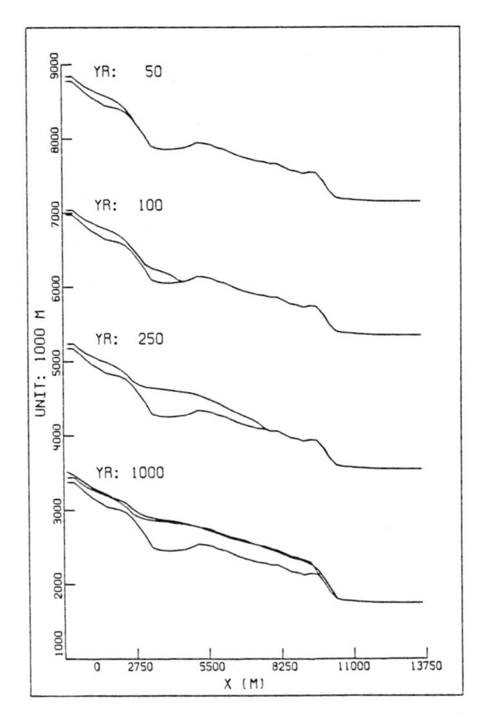

Figure 5a. Evolution of the model glacier in an experiment with no ice initially and a constant equilibrium line altitude of 2875 m. In the lower part the 1969-observed profile is also shown.

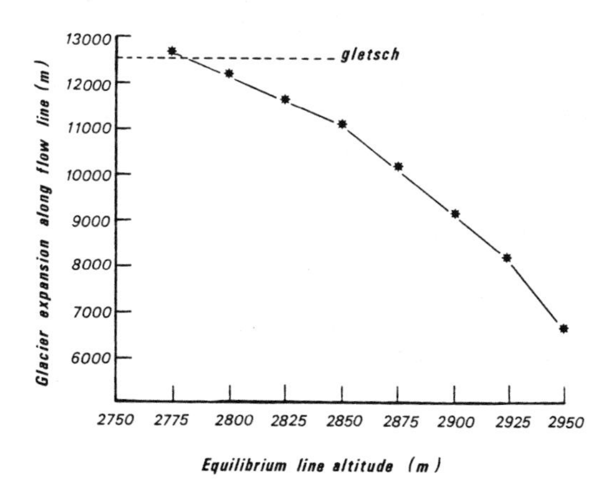

Figure 5b. Calculated steady-state length of the Rhône Glacier as a function of the equilibrium line altitude.

length and ice thickness distribution correspond fairly well to the shape of the Rhône glacier in 1969 (Fig. 5a). Although one cannot be sure that the observed 1969 profile was close to equilibrium, it can be concluded that the model gives a satisfactory description of the glacier. Figure 5b shows how the equilibrium length depends on E. For a 25 m change in E, the glacier reacts by changing its length between 500 and 1000 m. The larger values occur in warmer climates (E larger than 2875 m, say). This is mainly a geometric effect. It also appeared that the extreme glacier stands (in 1602 and 1969), when considered as steady states, correspond to a difference of 75 m in equilibrium-line altitude.

Depending on E, it takes 250 to 750 yr before a steady state is reached (starting with zero ice volume). Although this growth time is certainly larger than the relaxation time, the result nevertheless suggests that a simulation of the historic record should start with a carefully chosen initial condition, or preferably, with an initial time well before the beginning of the observed record. The (90%) response time was investigated more thoroughly by perturbing all calculated equilibrium states by a 25 m change in E. Response times than generally vary between 200 and 300 yrs. However, when the glacier tongue is on the valley floor (at $x > 11$ km), the response time is shorter, namely about 120 yr. This situation arises when E decreases from 2850 to 2825 m, and from 2825 to 2800 m and reversed.

We have compared measured surface velocities with calculated values. For this purpose the model was run to the 1890-profile (by adjusting E), because Mercanton's (1916) measurements are for the period 1885-1910. The comparison is made in Table I. The model values have been multiplied by 1.5 to take into account the fact that model velocities are vertically averaged.

Table I. A comparison of the surface velocities calculated by the model and the measurements of Mercanton for different grid points.

i	V calculated (m/y)	V Mercanton (m/y)
20-38	65	100
39-42	225	250
43-47	38	25

Calculated velocities appear to differ by some tens of percent from measured values. This is probably due to the non-equilibrium state of the model glacier, and may also be related to underestimation of the sliding component in the model. Also, the velocities given by Mercanton refer to the middle of the glacier, while model values represent average velocities over the width of the glacier. Nevertheless, the model seems to give a good overall description of the Rhone glacier.

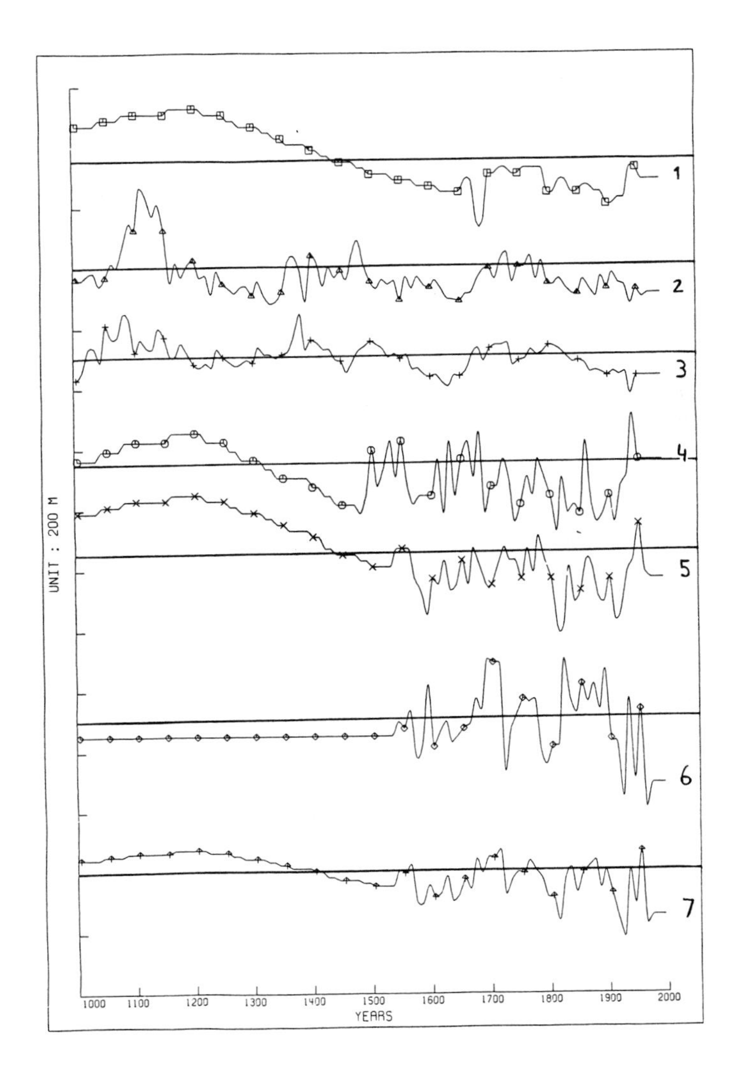

Figure 6. Forcing functions for the simulation of historical front variations, in terms of variations in E for optimized coefficients (horizontal lines are the mean equilibrium line altitudes). In which: (1) Central England summer temperature series; (2) tree-ring width series from Trier; (3) tree-ring width series from Spessart; (4) extended 'Basler Temperaturreihe'; (5) Pfister's summer temperature series; (6) Pfister's winter precipitation series, and (7) a combination of (5) and (6).

4. SIMULATION OF FRONT VARIATIONS

To simulate a record of historic front variations, a proper forcing function is required. It should preferably consist of mass-balance data, but existing series are far too short. This also applies to meteorological parameters, although series are longer. So the only way is to use proxy data. Another point to note is that, due to the large response time of the glacier, the integration should start well before the observed record of front variations begins, preferably several centuries. All integrations discussed here start in 1000 AD with zero ice thickness.

Several different forcing series were applied, see Fig. 6. The first two test were done with series of tree-ring width from Germany (from Trier and Spessart; Lamb, 1977). It is generally known that temperature and soil moisture determine differences in tree-ring width, provided that the tree is a stress-sensitive species. According to Lamarche and Fritts (1971), annual ring width from trees that lived close to the timberline (i.e. in marginal conditions) are related to glacier fluctuations. Although the series mentioned above are from trees from a lower altitude, we nevertheless made an attempt to use them as forcing function. The attempt is not succesful, as is obvious from Fig. 7a. There is no correlation between simulated and observed front positions. In particular the result, that the Rhône glacier should have its maximum extent right now, is curious. We believe that in this case the correlation between tree ring width and glacier mass balance is too weak. This is somewhat in contrast with the findings in Oerlemans (1986), where a simulation of front variations of Nigardsbreen (Norway) works best with a tree ring series from northern Sweden. However, in this region conditions are harder, and trees are much more sensitive to summer temperature.

In a third experiment 10-year mean values of central England summer temperature from Lamb (1977) were used. The result is shown in Fig. 7b. Again, there is not much similarity between observations and calculations. The same was found for Nigardsbreen (Oerlemans, 1986). We have serious doubts about the representativity of the central England temperature for central and northern Europe. The 'Basler Temperaturreihe' (Bider et al., 1959; Burkhardt and Hense, 1985; here we used 10-yr mean values) does only a slightly better job (Fig. 7c). It should be noted that the Basler Temperaturreihe was extended back to 1484 by using French wine harvest data. For the period 1000-1484 we matched it with the central England series. According to Fig. 7c, the result is reasonable for 1600-1830, but poor afterwards. A direct comparison of the two temperature records shows that the little ice-age cooling appeared somewhat earlier on the continent than in England.

Finally, runs were carried out with series based on Pfister's (1984) work, using his seasonal values of precipitation and temperature (taken to be equally important) for Switzerland from 1525. After some experimentation it turned out that a combination of summer temperature and winter precipitation works best. The corresponding result is shown in Fig. 7d. It is reasonable in the sense that the maxima of both 1602 and 1850 show up in the calculations. However, the predicted trend in the 20th century is opposite again to the observation.

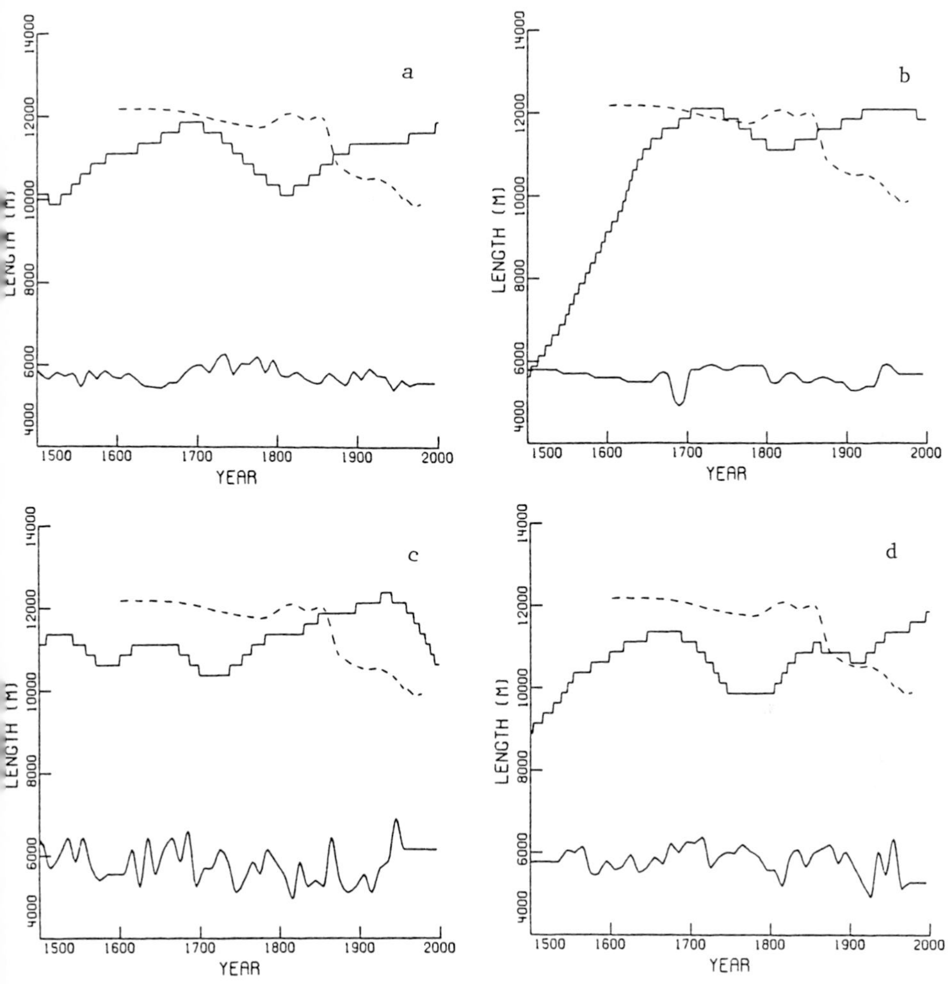

Figure 7. Simulation of historical front variations. In all cases, the lower curve represents the forcing function (arbitrary scale). The observed front positions are shown by the dashed line. Forcing functions: (a) tree-ring width from Trier/Spessart; (b) Central England summer temperature; (c) 'Basler Temperaturreihe', and (d) precipitation and temperature as reconstructed for Switzerland by Pfister.

5. DISCUSSION

An advance of the model glacier in the 20th century is common to all simulations and a puzzling result. We believe that it is not related to basic shortcomings of the numerical model, but rather to inadequate formulation of the mass balance. It seems that the Rhône glacier (and other glaciers in the Alps) rather follows the mean Northern Hemisphere temperature trend than the local climatological conditions. One reason could be that the mass balance of a glacier reacts more directly to radiation variations (total melt is roughly proportional to the summer radiation budget) than local meteorological variables like air temperature and precipitation. So when climatological trends during the last centuries were initiated by global radiative changes, and indications for this exist (e.g. Gilliland, 1982), our results would be more understandable.

To see how large the change in mass balance has to be to simulate the recent retreat of the Rhône glacier, some additional calculation were done. A fairly good result is obtained when a sudden 85 m (!) increase in equilibrium-line elevation is imposed from 1840 onwards (Fig. 8). In this case the same forcing as in Fig. 7d was used. However, such a perturbation of the mass balance is dramatically large. It seems that we are not going to understand the historic glacier variations without a renewed deep and careful investigation of how the mass balance of glaciers is related to climatic conditions.

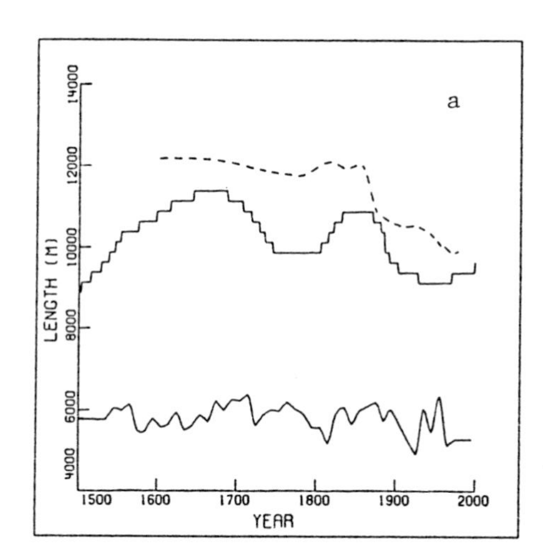

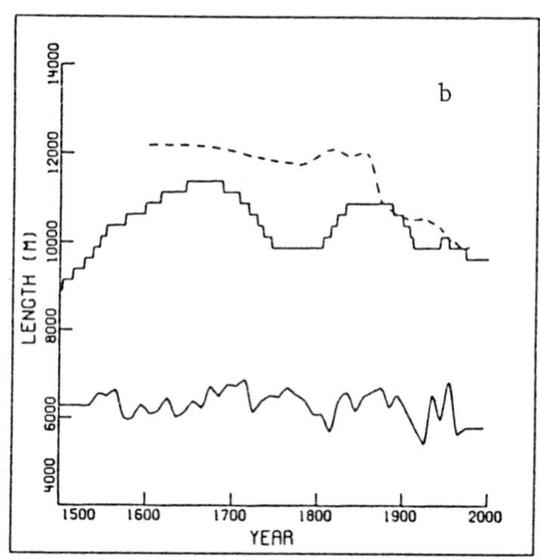

Figure 8a. As in Fig. 7d, but with an instanteneously imposed 85 m increase in E from 1840 onwards. Figure 8b. As in Fig. 7d, but with a linear shift of 1 m increase in E per year from 1840 onwards.

APPENDIX

The following values were used for the height of the surface h_s, height of the bedrock b(x), ice thickness H, steepness of the valley walls tanγ, and valley width at the surface.

i	h_s (masl)	b(x) (masl)	H (m)	tanγ	W' (km)
1	3520	3470	50	0	1.40
2	3480	3380	100	0	1.68
3	3420	3310	110	0	1.92
4	3360	3230	130	0	2.40
5	3300	3160	140	0	2.64
6	3260	3110	150	0	2.84
7	3220	3050	170	0	2.52
8	3170	3030	140	0	2.04
9	3140	3010	130	0	2.56
10	3100	2970	130	0	4.80
11	3020	2890	130	0	6.28
12	2960	2770	190	0	5.32
13	2920	2650	270	0	2.08
14	2900	2500	400	0	1.96
15	2880	2470	410	0	1.76
16	2870	2460	410	0	1.72
17	2860	2460	400	0	1.12
18	2845	2470	375	0	1.24
19	2825	2480	345	0.2	1.52
20	2800	2510	290	0.4	1.32
21	2780	2550	230	0.5	0.88
22	2765	2540	225	0.7	0.92
23	2740	2520	220	0.9	1.52
24	2700	2480	220	1.1	1.16
25	2670	2460	210	1.2	1.24
26	2640	2420	220	1.4	1.96
27	2620	2380	240	1.6	1.88
28	2600	2350	250	1.7	2.12
29	2560	2325	235	1.7	1.40
30	2535	2300	235	1.6	0.80
31	2520	2290	230	1.6	0.76
32	2490	2265	225	1.9	0.96
33	2460	2265	195	1.8	1.04
34	2420	2220	200	1.7	0.84
35	2380	2175	205	1.6	1.20
36	2360	2160	200	1.6	1.32
37	2330	2130	200	1.2	0.56
38	2300	2150	150	0.8	0.52
39	2170	2140	30	0.4	0.48
40	2035	2035	0	0	
41	1900	1900		0	
42	1810	1810		0.9	

i	h_s (masl)	b(x) (masl)	tanγ
43	1790	1790	1.8
44	1785	1785	1.7
45	1775	1775	1.6
46	1770	1770	1.5
47	1765	1765	1.5
48	1760	1760	1.5
49	1760	1760	1.5
50	1760	1760	1.5

REFERENCES

Aellen, M. 1981. Neuzeitliche Gletscher veränderungen, in Die Schweiz und ihre Gletscher. Kümmerly en Frey, Bern, 70-90.

Aubert, D. 1980. Les stades de retrait du Haut-Valais, Bull. Murithienne 97.

Bider, M., Schüepp, M. and Von Rudloff, H. 1959. Die Reduktion der 200-jährige Basler Temperaturreihe. Arch. Met. Geoph. und Biocl., serie 6d9, 360-412.

Bindschadler, R. 1980. The predicted behaviour of Griesgletscher, Wallis, Switzerland, and its possible threat to a nearby dam. Zeitschrift für Gletscherkunde und Glazialgeologie bd. 16, (1), 45-59.

Budd, W.F. and Jenssen, D. 1975. Numerical modelling of glacier systems. IAHS 104, 257-291.

Budd, W.F., Keage, P.L. and Blundy, N.A. 1979. Empirical studies of ice sliding. J. Glaciol. 23 (89), 157-170.

Burkhardt, Th. and Hense, A. 1985. On the reconstruction of temperature records from proxy data in Mid-Europe. Arch. Met. Geoph. Biocl. Ser. B, 35 (4), 341-359.

Furbish, D.J. and Andrews, J.T. 1984. The use of hypsometry to indicate long-term stability and response of valley glaciers to changes in mass-transfer. J. Glaciol. 30 (105). 199-211.

Gilliland, R.G. 1982. Solar volcanic and CO_2 forcing of recent climatic changes. Climatic Change 4, 111-131.

Hooke, R. LeB. 1981. Flow law for polycristalline ice in glaciers: comparison of theoretical predictions, laboratory data and field measurements. Reviews of Geophysics and Space Physics 19 (4-81), 664-672.

Huybrechts Ph., de Nooze, P. and Decleir, H. 1988. Numerical modelling of Glacier d'Argentière and its historic front variations. This volume.

Jost, W. 1936. Die seismischen Eisdickenmessungen am Rhônegletscher 1931. Denkschriften der SNG 71 (2), 26-42.

Kruss, P.D. 1984. Terminus response of Lewis glacier, Mount-Kenya, Kenya, to sinusoïdal net-balance forcing. J. Glaciol. 30 (105), 212-217.

Lamarche, J.R.V.C. and Fritts, H.C. 1971. Tree rings, glacial advance, and climate in the Alps. Zeitschrift für Gletscherkunde und Glazialgeologie 7, (1-2), 125-131.

Lamb, H.H. 1977. Climate, Present, Past and Future. Climatic history and the future, vol. 2, Methuen, London.

Mercanton, P.L. 1916. Vermessungen am Rhônegletscher 1874-1915. Neue Denkschriften der SNG 52, 190 pp.

Müller, R., Ohmura, A, Schroff, K., Funk, M., Pfirter, K., Bernath, A. and Steffen. K. 1980. Combined ice water and energy balances of a glacierized basin of the Swiss Alps - the Rhônegletscher project. In: F. Müller (Ed.) Geography in Switzerland, 57-69.

Oerlemans, J. 1986. An attempt to simulate historic glacier front variations of Nigardsbreen, Norway. Theor. and Appl. Climatology 37, 126-135.

Oerlemans, J. and Van der Veen, C.J. 1984. Ice Sheets and Climate, Reidel, Dordrecht, 217 pp.

Paterson, W.S.B. 1981. The physics of glacier, 2nd ed., Pergamon Press, Oxford, 380 pp.

Pfister, C. 1984. Klimageschichte der Schweiz 1525-1860. Academica Helvetica 6 (1), 184 pp.

Wächter, H.P. 1983. Eisdickenmessungen auf Alpengletscher. In: Das Gelbe Heft, Zürich, 21-22.

THE RESPONSE OF THE EQUILIBRIUM LINE ALTITUDE TO CLIMATE FLUCTUATIONS: THEORY AND OBSERVATIONS

M. Kuhn
Institute of Meteorology and Geophysics
University of Innsbruck
Austria

ABSTRACT

With the aim of establishing a quantitative relation between equilibrium line altitude and fluctuation of climatic elements, this paper first formulates mass balance and energy balance components in a compatible form. The model is designed for seasonal averages, uses air temperature, humidity and net radiation as independent variables and needs specified turbulent transfer coefficients in order to calculate the melt rate.

Given cumulative accumulation at the equilibrium line as a further independent variable the model is used to predict the adjustment of the equilibrium line altitude following a small disturbance in one of the variables. The altitudinal mass and energy balance gradients needed for that purpose are taken from alpine observations. Model predictions are compared with 30 years observations at Hintereisferner.

1. INTRODUCTION

Glacier length, mean specific mass balance and equilibrium line altitude are three parameters frequently used to characterize the state of a glacier. Their interannual variations or deviations from a long term mean value may be interpreted as consequences of climatic variations.

This kind of interpretation is complicated by the dynamics of ice flow. Unless the climate signal lasts long compared with the reaction time of a glacier, its length will be in a transient state. Even the mean specific mass balance, although determined on an annual basis, may contain a small transient effect, since it is averaged over the entire glacier surface the lowest part of which may not be in equilibrium with present climatic conditions.

Thus, the mass balance/altitude profile or the altitude of the equilibrium line are the parameters most suitable to interprete the glacier's response to interannual or short-term climatic fluctuations.

In the folowing, the mass balance is formulated in terms of

J. Oerlemans (ed.), Glacier Fluctuations and Climatic Change, 407–417.

seasonal values of accumulation, radiation balance, temperature and humidity of the air and surface, using the respective altitude derivatives in order to describe the establishment of a new equilibrium line altitude following minor disturbances of any of the variables enumerated. The results of such computations are then compared to the 30-year series of climatic and glaciologic data at Hintereisferner.

2. FORMULATION OF THE MASS AND ENERGY BALANCE

If $\dot{b}$ denotes the rate of change of mass per unit horizontal surface of a glacier, the specific net accumulation may be expressed as

$$c = \int_0^{\tau'} \dot{b}\, dt \qquad [\mathrm{kg\ m^{-2}}] \tag{1}$$

where τ' is the duration of the accumulation period in days. Similarly, specific net ablation is

$$a = -\int_{\tau'}^{\tau'+\tau} \dot{b}\, dt \tag{2}$$

where τ is the duration of the ablation period and $\tau'+\tau$ is the balance year. Thus the specific net balance is

$$b = c - a \tag{3}$$

We shall assume that in any year c is directly proportional to precipitation in the period $0 - \tau'$, which implies that there is no interannual change of wind drift.

Let us further assume that at the equilibrium line all accumulation melts first, thereby consuming the specific energy

$$-Q_M = \dot{m}\, L_M, \text{ or}$$
$$-\tau\overline{Q_M} = m\, L_M = c\, L_M \tag{4}$$

where $\dot{m}$ is the melt rate in kg m^{-2} d^{-1}, and L_M = 0.33 MJ kg^{-1}.
This assumption implies that there is no direct sublimation from the solid phase but only evaporation from the liquid phase, and that ablation by snow drift is excluded.
The meltwater has then several options:

1. It runs off, which needs no further comment.
2. It evaporates, which is taken into account by the latent heat flux

$$Q_L = -\dot{e}\, L_v \tag{5}$$

   (where L_v = 2.5 MJ kg^{-1}) but does not further affect mass balance. Note, however, that each kg of water that evaporates prevents L_v/L_M = 7.5 kg of snow from being melted.
3. It percolates and refreezes in the cold snow pack, thereby transporting latent heat of melting, which is used for heating the

snow below the surface. The heat flux density due to refreezing is smaller than the average total heat flux density $\overline{Q_S}$ in the snow pack which is related to the snow temperature T_m at the beginning of the ablation season:

$$\overline{Q_S} = -c_1 \int_Z^0 \rho_s \frac{\partial T_s}{t} dz = \frac{c_1}{\tau} \int_Z^0 \left[(\rho_s T_s)_{t=0} - (\rho_s T_s)_\tau\right] dz \quad (6)$$

where c_1 is the specific heat of ice ($\approx$ 2000 J kg^{-1} K^{-1}), ρ_m is snow density and Z the depth to which temperature is observed to change. Generaly, Q_S in a porous snow pack will contain contributions due to latent heat of melting, evaporation, radiation, sensible heat exchange and genuine conduction.

4. Finally, meltwater may be stored in the snow matrix without freezing during the ablation season, but may freeze in the following winter.

The meltwater balance then is

$$m = r + e + f + s \qquad (kg\ m^{-2}) \quad (7)$$

where the right hand terms refer to the four paragraphs above. The sum of f + s is called internal accumulation.
For the computation of the energy balance we set m = a, express e by Q_L (eq. 5) and include f and s in Q_S (eq. 6), so that there is no need to know the amount of refreezing explicitly. The energy balance at the equilibrium line can then be initialized by values of C and s T_S dz at t = 0. An alternative approach was used by Ambach (1985) who included all thermal effects of superimposed ice by multiplying c by a factor that is specific to a given location, e.g. 5/3 at the equilibrium line in the EGIG-profile in Greenland.

For mass balance calculations, however, f and s must be known explicitly. They can reach significant proportions of net accumulation in Alaskan glaciers (Trabant and Mayo, 1985), are of the order of 1% of c on the partly cold tongue of Storglaciären (Östling and Hooke, 1986) and are likely to be negligible on temperate glaciers.

The surface energy balance can be expressed as

$$Q_R + Q_H + Q_L + Q_S + Q_M = 0 \quad (8)$$

where Q_H and Q_L are the turbulent sensible and latent heat flux densities, respectively, and Q_R is net radiation. The latter reads

$$Q_R = (1 - a)\ G + E\downarrow - E\uparrow \quad (9)$$

where a is albedo, G global irradiance, E↓ longwave irradiance and E↑ longwave emittance.

The signs of the terms in Eq. (8) are positive if the flux is directed toward the surface, melting implies a negative value of Q_M.

The following of these variables can be formulated in a general way:

$$E\uparrow = \sigma T_o^4 \tag{10}$$

with $\sigma = 5.67 \cdot 10^{-8}$ W m^{-2} K^{-4}, T_o being the surface temperature. $E\uparrow$ = 315 Wm^{-2} for T_o= 0°C.

$$Q_H = \rho_a C_p r_H{}^{-1} (Ta - TO) = \alpha_H (Ta - TO) \tag{11}$$

where air density $\rho_a \approx 1$ kg m^{-3}, specific heat C_p = 1005 J kg^{-1} K^{-1}.

Values of the resistance to heat transfer r_H, or the bulk transfer coefficient α in eq. (11) have not often been published explicitly, but they can be related to the parameters of the logarithmic profiles in a straightforward manner (Kuhn, 1979). In neutral stratification

$$\alpha_H = \rho_a c_p r_H{}^{-1} = \rho_a c_p \kappa u_* / \ln(z/z_0) \tag{12}$$

Here u_* is the friction velocity, a direct indicator of the state of turbulence of the boundary layer:

$$u_* = \frac{\partial u}{\partial z} \kappa z \tag{13}$$

z_0 is the roughness parameter which increases with physical roughness of the surface, but cannot be quantitatively predicted from inspection of the surface. It is obtained from

$$z_0 = z \exp(- \kappa u(z)/u_*) \tag{14}$$

The value of α_H thus depends on u_*, z_0 and the reference height z which generally is taken as z = 2 m for practical reasons.

Table II in Kuhn (1979) summarizes wind profile measurments on various glaciers. If these are evaluated according to eq. (12) to (14) one finds

$$0.09 < u_* < 0.46 \ ms^{-1}$$
$$1.8 < z_0 < 9.0 \ mm$$
$$32 < r_H < 170 \ sm^{-1}$$
$$6 < \alpha_H < 31 \ Wm^{-2} K^{-1}$$
$$0.5 < \alpha_H < 2.7 \ MJ\ m^{-2} d^{-1} K^{-1}$$

where the individual extremes do not pertain to identical cases. For α_H, Tanzer (1987) found a long term average of 0.5 MJ $m^{-2}d^{-1}K^{-1}$ at Hintereisferner, Funk (1985) found 0.9 - 1.5 MJ m^{-2} d^{-1} K^{-1} for weekly averages at the Rhône glacier, Schug (1987) evaluated α_H = 0.8 MJ m^{-2} d^{-1} K^{-1} for a three day period at Schwarzmilzferner. These results differ by a factor of 3 which means a relative uncertainty of turbulent fluxes comparable to an albedo range from 0.25 to 0.75 (= 0.75 to 0.25 absorbed).

Very similar considerations apply to the turbulent transfer of latent heat

$$Q_L = -\dot{e} L_v = L_v r_v{}^{-1} (\rho_{va} - \rho_{vo}) = \alpha_v (\rho_{va} - \rho_{vo}) \tag{15}$$

where $\dot{e}$ is the rate of evaporation, $L_V = 2.5$ MJ kg^{-1}, $r_V \approx r_H$, ρ_{va} and ρ_{vo} are water vapor density of the air and at the surface, respectively, and α_V is the bulk transfer coefficient for latent heat.

From data published by Markl and Wagner (1977) and similar measurements on alpine glaciers it appears that τS (eq. (6)) is typically of the order of -10 MJ M^{-2}, equivalent to 30 kg m^{-2} ablation, so that it will be neglected in the following discussions.

Equation (8) is now rewritten as:

$$(1 - a)\ G + E\downarrow - \sigma T_o{}^4 + \alpha_H\ (T_a - T_o) + \\ + \alpha_v\ (\rho_{va} - \rho_{vo}) + Q_M = 0 \qquad (16)$$

or

$$Q_R + \alpha_H\ \Delta T + \alpha_v\ \Delta\rho_v + Q_M = 0. \qquad (17)$$

In Eqs. (16) and (17) Q_M is the dependent variable. There are truly independent variables like G, $E\downarrow$, ρT_a, ρ_{va}; variables that have a limited range like a, T_o and ρ_{vo}, and the transfer coefficients which increase with windspeed or surface roughness.

3. THE BALANCE AT THE EQUILIBRIUM LINE

By definition, at the equilibrium line accumulation equals ablation

$$b = c - a = 0 \qquad \text{and } c = a = m \qquad (18)$$

In order to establish a link between climate and mass balance, we shall treat c as an independent variable and express a in terms of the energy balance. For this purpose cumulative rather than net values of c have to be used, i.e. the winter balance plus summer snow falls.

From eq. (4) we express Q_M and insert into (8), again neglecting Q_S:

$$-\overline{Q_M} = L_M\ \tau^{-1}\ c = \overline{Q_R} + \overline{Q_H} + \overline{Q_L} \qquad (19)$$

In the following we shall drop time average bars and express variables as function of altitude z. So

$$L_M\ c(z) \approx \tau\ (z)\ [Q_R(z) + Q_H(z) + Q_L(z)] + L_M\ b(z)$$

or

$$L_M\ c(z) \approx \tau\ (z)\ [Q_R(z) + \alpha_H(T_a(z) - T_o(z)) + \\ + \alpha_v\ (\rho_{va}(z) - \rho_{vo}(z))] + L_M\ b(z) \qquad (20)$$

where the last term on the right side of eq. (20) makes it valid for z other than h, the equilibrium-line altitude.

4. RE-ESTABLISHED BALANCE AFTER A CLIMATIC DISTURBANCE

Equilibrium (b = 0) at an initial altitude h_1 can be disturbed by minor changes δT_a, δc, δQ_R and so on. These will lead to a new equilibrium at $h_2 = h_1 + \Delta h$ which depends on the altitude gradients of all variables in eq. 20. Note that $\partial/\partial z$ is the change with altitude and not the change with height above the surface. At the new equilibrium-line altitude, the variables will have the following values

$$c(h_2) = c(h_1) + \partial c/\partial z \; \Delta h + \delta c$$

$$\tau Q_R(h_2) = \tau Q_R(h_1) + \partial \tau Q_R/\partial z \; \Delta h + \delta \; \tau Q_R \quad (21)$$

etc.

Subtracting the old (h_1) from the new (h_2) equilibrium condition (Eqs. 20 and 21) leaves only terms containing Δh and δ. We obtain

$$L_M \tau^{-1} \; \partial c/\partial z \;\; \Delta h + L_M \tau^{-1} \; \delta c = \frac{\partial Q_R}{\partial z} \; \Delta h + \delta Q_R + $$

$$+ \alpha_H \frac{\partial T_a}{\partial z} \Delta h + \alpha_H \; \delta T_a - \alpha_H \frac{\partial T_0}{\partial z} \Delta h - \alpha_H \delta T_0 + \quad (22)$$

$$+ \alpha_v \frac{\partial \rho_{va}}{\partial z} \partial h + \alpha_v \; \delta \rho_{va} - $$

$$- \alpha_v \frac{\partial \rho_{v0}}{\partial z} \Delta h - \alpha_v \; \delta \rho_{0v}$$

Denoting the individual terms by Q_i we find a solution for Δh :

$$\Delta h = - \Sigma \delta Q_i / \Sigma \; (\partial Q_i / \partial z) \quad (23)$$

A number of implications of this algorithm need to be spelled out:

1. All disturbances are independent of altitude (Lliboutry's linear balance assumption, 1974).
2. $\partial Q_i/\delta z$ is independent of altitude and of time, i.e. the gradients do not change with changing climate conditions (see, however, the real case in Kuhn 1984).
3. Possible feedbacks are not included in these computations.
4. While the δQ_i can act collectively or individually, the altitude adjustment is always effected by the sum of all $\partial Q_i/\partial z$.

5. NUMERICAL VALUES

From the investigations of the Meteorological Institute of Innsbruck on Hintereisferner, the following typical values are found at the mean equilibrium line altitude (h = 2960 m). Values for Q_i are extrapolated from Wagner (1979, 1980) and Tanzer (1987) who evaluated measurements of July - September 1971. Q_M was increased by 10% compared to ablation measurements in order to close the balance. Q_M is equivalent to the cumulative ablation at the measuring site which was below equilibrium line altitude in 1971. Values of $\partial Q_i/\partial z$ have been evaluated from various investigations in the altitude range from 1900 to 3000 m.

Table 1

τ = 120 d (June-Sept)		
L_M = 0.33 MJ kg^{-1}		
Cumulative ablation		
a_c = 2700 kg m^{-2}	$\overline{\overline{Q}}_m$ = -7.5 MJ m^{-2}d^{-1}	$\partial Q_M/\partial z$ = -2.8 kJ m^{-2}d^{-1}m^{-1}
$\partial c/\partial z$ = 1 kg m^{-2}m^{-1}	$\overline{\overline{Q}}_R$ = 5.8	$\partial Q_R/\partial z \simeq 0$ (see text)
α_H = 5.8 W m^{-2}K^{-1}		
= 0.5 MJ m^{-2}d^{-1}K^{-1}		
ρ_a = 1 kg m^{-3}		
c_p = 1005 J kg^{-1}		
$r_H = r_v$ = 173 sm^{-1}		
$\overline{\overline{T}}_a$ = 1.9°C		
$\overline{\overline{T}}_0$ = -1.8°C	$\overline{\overline{Q}}_H$ = 1.9	
$\partial T_a/\partial z$ = -0.006 K m^{-1}		$\partial Q_H/\partial z$ = -3.0
$\partial T_0/\partial z \simeq 0$		
L_v = 2.5 MJ kg^{-1}		
α_v = 14.5 . 10^3 WM^{-2} (kg m^{-3})$^{-1}$		
= 1.25 . 10^3 MJ m^{-2}d^{-1} (kg m^{-3})$^{-1}$		
$\overline{\overline{\overline{\rho}}}_{va}$ = 4.2 . 10^{-3} kg m^{-3}		
$\overline{\overline{\overline{\rho}}}_{v0}$ = 4.3	$\overline{\overline{Q}}_L$ = -0.1	
$\partial \rho_{va}/\partial z$ = -1.5 . 10^{-6} kg m^{-3}m^{-1}		
$\partial \rho_{v0}/\partial z \approx 0$		$\partial Q_L/\partial z$ = -1.9
	$\overline{\overline{Q}}_S$ = -0.1	$\partial Q_S/\partial z \simeq 0$
	ΣQ_i = 0	$\Sigma(\partial Q_i/\partial z)$ =-7.7 kJ m^{-2}d^{-1}m^{-1}

In table 1 the change of radiation balance with altitude was assumed to be approximately zero. Inspecting the profiles of individual components of the radiation balance, however, one finds an increase in global irradiance and a decrease in downward longwave irradiance which nearly compensate each other, a minor decrease of longwave emittance and significant change of albedo which may range from as little as 0.15 on the debris-covered tongue to as much as 0.80 with fresh snow in the upper reaches of a glacier. On a clear summer day with G = 25 MJ $m^{-2}d^{-1}$, the absorbed fraction (1-a) G then may vary from 21 MJ $m^{-2}d^{-1}$ (2500 m) on the tongue to 5 MJ $m^{-2}d^{-1}$ (3500 m) in the highest parts. This is indeed a strong gradient, equal to about -1.6 MJ $m^{-2}d^{-1}$ $(100\ m)^{-1}$ and thus of comparable magnitude to the effect of temperature gradients ($\alpha_H\ \partial T_a/\partial z$) and yet it was not included in the computation of the equilibrium line shift. The reason for neglecting it is obvious: on the average, the albedo at the equilibrium line will remain the same as it is determined by the snow/ice transition, which is widely independent of the altitude at which it occurs. In other words, the albedo pattern (just as the surface temperature pattern) moves with the transient snow line and thus with the equilibrium line.

Using the values of table 1, equ. (23) can be solved for individual disturbances each acting alone. This yields

$\delta Q_R = \pm 1$ MJ $m^{-2}d^{-1}$ $\qquad \Delta h = \pm 130$ m

$\delta T_a = \pm 1$ °C $\qquad \Delta h = \pm 65$ m

$\delta \rho_{va} = \pm 1$ g m^{-3} $\qquad \Delta h = \pm 160$ m

$\delta c = \pm 1$ kg m^{-2} $\qquad \Delta h = \pm 0.35$ m

In order to compare these effects to one another the respective standard deviations are inserted as disturbances.

$\delta c = \sigma_c = \pm 540$ kg m^{-2} $\qquad \Delta h = \pm 190$ m

$\delta T_a = \sigma_t = \pm 0.6$ K $\qquad \Delta h = \pm 40$ m

$\delta \rho_{va} = \sigma_\rho = \pm 0.25$ g m^{-3} $\qquad \Delta h = \pm 40$ m

($\delta Q_R = \sigma_R \approx \pm 1$ MJ $m^{-2}d^{-1}$ $\qquad \Delta h \approx \pm 130$ m)

where σ_R is an estimate. The observed (30 year) standard deviation of h is ±120 m. Note that these values differ from earlier calculations (Kuhn, 1981).

5. POSSIBLE RELATION AMONG THE 'INDEPENDENT' VARIABLES

On the scale of an alpine glacier there is relatively little true feedback in the sense of an atmosphere-surface-atmosphere reaction. What we want to consider here is the probability of the various disturbances occurring simultaneously. Recalling that the energy balance at the equilibrium line

$$(1-a)G + E\downarrow - E\uparrow + \alpha_H (T_a - T_0) + \alpha_v(\rho_{va} - \rho_{v0}) + \tau^{-1} L_M c \approx 0 \quad (24)$$

and keeping T_0 constant we find a negative feedback in the turbulent transfer of sensible heat since an increase δT_a also increases stability and thereby decrease energy transfer.

There is an obvious interrelation between T_0 and $E\uparrow = \sigma . T_0^4$, and a less obvious one between T_a and $E\downarrow$. If we hypothesize that 70% of $E\downarrow$ is determined by the temperature of the lowest atmosphere T_a, then any δT_a leads to a simultaneous $\delta E\downarrow = 0{,}7 \cdot 4\, \sigma T_a^3 \delta T_a = 3.2\ \delta T_a\ (\mathrm{Wm}^{-2})$. On the other hand, $\delta T_a < 0$ means increased summer snowfall which leads to a disturbance $\delta Q_R < 0$ due to increased albedo. There is an empirical relation between global radiation and cloudiness, which compensates decreasing G by increasing $E\downarrow$. Finally, an increase of vapor density ρ_{va} with temperature T_a has been found which is of the order of $1\ \mathrm{gm}^{-3}\ \mathrm{K}^{-1}$.

7. TEST WITH THE 30-YEAR SERIES OF HINTEREISFERNER

The denominator of eq. (23) determines the sensitivity of the equilibrium line adjustment to climatic disturbances δQ_i. When $\Sigma(\partial Q_i/\partial z)$ goes to zero the equilibrium line is outside of the glacier and the sign of δQ_i determines whether the glacier completely becomes ablation or accumulation area, respectively. A large denominator dampens the reaction of Δh.

In the determination of $\Sigma(\partial Q_i/\partial z)$, the most uncertain quantities are α_H and $\partial c/\partial z$. Table 2 presents the effect of changing these quantities in a conceivable range.

Table 2: Values of $\Sigma(\partial Q_i/\partial z)$ for $\tau = 120$ d, $\partial T_a/\partial z = -0.6\mathrm{K\ km}^{-1}$, $\delta Q_{va}/\delta z = -1.5\ \mathrm{g\ m}^{-3}\mathrm{km}^{-1}$ and variable $\alpha_H(\alpha_V)$ and $\partial c/\partial z$

	α_H = 0.5	1.0	1.5	$\mathrm{MJ\ m}^{-2}\mathrm{d}^{-1}\mathrm{K}^{-1}$
	r_H = 172	86	58	$\mathrm{s\ m}^{-1}$
$\partial c/\partial z$ ($\mathrm{kg\ m}^{-2}\mathrm{m}^{-1}$)				
1.0	7.7	12.6	17.5	
2.0	10.7	15.4	20.3	

Since table 2 gives a possible range of $\Sigma(\partial Q_i/\partial z)$ from 7.7 to 20.3 one hopes to narrow this range by comparison of computed to observed values of Δh. The annual mass balance of Hintereisferner has been determined since 1952/53 by the direct method, which also furnishes the altitude of the equilibrium line. Simultaneous climatic records at the valley station Vent (1900 m) have been used to compute h, considering δT_a (1 May - 30 Sept.), $\delta(\rho_{va}/\bar{\rho}_{va})$ (1 May - 30 Sept.) and δc from winter precipitation plus summer snowfalls.

Table 3: Comparison of calculated and observed Δh on Hintereisferner, deviations from the mean of 1953 to 1982.

	$\bar{h}$(m)	σ_h(m)	correlation with observed values	
observed	2960	±118		
calculated		±148	0.73	using values of table 1
		±174	0.80	doubling δT_a
		±266	0.22	doubling δc

In table 3 observed values are compared with calculations according to equation (3), using values of table 1. Larger values of $\Sigma(\partial Q_i/\partial z)$ would decrease the standard deviation of the calculated Δh without changing the correlation coefficient.

An attempt was made to incorporate the simultaneous effect of $\delta E\downarrow$ due to δT_a. Since the radiative transfer coefficient α_S

$$\alpha_S = \frac{dE\downarrow}{dT_a} = 4\ \varepsilon\ \sigma\ T_a^3 \tag{2}$$

is of comparable magnitude to α_H ($\alpha_S = 4.6\ Wm^{-2}K^{-1}$ at 0°C for $\varepsilon = 1$) this effect was simulated by doubling δT_a.

The resulting standard deviation is higher, but correlation is improved from 0.73 to 0.80.

Doubling the effect of accumulation did not yield any improvement.

Judging from these results it seems that larger transfer coefficients α_H and α_V as well as incorporation of the interdependence between various δ would improve the agreement of calculated and observed values. While this is a question of refining the model structure I believe that the model is finally limited by the natural variability of coefficients and gradients which were treated as constants here but change in reality with the weather situation prevailing in individual years.

Acknowledgement: This investigation was sponsored by the Geophysical Commission of the Austrian Academy of Sciences.

REFERENCES

Ambach, W. 1985. Characteristics of the heat balance of the Greenland ice sheet for modelling. Journal of Glaciology, vol. 31, 107, 3-12.

Funk, M. 1985. Räumliche Verteilung der Massenbilanz auf dem Rhonegletscher und ihre Beziehung zur Klimaelementen. Geographisches Institut der Eidg. Technischen Hochschule Zürich, Heft 24, 183pp.

Kuhn, M. 1979. On the computation of heat transfer coefficients from energy-balance gradients on a glacier. Journal of Glaciology, vol. 22, 87, 263-272

Kuhn, M. 1981. Climate and glaciers. Proceedings of the Canberra symposium on sea level, ice and climatic change. IAHS publ., no. 131, 3-20.

Kuhn, M. 1984. Mass budget imbalances as criterion for a climatic

classification of glaciers. Geografiska Annaler, 66A, 3, 229-238.
Lliboutry, L. 1974. Multivariate statistical analysis of glacier annual balances. Journal of Glaciology, vol. 13, 69, 371-392.
Markl, G., H.P. Wagner 1977. Messungen von Eis- und Firntemperaturen am Intereisferner (Ötztaler Alpen). Zeitschrift f. Gletscherkunde und Glazialgeologie 13, 1-2, 261-265.
Östling, M., R.L. Hooke 1986. Water storage in Storglaciären, Kebnekaise, Sweden. Geografiska Annaler, 68A, 4, 279-290.
Schug, H.-J. 1987. Der Schwarzmilzferner-Meteorologisch-glaziologische Untersuchung an einem Kleingletscher in den Allgäuer Alpen. Diplomarbeit, Universität Innsbruck, 74pp.
Tanzer, G. 1986. Berechnung des Wärmehaushalts an der Gleichgewichtslinie des Hintereisferners. Diplomarbeit, Universität Innsbruck, 103pp.
Trabant, D.C., L.R. Mayo 1985. Estimation and effects of internal accumulation on five glaciers in Alaska. Annals of Glaciology, 6, 113-117.
Wagner, H.P. 1979/1980. Strahlungshaushaltsuntersuchungen an einem Ostalpengletscher während der Hauptablationsperiode. Teil 1: Kurzwellige Strahlung, Teil 2: Langwellige Strahlung und Strahlungsbilanz. Archiv f. Meteorologie, Geophysik und Bioklimatologie, B, 27, 297-324 and 28, 41-62.

GLACIOLOGY AND QUATERNARY GEOLOGY

1. V. V. Bogorodsky, C. R. Bentley and P. E. Gudmandsen, Radioglaciology. 1985. ISBN 90–277–1893–8.
2. I. A. Zotikov, The Thermophysics of Glaciers. 1986. ISBN 90–277–2163–7.
3. V. V. Bogorodsky, V. P. Gavrilo and O. A. Nedoshivin, Ice Destruction. 1987. ISBN 90–277–2229–3.
4. C. J. van der Veen and J. Oerlemans (eds.), Dynamics of the West Antarctic Ice Sheet. 1987. ISBN 90–277–2370–2.
5. J. S. Aber, D. G. Croot and M. M. Fenton, Glaciotectonic Landforms and Structures. 1989. ISBN 0–7923–0100–5.
6. J. Oerlemans (ed.), Glacier Fluctuations and Climatic Change. 1989. ISBN 0–7923–0110–2.